Tides of Change on Grand Manan Island

Tides of Change
on Grand Manan Island

Culture and Belonging in a Fishing Community

Joan Marshall

McGILL-QUEEN'S UNIVERSITY PRESS

Montreal & Kingston · London · Ithaca

© McGill-Queen's University Press 2009

ISBN 978-0-7735-3475-9 (cloth)
ISBN 978-0-7735-3476-6 (paper)

Legal deposit first quarter 2009
Bibliothèque nationale du Québec

Printed in Canada on acid-free paper that is 100%
ancient forest free (100% post-consumer recycled),
processed chlorine free.

This book has been published with the help of a
grant from the Canadian Federation for the
Humanitites and Social Sciences, through the Aid
to Scholarly Publications Programme, using funds
provided by the Social Sciences and Humanities
Research Council of Canada.

McGill-Queen's University Press acknowledges the
support of the Canada Council for the Arts for our
publishing program. We also acknowledge the financial support of the Government of Canada through
the Book Publishing Industry Development Program
(BPIDP) for our publishing activities.

Library and Archives Canada Cataloguing
in Publication

Marshall, Joan, 1943–
Tides of change on Grand Manan Island : culture
and belonging in a fishing community / Joan
Marshall.

Includes bibliographical references and index.
ISBN 978-0-7735-3475-9 (cloth)
ISBN 978-0-7735-3476-6 (pbk)

1. Grand Manan Island (N.B.)—Social conditions.
2. Grand Manan Island (N.B.)—Economic conditions. 3. Grand Manan Island (N.B.)—Social life
and customs. 4. Fisheries—Social aspects—New
Brunswick—Grand Manan Island. 5. Fisheries—
Economic aspects—New Brunswick—Grand Manan
Island. I. Title.

HN110.G715M37 2009
971.5'33
C2008-905689-2

This book was designed and typeset by studio
oneonone in Adobe Garamond 11/14

For the people of Grand Manan

Contents

Maps, Figures, and Tables

Maps

Figures

Tables

Acronyms

ABMA	Aquaculture Bay Management Area
APC	Atlantic Policy Congress (of First Nations Chiefs)
ATV	all-terrain vehicle
CAFSAC	Canadian Atlantic Fisheries Scientific Advisory Committee
DAFA NB	Department of Agriculture, Fisheries and Aquaculture, New Brunswick
DFA	Department of Foreign Affairs (Canada)
DFA NB	Department of Fisheries and Agriculture, New Brunswick
DFO	Department of Fisheries and Oceans (Canada)
DIAND	Department of Indian Affairs and Northern Development (Canada)
EEZ	exclusive economic zone
EI	Employment Insurance
FCMA	Fisheries Conservation and Management Act (USA)
FCR	feed conversion ratio
FERC	Federal Energy Regulatory Commission (USA)
GMFA	Grand Manan Fishermen's Association
GPS	global positioning system
ICJ	International Court of Justice
ICNAF	International Commission for the North Atlantic Fisheries
ISA	infectious salmon anaemia

ITQS   individual transferable quotas
LAPPS   limited access privilege programs
LNG   liquefied natural gas
MOU   Memorandum of Understanding
NFB   National Film Board
NGOS   non-governmental organizations
PSP   Parents Supporting Parents
RAP   Regional Assessment Process
SEC   Special Eligibility Criteria
UNCLOS United Nations Convention on the Law of the Sea

Acknowledgments

Even trying to articulate my appreciation to the people of Grand Manan for their understanding, cooperation, tolerance, and friendship over the twelve-year period in which I was involved in research in their community seems somewhat inadequate. Words cannot express the feelings of gratitude I have for so many who have shared their lives with me in ways that not only contributed to my understanding of their struggles and successes as their lives were transformed but that also enriched my own life. The respect and admiration I have for the strengths of the women who kept together home and community while their husbands were away fishing has tempered my own complaints, which seem trivial in comparison. But, for a few, their equally devastating attempts to cope with the patriarchal constraints of their island culture became a source of personal angst as I debated the ethics of intervention or tangible support. For me as a researcher, the ongoing balancing of participatory action and observation with active contributions to community lives frequently presented dilemmas that could not be easily resolved. Shared confidences had to be carefully and sensitively entered as "data," even as I felt the weight of information that seemed to demand action. At the same time I watched the men manoeuvre boats through treacherous straits against surging tides, carefully peeling and pointing weir stakes, trimming top poles that had been gathered in the woods, repairing engines, and rescuing trapped porpoises. The wide range of skills and different types of knowledge that were required in their daily activities as fishers was a constant source of admiration. These men knew how to work, how to innovate, and how to survive in a world defined essentially by the natural environment. They, too, like the women, shared parts of their lives with me, usually on the boats and wharves rather than in the kitchens, but always with a generosity that belied their tough-talking exteriors.

The research began in 1995, and I will be forever grateful to the Royal Canadian Geographical Society for funding the first full summer of research in 1996, when I was unemployed. Once I was committed, there was no changing course as I became enthralled by the culture of a community that seemed to be on the brink of such shattering changes that I felt the documentation would be enormously important – for academics, policy decision makers in governments, and, of course, the Grand Manan community itself. And so I continued for several years with my own resources, eventually (thankfully) receiving funding from the Social Sciences and Humanities Research Council of Canada from 1999 until 2006. That continuing support was essential and is very much appreciated.

Everyone at McGill-Queen's University Press has also been enormously supportive and helpful, including Philip Cercone, whose warm welcome when I first mentioned the book encouraged me to continue. I am very appreciative of the input provided by two anonymous reviewers, whose volunteer contribution of so many hours carefully reading the original manuscript resulted in a much improved book, especially in relation to the reorganization of the material and the reorientation of major themes. The editors, Joan McGilvray and Joanne Richardson, who read the manuscript with the detail that inevitably resulted in more work for me, but also a much improved text, are sincerely thanked.

During the course of the research I needed the help of several technical people whose skills in map-making, poster design, compilation of statistical data, and bibliographic research were invaluable. Among them, Christine Earl at Carleton University; Pete Barry at McGill University; graduate students Monica Kollerstrom, Sarah Shiff, and Natalie Foster; and design artist Christian Lambert were especially helpful. I am very grateful to Eric Allaby for so generously providing his original drawings of weirs that are included as figures 2.1 and 2.2. Also thanks are extended to Nicole Green Wolfe for use of one of her photos of her grandfather, Smiles Green. All other photographs are my own.

Unquestionably, however, it was the people of Grand Manan who made this research possible, whose generous sharing of their lives has opened up for others the realities of rural lives in transition. The first people I met were Cecelia Bowden and Gene Gillies, at whose cottages I often stayed, especially during the winter months. Not only were they gracious hosts, but they generously gave me special rates, which meant my research funding could be extended. In the first autumn, when I was preparing a research proposal, I stayed at the bed and breakfast operated by Hazel and Alan Zwicker, who became good friends and were initially a source of contacts that helped introduce me to other islanders. Early interviews with women, both as individuals and collectively at their Bible study groups or their weekly ramoli nights, meant that I was able to begin the process of untangling the gender relations that seemed so central to islander identities. Not wanting to divulge too many names of people who would prefer to remain anonymous, I nevertheless want to extend my appreciation to them for their openness and willingness to be part of this long project. Many continue to be friends today, and I only hope that publication of this book will not change that! I was often invited to accompany them to events such as church suppers or bingo nights, all of which provided an opportunity to gain insights into the social worlds of the island in ways that formal interviews never could.

Some individuals were especially helpful in a variety of ways, such as Charles Jensen at the Boys and Girls Club, whose insights into youth culture provided invaluable information. Working with him on the development of the film project was particularly rewarding. Robbie Griffin's help in sorting through the information he had collected in preparation for his school reunion of 2000 and the compilations of school data prepared for me by Janice Naves were essential to eliciting an accurate understanding of the migration patterns of young people. To the several professionals in social services who granted me interviews, I owe a particular debt of gratitude. As well, I appreciated the willingness of Gail Miller, Hayley Outhouse, and Linda Griffin to share their time and personal experiences in the dissemination of study results at two conferences – one in Halifax (2003) and one in Montreal (1997). Invitations to accompany the men to the weirs or on their lobster boats were always appreciated, resulting in hundreds of photographs and impromptu interviews that gave me insights into the working lives of islanders. Bud Brown, Neil Morse, Lester Bass, Allison Naves, Allison Monroe, Harry Stanley, Wayne Ingalls, Herbie Lambert, Rod and Bradley Small, Paul Brown, David and Stephen Bass, and Rodger and Andrew Maker all welcomed me aboard their boats, despite the traditional belief that women were bad luck.

The support and friendship of people such as Janet Dexter Ingersoll, Joan Barberis, Helen and Lester Bass, Ruth and Bill Bass, Smiles and Alice Green, Joey Green, Len and June Brierly, Cecelia Bowden and Gene Gillies, Adian and

Ruth Green, Bob and Judy Stone, Carmen and Pete Roberts, Lydia Parker, Betty and George Brown, Nathan and Karen Bass, Arthur Brown, Anneka Gichuru, Erleen Christensen, Wendy Dathan, and Margery Small meant that extended stays on the island were like homecomings. Their initial help in providing interviews or information related to the changes on the island merged into visits and casual chats as friends. Unhappily, many of these friends have died, and I miss them all. The island is changing and their passing is an important part of that transformation. The many generous gestures of friendship included George and Betty Brown bringing me wood for the fireplace during an inclement May; seventy-five-year-old Carolyn Flagg, tired from dulsing, sharing her lunch with me at her beloved camp on the seawall at Dark Harbour; Philip Russell's unsolicited gift of clams while I was grocery shopping; and Rodger Maker's dory shuttles out to Wood Island. Islanders have been both generous and welcoming, and I shall always treasure the enrichment they brought to my life during the years I have devoted to trying to understand the forces of change that were affecting their lives. Special thanks are extended to the three island couples who so willingly and graciously agreed to read the manuscript prior to final submission. In asking them about the "rightness" and "fairness" of interpretations I acknowledged that we might not agree on some issues, but I needed to feel that I had achieved balance in the presentation of many very difficult situations. To Bob and Judy Stone, Allison and Janice Naves, and Charles Jensen I owe a particular debt of gratitude for that undertaking, as I do to Wendy Dathan and Erleen Christensen, who diligently, helpfully, and thoughtfully read and commented on the manuscript. The friendship of Alison Deming and her sensitivity to and understanding of the many complexities of island relationships provided valuable support to me through the years.

Finally, I extend my fond appreciation to my family, especially to Hugh, whose untimely death in 2006 meant that he could not reap the satisfaction of seeing this project to completion. Hugh never complained that I was absent for so many weeks each year (about eight to ten), and he always sent me off at 5:30 AM with a two-egg breakfast to sustain me during the twelve-hour journey to the Maritimes.

To all of these friends, and more, I wish to say "thank you" and to emphasize that this book could not have been written without you.

Tides of Change on Grand Manan Island

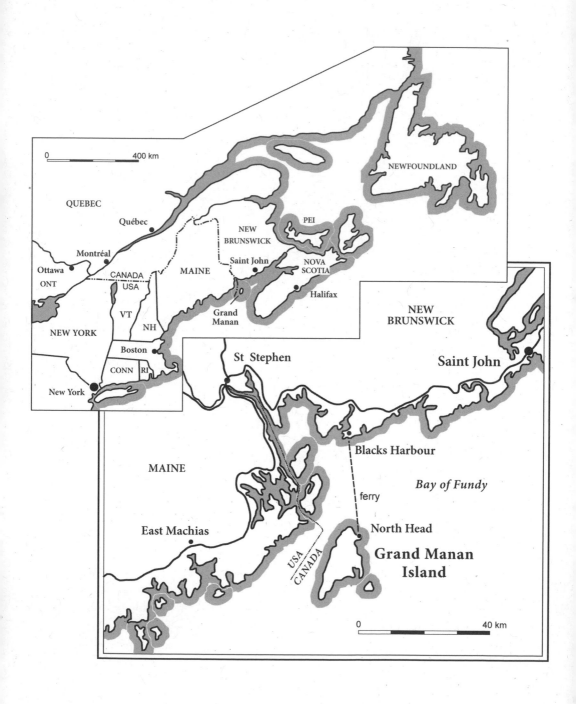

Introducing Grand Manan

I first came to Grand Manan in 1989 as a visitor for a few days. Captivated by the working landscapes of the small fishing villages on the island, and fascinated by the traditional activities of smoking herring and gathering dulse, I returned again in 1991 and 1993. Even then it was apparent that changes were beginning to encroach on the economic livelihoods that had defined the community for almost two hundred years, and I wondered about the social and cultural implications. Several aquaculture sites had been established by 1993, and only two smokehouses continued to produce the lovely molten-gold herring that would be shipped out in large wooden boxes. Having just completed my doctoral research and seen it published as a book, in 1995 I visited again with a view to looking more deeply into the emerging changes and the possibility of developing a long-term research program.

Initially, I think my expectation was a four- to five-year commitment that could effectively document and describe changes that seemed inevitable and that I thought might even overwhelm the island community, especially the women. It was quickly apparent that a focus on women alone, without understanding the men, the children, and the lives of the elderly, would be unrealistic and, ultimately, unsuccessful. Also, gradually I came to realize that four to five years represented only a beginning: the complexity of social and economic patterns of change within the community simply demanded a longer time period in which to explore and document it. The study became a twelve-year love affair that was intermittently tinged with sadness and anger as various episodes and inevitable tensions within the community became prominent. The initial perceptions I had of an indomitable culture, with strengths, understandings, and survival strategies in the face of unpredictable harvests and risks at sea, became more nuanced, incorporating ambiguities and contradictions that precluded a romantic view of this traditional fishery-based island. I recall feeling almost betrayed during one particularly memorable March week; suddenly I was being forced to acknowledge some of the dark undersides of community relationships. The vulnerabilities of some who seemed to have lost their way were juxtaposed with the fierce struggles of others who resisted traditional definitions of roles. There were many questions, many imponderables, that argued for a depth of exploration that was quite daunting. The romantic ideal was no more.

During the 1990s and into the beginning of the new millennium, Grand Mananers experienced a change that went beyond "enormous" or "rapid." The changes that came to this small island were transformative because they were qualitatively different from any that had affected their community in the past and because they were largely out of the control of islanders themselves. The years of change that I describe, between 1995 and 2006,[1] touched every person and every family whose lives had depended upon the fishery for two hundred years. This period of significant and irreversible change affected lives and relationships, both social and economic, as well as the very meaning of "community" and what it is to be a Grand Mananer. Historical personal and collective identities that had been inextricably woven into a profound sense of place were being challenged and transformed by events and circumstances that islanders scarcely understood and over which they had very limited control.

The ethnographic research upon which this study is based offered particular challenges. In the beginning, I was fearful and uncertain about how to proceed, how to meet people, how to ask for interviews, and even how to know what questions I wanted to ask. Books on research methodologies had, of course, been consulted. But the reality of entering a strange new community, with a culture very different from my own, was intimidating. One of my most fortunate decisions was to spend the first week of my exploratory visit at a bed-and-breakfast run by an island couple who had themselves had a varied experience of the fishery and their church. They were well known on the island and were able to provide me with an initial list of people who might be willing to talk with me. Using their names as references, I was indeed able to begin meeting islanders who agreed to talk about their lives, their hopes, their expectations, and their tragedies. As well, I did a lot of wandering about on the wharves, which often led to interesting and fruitful conversations that provided further leads and interview possibilities. At church as well, people who welcomed me were interested in hearing about my purpose on the island, and often I was invited for tea or simply to "drop in anytime." Invariably, it was the casual encounters rather than the more formal telephone-requested interviews that produced the most rich and nuanced information about island lives and landscapes.

One of the most challenging and sensitive problems involved how to deal with the necessity of obtaining informed consent. One simply cannot hand out a consent form to a fisher on the wharf and expect anything approaching an open conversation. Any communication would end right there. On the other hand, immediate acknowledgment of my work and its eventual goal (a book, definitely not a report to Revenue Canada!) was frequently enough to assure people of the ethical treatment of any information that they might impart. Formal interviews, in homes or offices, always included a request for a consent form signature. Even this provided some tricky moments, and, inevitably, adult islanders refused to be identified. The contrast between adults and teenagers was striking. Without exception, when I interviewed the latter and asked whether I could record the interview there was never any objection. Indeed, they seemed pleased that I valued their comments enough to want to record them. Obtaining their signatures was never a problem. Frequently, parents stayed in the room for the interview or were in the next room, which certainly affected some of the results.

In addition to interviews, both formal and informal, there were meetings (the municipal council, the Rotary Club, the transport commission, the Boys and Girls Club of Grand Manan Island [hereafter Boys and Girls Club], the tourism association, and many more), school graduations, social events such as bingo and parades, and church services, all of which provided insights into the evolving social worlds and community relationships between 1995 and 2006. As more people began to ask "who are you?" I put up a notice in one of the gas stations (at a time when there were still two on the island) that described my study and provided my coordinates in case someone wanted to contact me. As I became more and more accepted within the community, I would occasionally be asked to provide input related to social services or to sit on boards that were trying to find solutions to youth problems. I was able to contribute in various other ways as well, such as donating my photographs for the production of greeting cards as a fundraiser for the Boys and Girls Club, and designing a media project that encouraged students to learn about television production and film techniques. For me there was inestimable satisfaction in being allowed to participate in these community projects, partly because they offered me a way of legitimizing my work that went beyond personal accomplishment.

Unquestionably, these small contributions were possible because of the openness and encouragement of the then executive director of the Boys and Girls Club, Charles Jensen. His clear vision of the potential that the club offered island youth and his unwavering commitment to developing programs that would attract and support youth in a variety of activities provided an exceptional foundation for the future. His departure from the island after an intense twelve years of service was an enormous loss for the community. I was fortunate to have been able to work with him in ways that benefited both of our goals. Not only did he provide help with introductions, but his views and perceptions of events as they unfolded were invariably based upon conversations with people on all sides of any issue. He understood the dynamics and tensions of community relationships and was able to hold onto his own values when positions had to be taken. I learned a lot from him. Asking people to open their lives to you is, in one sense, quite arrogant and invasive, and I never lost the feeling that I was extremely privileged to be able to share in the lives of so many islanders.

Using both the "hard data" of demographic and economic statistics and the personal stories of individual lives, gathered over a twelve-year period, I have tried to present a narrative that is both complex and clear, both explanatory and understanding. In writing of these stories, experiences, and perceptions I have occasionally used fictitious names in order to protect the identities of those who would not wish them to be known. Some identities are revealed, but only under specific circumstances: when permission has been granted, when statements were made in public venues or published in news stories, and when the situation is clearly non-threatening or complimentary. The interpretations that I give to events, behaviours, and outcomes are, of course, tempered by my own life experiences and education. Any "objectivity," as such, is impossible. The impact of my middle-age status and female gender certainly affected many aspects of the research process and outcomes. All I can hope is that my interpretations fit equably into the theoretical frameworks that I propose and that the community itself recognizes the plausibility of my rationales.

As Anne Buttimer (1980) warned so many years ago, the subjects of our research do not necessarily see the world as we do, which is not to say that our interpretations are wrong. Every time I sent copies of my articles to islanders for comments, I did so with some trepidation. Usually the responses were positive, and, indeed, as an ongoing practice it served me well in terms of credibility and continuing cooperation and trust. One especially memorable comment came from a young woman who had been working on a local committee to find solutions to the youth drug problem. Having read my article, she said: "It's so nice to see it all set down like this; and to realize how my experience is all part of life on this island and affects everyone." Clearly, the lives of Grand Mananers are important, and I hope that those from urban Canada who read about their lives here will understand my frustration with the opinion of John Ibbitson, who wrote: "The rural will become a boutique lifestyle, a refuge for the countercultural, the retired, and for those who simply love the land. They'll make do" (*Globe and Mail*, 29 March 2007, A4). We cannot write off rural communities in such a cavalier way.

The challenges of observing without judging, of participating by following, of recording without excluding, and of interpreting while providing ethical balance have all been persistently at the forefront of my concerns, both in daily practice on the island and as I write. Again, I have changed the names of some to protect their identities, while I have correctly named others. In listening to and transcribing the voices and stories of so many islanders, I have been enriched and humbled by their strengths, foibles, and struggles. Life in a fishing village is never easy; and the lives of Grand Mananers have been especially challenged by the relative isolation that has inhibited possibilities for adapting to a rapidly changing world. Intertwined and enmeshed in the changing technologies of fishing and in the introduction of salmon aquaculture are new regulatory regimes, government legislation, institutional frameworks for health and education, and changing social attitudes and values that affect the historic roles of religion and marriage. In a decade, all of these have been affected and, in some cases, radically altered. It is probably not an exaggeration to suggest that the island of Grand Manan has changed more in this decade than in any other during its two-hundred-year history.

What follows is in many ways as much a personal journey as it is a complex weaving of stories and events that have transformed Grand Manan. In the documentation and questioning of others, I was forced to consider my own values and assumptions. Standing on the wharf one day in July, watching a tragedy unfold as the tail of a whale emerged from the nets of a seiner, I had to make split-second choices about photography, interviews, and disclosure. My cultural assumptions and academic "objectivity" did not necessarily serve me well in my relationships within this historic fishing community. No longer dependent upon the wild fishery, islanders have been faced with new choices and difficult transitions, most of which have been largely outside their control. Men and women both, unable to continue in their traditional jobs and roles, have had to redefine their identities, means of livelihood, and personal and community relationships. Grand Manan may still appear to be a tourist's dream getaway, but the reality for islanders is much different. This is their story insofar as I have been able, with their help and friendship, to tell it.

1

Introducing Grand Manan

Grand Manan is a physical riddle. Though it may have been larger at the time of creation, and the little isles and islets, some no more than ledge tops, that go to make up the archipelago seem to indicate that it was, it has existed within the histories period [sic] as it appears today. ~ *Grand Manan Historian*

Archipelago and Island

Islands hold a fascination for people that transcends simple explanation. Islands are bounded geographically by water, but they can also be circumscribed by boundaries of cultural difference and social distinctiveness. Depending upon historical, political, and geographical circumstances, islands' have fascinating stories to tell and cautionary tales to offer. Grand Manan, an island in the Bay of Fundy on Canada's east coast, has an especially rich and tumultuous story that contains many lessons and just as many dilemmas. While Grand Manan is properly defined as an archipelago, it is usually referred to as Grand Manan Island, New Brunswick, its length and width being approximately seventeen and seven miles, respectively. Nonetheless, until the recent past, its outer islands (including Kent, Long, Nantucket, Outer Wood, Wood, Cheney, and Ross) were inhabited as bases of the wild fishery and important commercial activities, notably the production of smoked herring. Today, only White Head and the main island of Grand Manan (often referred to as "the Main") continue to be homes to the 2,500 residents (950 households), most of whom claim a lineage that goes back many generations. Almost twenty-five miles from the mainland – ninety minutes by ferry – Grand Manan is overwhelmingly defined by the sea. Its shorelines and landscapes, villages and woods, are all templates for the many activities and livelihoods that reflect a historic dependence upon the sea.

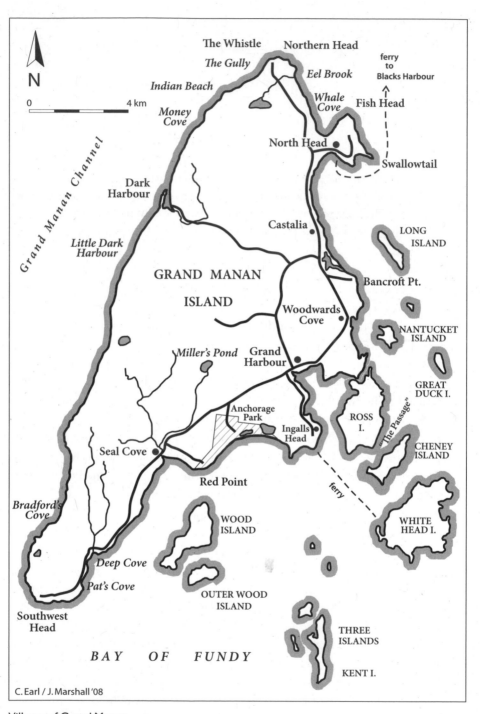

N

0   4 km

The Whistle

The Gully

Indian Beach

Money Cove

Northern Head

Eel Brook

Whale Cove

ferry
to
Blacks Harbour

Fish Head

North Head

Swallowtail

Grand Manan Channel

Dark
Harbour

*Little Dark
Harbour*

Castalia

LONG
ISLAND

GRAND MANAN

ISLAND

Bancroft Pt.

Woodwards
Cove

NANTUCKET
ISLAND

Miller's Pond

Grand
Harbour

GREAT
DUCK I.

Anchorage
Park

Ingalls
Head

ROSS
I.

"The Passage"

CHENEY
ISLAND

Seal Cove

Red Point

*Bradford's
Cove*

WOOD
ISLAND

ferry

WHITE
HEAD I.

Deep Cove

Pat's Cove

Southwest
Head

OUTER WOOD
ISLAND

THREE
ISLANDS

BAY   OF   FUNDY

KENT I.

C. Earl / J. Marshall '08

Villages of Grand Manan

The physical characteristics of the main island clearly define the historic settlement pattern that is only today beginning to free itself from constraints on mobility (through the arrival of paved roads and cars) and to reflect common features of mainland real estate-induced changes, such as cliff-top homes where views enhance land values. While the east side of the island, underlain by old sedimentary formations, is at sea level and provides many opportunities for harbours and access to beaches, the western side is dominated by high, three-hundred-foot cliffs that have inhibited settlement except for fishing and hunting camps. Formed during the Triassic period, these cliffs are relatively young igneous formations, volcanic in origin, with striking columnar walls of basalt or bedded layers of rock that rise steeply out of the ocean. A fault line extending from Red Point across the island to Whale Cove separates the two zones of geological formation. Along the western side of Grand Manan, known by locals as the "back of the island," the most distinctive natural formations are long gravel bars that enclose brackish ponds, providing both a safe haven against storms and landing possibilities for dories or skiffs. These gravel bars are about sixty feet high and over 120 feet in width, well above the high-tide line (averaging twenty-five feet), and have been favourite sites for camps that served as temporary homes for previous generations of fishers who went there during the week to fish or gather dulse. Today they are more appreciated as recreational weekend "getaways."

On the eastern side the much older rock, dating to Precambrian times, is ancient sedimentary and metamorphic rock that is interspersed by jagged outcroppings of resistant volcanic rock, extremely treacherous for boats. The harbours along the eastern side, however, are well located for protection against prevailing westerly winds, although quite exposed to the more violent northeasterly storms. North Head, Ingalls Head, and Seal Cove all offer excellent harbour facilities. Woodwards Cove, on the other hand, is accessible only at high tide and, since about 1990, has ceased to be used for the traditional fishery. It is now a main centre of aquaculture because of its centrality within the archipelago of small protective islands, including Nantucket, Long Island, High Duck Island, Low Duck Island, and Great Duck Island.

The existence of good anchorages during the fair weather westerly flow of air was not always enough to ensure adequate protection during seasons of storms and high tides. Despite the harbours at Flagg's Cove (North Head), Grand Harbour, and Seal Cove, in the mid-nineteenth century, lacking the "breakwaters, as we know them today, the coastal fishing fleet for the most part consisted of boats small enough to carry, or haul to shore shelter often over long, or extensive tidal areas. A special type of boat was built here, adapted to these conditions; they were sturdy little vessels but restricted the cruising range of owner or master. For decades, Grand Mananers were not only inshore fishermen, they were hardly more than shoreline operators. That a people continued to exist and to gradually surmount these natural handicaps is the wonder of it all."[1]

Indeed, early contacts with Indians[2] and English explorers record their negative reactions to the potential protection these harbours might offer. Grand Manan was sighted by Jean Cabot in 1498 and by Gaspar Cortoreal in 1501, but the earliest map of this island was only published in 1558. On it is shown a cape at the mouth of the Bay of Fundy called *C. de las muchas isles* (Cape of Many Islands), which included Grand Manan (Scherman 1971, 11). The first written reference to the island is contained in the reports of Pierre du Guast, Sieur de Monts, and Samuel de Champlain in 1606. Following decades of negotiations between Britain and France, the land and seas of eastern Canada and the United States were being constantly exchanged. Then a circumnavigation by Captain William Owen in 1770 provided a detailed description of the merits of Grand Manan's many harbours and inlets. After his first landfall at Castalia, he suggested that it might provide shelter in a crisis, a "tolerable asylum." Continuing around to Ross Island and Cheney Passage, where he became grounded during low tide, he was not impressed with any real possibilities for good harbours. He felt there might be excellent anchorage at Grand Harbour until the tide swept out, leaving only a shallow depth at the tip of what is today Ingalls Head. He was apparently more enthusiastic about the prospects for Seal Cove, except that heavy fog and gales drove him towards the mainland (15).

Distinct geological zones and harbour characteristics are not the only distinguishing features that contribute to the physical variety and mystery of the landscapes on this small island. Tidal and intertidal zones contribute to variable habitats for sea weeds and crustaceans that are harvested by islanders. The strong tides of the Bay of Fundy, known as the highest in the world, vary on average even between northern and southern ends of the island, from up to twenty-eight feet at North Head to about twenty-five feet in the south. As well, the great tidal range results in powerful currents throughout the spring as well as neap tides that create dangerous and unpredictable conditions along the many ledges and islets that define the archipelago, a direct cause of hundreds of wrecks and much loss of life. A lifelong interest of Eric Allaby, islander and former member of the New Brunswick Legislative Assembly, these wrecks were the subject of his extensive, decades-long research. As one might expect, by far the majority of the shipwrecks occurred prior to the development of radar and global positioning system (GPS) technologies. The strong tides and tricky cross-currents were the most responsible factors "for the heavy loss of shipping about Grand Manan" (Allaby 1984, 18). As he describes, being at the entrance to the Bay of Fundy during the period of settlement growth, especially following the American Revolution, Grand Manan and its treacherous reefs and ledges created challenging obstacles for the commercial traffic that plied the waters between the emerging colonial states and Britain: "The ships and schooners of yesteryear were the trains and trailer-trucks of today; the cargoes they carried reflected the commerce of the time. Outward bound shipping typically carried lumber for England, salt fish for the West Indies or gypsum for New York. Inward bound ships brought British clothing and household goods, or salt, sugar and rum from the West Indies" (18).

Among the stories and family histories told by islanders today are the exploits and heroic life-saving missions of lighthouse keepers and early settlers scattered along the island coastline. The most oft-told story, perhaps, is that of the *Lord Ashburton*, which, in 1857, foundered off the northern cliffs of the same name. Indeed, the treacherous seas around Grand Manan meant that many islanders were able to offer their pilotage skills to American fishers who ventured into these waters: "Those in want of employment readily procure it by engaging themselves in U.S. fishing vessels, for which they are exceedingly well paid; and as they are generally good fishermen and well acquainted with Pilotage of the Bay, they are invaluable."[3] The wrecking of the schooner *Velma* in 1900 led to a valiant rescue of several men by Captain Cheney and his two sons – heroism that was recognized with the presentation of medals by both Canada and the United States. According to Allaby, the greatest recorded loss of life occurred in 1909, when the steamer *Hestia* was wrecked upon the Old Proprietor Ledge, costing the lives of thirty-five men. In his research he has documented over three hundred wrecks, mainly occurring in the first two hundred years of recorded Grand Manan history, beginning in 1720, but especially after the Revolutionary War in 1776. The most recent feat of heroism, described in matter-of-fact detail by the central hero, Vernon Bagley, occurred in 1963. As game warden at the time, Bagley was called by Sydney Guptill, keeper of the Southern Head lighthouse, to help in the rescue of two men whose small boat was floundering without a working engine off Southwest Head during a midnight storm. Bagley volunteered to be lowered by rope down the cliff face to rescue one of the men, and his ninety-minute ordeal of securing the safety of a man many pounds heavier than he was became a legend that was picked up by news services across the continent. He was awarded the Carnegie Silver Medal for bravery, while Guptill received the bronze medal. "The only way I could've got the gold," he told me with a wry twinkle, "would have been to have died!"

Lighthouses, fog horns, and bell buoys around the island testify to the dangers of navigation. Beginning with the "first lighting of the lantern at Gannet Rock Lighthouse on Christmas Eve 1831, and the construction of a light station at Machias Seal Island the following year, there dawned a part of Grand Manan history and a facet of local society which grew to mean life long careers for scores of island natives over the past century and a half" (Allaby 1984, 22). With the advent of automation and the advanced technologies that have become the "eyes" for the boat captains, these careers have become part of the family lore that continues to define identities and senses of belonging, despite the fact that the possibilities for wrecks and the need for land-based lighthouses has greatly diminished.

Landward, the vegetation on the island is dominated by grey birch, maple, tamarack (also known as "hackmatack"), fir, and spruce. Because of the salt spray the wood is extremely dense, making it excellent for high-quality paper, and is therefore highly prized by mainland pulp companies. It was also the source of timber for the early boat-building enterprises that engaged many skilled boat builders during the nineteenth century. As well as dense forests, bogs and saltwater marshes define large areas of the vegetation zones of Grand Manan. The natural history of the islands includes a fascinating variety of habitats, from shoreline tidal zones to broad swaths of dense marsh grasses and sea lavender as well as bogs dotted by delicate orchids, soft cottongrass flowers, pitcher plants, and bakeapple berries. Surrounding wreaths of gooseberry and raspberry bushes and dunes that support edges of lamb's quarters and pea vines complete a picture of lush and diverse habitats. Grand Manan is a veritable garden, offering a feast of wild berries and edible plants that span all seasons. No one could starve on this rich island.

Interestingly, as small as it is, Grand Manan has distinctive microclimates that affect daily lives and even people's choices of where to live. While the north end of the island at North Head and Castalia is noted for early springs and sunny clear days throughout the summer months, the southern end of the island is known for its fog. At first a visitor might be tempted to brush off the derisive comments of North Headers about the villages of Seal Cove and Deep Cove having too much fog; but any length of time spent there in May and June will soon give credence to these observations. Indeed, the iridescent lupines of June bloom two weeks later in Seal Cove, and their blue, purple, and pink jewel-like colours continue to tempt photographers even into July, long after they have died back in North Head. Similarly, vegetable gardens produce lettuce and strawberries several weeks earlier in North Head. The fog seems to hold in against the bank along Red Point Road in Seal Cove, always the last place to enjoy sun during the late spring days of inclement weather. In winter, even when snow melts as quickly as it falls on North Head, it can blanket the hills leading into Seal Cove for many hours, acting as a deterrent to islanders wanting to drive up the island during the winter months.

It is generally acknowledged that Grand Manan – visited by Champlain in 1605 and sporadically by Indians until as recently as the mid-twentieth century – did not have permanent settlement until the mid-eighteenth century. Fleeing the American Revolution, many of the early settlers had arrived from the American states of the eastern seaboard.[4] Nonetheless, some have argued that, despite the lack of artifactual evidence, the Passamaquoddy tribes who had an encampment at Pleasant Point in Maine may have had the first permanent camps on Grand Manan. They are known to have had seasonal camps at Indian Beach between Money Cove and Long Eddy, and at Indian Camp Point on Ross Island; but, according to most sources, these were only for summer sealing and fishing activities.[5] "When Joel Bonney and his party arrived at Grand Manan in 1779, the Passamaquoddy Indians were incited by Revolutionists in Machias to assert a claim to the island. The Bonney family sent a representative to the mainland to discuss the matter with the tribal chieftains when it was agreed, on payment of 10 dollars and a heifer, they might spend the winter here, but would have to leave as soon as possible the following spring."[6] That winter the first white child was born on the island, and the next spring the Indians arrived to make sure that the white people left their temporary home.

A few years later, however, land grants were being offered to settlers who would agree to make improvements and to encourage other settlement. The earliest permanent settlers, Captain Thomas Ross and Moses Gerrish, had "received a License of Operation to all of Grand Manan,"[7] moving to Ross Island in 1784. The earliest known headstone can be found on Ross Island, marking the burial of William Ross, son of the original settler Captain Thomas Ross, who drowned in 1828. Ross and Gerrish were soon followed by William Cheney from Massachusetts, who settled on Cheney Island in 1784, and Dr John Faxon, who established a fishing and shipbuilding centre at Seal Cove.

The earliest arrival on White Head Island was William Frankland, who first cleared land in Grand Harbour on the main island in 1801, before moving to White Head in 1805. Early nineteenth-century accounts describe the intense relationships of islanders with the sea and with their American neighbours to the south.

> The people of Grand Manan are active, industrious, and hardworking, capable of enduring great hardship and fatigue. The young men, from lack of employment at home, engage on American fishing vessels; they get good wages, because they are active, hardy sailors, excellent fishermen, and admirable pilots for the Bay. The Americans say "there is no better man on board a fishing vessel than a native of Grand Manan, if you take him away from his own island." (Perley 1852, 107)

Incorporated into these accounts are references to families who arrived in the early 1800s and who still live on Grand Manan, such as Nathaniel Doggett (now Daggett), who had a curing establishment west of Cameron's (Pettes) Cove; Joel Ingersoll, who received a grant in Woodwards Cove; and Josiah Flag, Nehemiah Small, and Francis Gubtail in North Head. This long lineage of family connections and dependence upon the sea has been both a strength (as the community learned how to survive the unpredictable vagaries of the wild fishery) and a weakness (as it struggles in the present world of rapid change). The population in 1805 was about 200 persons; by 1818 it was shown as 384; in 1824 as 1,003; in 1851 as 1,187; and in 1861, the last census before Confederation, as 1,535. Of these, only 354 were not natives of the island. Twenty years later, by 1881, the population had grown to over 2,500, showing an increase of 749 inhabitants (43 percent) in just two decades.[8] Three years later, the first ferry service was inaugurated with the formation of the Grand Manan Steam Boat Company; and Captain John Pettes opened the Marble Ridge Inn at North Head, the first tourist establishment and one that is still in operation as the Marathon Inn.

Historically, the land-based political boundary between New Brunswick and Maine evolved in the period between 1798 and 1842 (Gentilcore 1993, pl. 21). The settlement, which began from about 1784 onwards, followed the 1783 Treaty of Paris, which defined the St Croix River as the boundary between Britain and the United States. Grand Manan had been excluded from early eighteenth-century maps because, as one 1786 British observer noted, there was no conclusion as to which country could rightfully claim the island (Small 1997, 18). It seemed to be "shrouded in the impenetrable fogs of the Bay of Fundy" until it was resolved in 1817 (25). The sea boundary, on the other hand, as I describe later in the book, remains unresolved and is an ongoing source of tension and conflict for the fishers of Grand Manan.

This brief summary of how the islanders' origins are tied to the United States suggests why there is an ongoing question of borders, cultures, and identities. For generations Grand Mananers' ties to their American neighbours, beginning with the eighteenth-century Loyalists, have been stronger than their ties to Canadian mainland institutions. Indeed, the early existence of a rail-ferry link to Boston greatly facilitated travel between the American mainland and Grand Manan. Directly affecting religious values and adherence, nineteenth-century art, contemporary shopping habits and vacation plans, and the choice of summer residence, the proximity and influence of the United States continue to be important elements of Grand Manan culture and identity. Referring to the establishment of the early churches on the island, Wayne Reppert, writing in the *Grand Manan Historian*, describes a hybridity of border culture between "largely poor and unschooled fishermen and farmers of Grand Manan … [and] the Loyalist establishment," and he comments on the "civilizing influence" of a "fervent Baptist revival."[9] Moving up the eastern seaboard from the United States, the Baptist Church's arrival on Grand Manan in the late nineteenth century has had a profound impact upon the culture of the island – one that continues today, with many young people opting to attend American bible colleges.

Artists who visited Grand Manan in the nineteenth century used the fluid boundary between New England and Grand Manan to interpret landscape and fishing cultures in ways that illuminated the ambiguity of "place." In her exploration of representations of Grand Manan in nineteenth-century painting, Doreen Small (1997, 25) has shown how the "nebulous political status of the island, and its relative anonymity, provided the basis for Grand Manan to be read in terms of a New England regional landscape and as a site of Canadian landscape iconography." Moreover, the New England imagery that was reflected in paintings revealed the strong religious overtones that reflected the prevailing culture of the period. The moral influence of nature was represented in soaring cliffs reminiscent of cathedrals, and the ocean was seen as being related to the "original biblical meaning of the wilderness-as-void" (34). For writers, too, the landscapes and boundaries of islands were seen as protective of religious values and of the purity of Puritan households. In her novels, Beecher-Stowe extolled the virtues of holding to the Sabbath, a dominant aspect of the norms of coastal communities.

Journalists who visited the island felt that its isolation from the mainland affected islander sensibilities. On Grand Manan "there remains a primitive type of people unspoiled by the inroads of visitors or the acquired smartness resulting from constant intercourse with city boarders. The fishermen have that lordly indifference to the outside world which is only found among folk by the sea" (A. Brown, quoted in Small 1997, 47). This "lordly indifference" is still apparent in conversations today, as islanders refer to their special culture of "difference" and complain about interference from the "outside." The boundary with the United States, the fact of being an island, the history of the Baptist Church, and Loyalist origins all created an environment within which ambiguity and contradiction had an inevitable impact upon social and cultural formation.

Any exploration of island changes from 1995 to 2006 must take into account the heritage of generations of American ties, boundary discussions, and isolation from the mainland, all of which inform the entire network of personal, social, and economic relationships. While it is apparent that the diverse traditional fishery provides the crucial foundation for island identities, in fact a complex interplay of historical and cultural factors have contributed to the outcomes that are today presenting the challenges that have come with rapid change. There are also important spatial characteristics related to the clusters of homes in small villages along the eastern shore that have contributed to particular identities over generations and to distinctive patterns of movement across island territories. The five main villages are North Head, Castalia, Woodwards Cove, Grand Harbour, and Seal Cove. As well, residents of Ingalls Head see their homes as distinct from those of Grand Harbour, and certainly the unaffiliated administrative district of White Head Island (population 250) is in no doubt about its difference from the main island. As I discuss in chapter 8, the historical focus of shelter, home, and work in these tiny nucleated settlements has been imprinted upon the culture of Grand Manan and is only recently gradually being blurred due to the forces of centralization and cross-island mobility. Spatial boundaries are melting, along with social and cultural ones. Still, one is constantly aware of the existential reality that Grand Manan's island situation overwhelms everything else. As Nicolson (2001, 141) comments in his beautiful description of three islands in the Hebrides: "Something of the sense of holiness on islands comes, I think, from this strange, elastic geography. Islands are made larger, paradoxically, by the scale of the sea that surrounds them … The sea elevates these few acres into something they would never be if hidden in the mass of the mainland."

Inklings of Change

In the more than twelve years I spent documenting change on Grand Manan, (notably from 1995 to 2006), I was constantly taken aback by the never-ending dynamism and fluidity that was obvious during each trip, even when spaced only a few months apart. Change was a constant: in the details of alterations in the landscape, in the nature of businesses, and in the daily lives of the people. Rarely was I away from the island for a period exceeding three months, and inevitably, upon every visit I encountered new buildings, new businesses, bankruptcies, and, of course, marriages, births, and deaths. Indeed, the value of frequent trips as well as extended stays became obvious to me very early on, contributing to both my own sense and to the community's that I was not merely a "visitor" but, in some small way, part of the community. On average, I spent ten weeks per year on Grand Manan, in all seasons, in every phase of work and recreational activities. The constant ebb and flow of new businesses and failed enterprises demonstrated to me the islanders' ongoing resilience, hopefulness, naivety, and creativity as they constantly strove to improve their lives. New restaurants might change ownership four times over six years and then close; craft shops were relocated; convenience stores expanded, changed functions, and opened new outlets. There is a pervasive culture of entrepreneurship and creativity that is intrinsically bound up with individualism and competitiveness: this has been both essential for survival and problematic for adaptation. All of this describes the invincible determination and flexibility that have defined island life. It is in taking the long view that significant changes can be appreciated not only in the physical landscape and the structure of the economy but also in the attitudes and values of islanders themselves.

During my first visit to the island in 1989, I had seen a vibrant smoked herring industry along several wharves, occupying both men and women throughout the summer months. A photo taken that year shows a herring carrier beside the wharf at Woodwards Cove (a wharf dependent upon high tides for access), probably the last year that the carriers were in there. By 1993, there were only two sites for smoking herring that were still active on Grand Manan: the Inger-soll smokehouses at Woodwards Cove and the Bensons and Ingalls sheds at Seal Cove. It was obvious that change was imminent. People talked about the new ferry, launched in 1990, with a sense that it would bring new economic opportunities, even though the markets for smoked herring were declining. In formulating a research strategy in 1995, my earlier introduction to the commu-nity provided a basis for speculating about the nature of social and economic change and for designing a methodology that could effectively capture the complexity of that change over the coming decade. I knew it would be a long-term study, although in 1995 I certainly did not envision twelve years. Of course, 2006 is not an "end." But the events themselves in that summer of 2006 do provide a "bookend" of sorts that, in ways both tragic and positive, suggest a way forward in a newly defined world of globalized food production and mar-keting systems. Far from the community-based smoked herring sheds, the aquaculture sites are embedded in a global agro-food industry that depends upon government support and high-risk, large-investment capital inputs. Everything has changed.

It was not only economic structures that were being affected. Fundamental values associated with views on religion, alcohol, marriage, and education all evolved in the years from 1995 to 2006 to an extent that, even though I docu-mented it, I still find astonishing. The increasing problem of drug use and the deaths of young men was only one sign of a particularly tumultuous period and of a community in transition. That the transition has been as difficult as it has been is, I believe, mainly related to two crucial factors: Grand Manan's relative geographic isolation and its low rates of in-migration. Together, these factors have contributed to intense differences, demarcations that have been both a strength and a weakness as the community struggles to adapt to modern rela-tions with government, corporations, and new technologies. While forces of globalization have begun inexorably to permeate the structures and networks of Grand Manan, there has been both conscious and unconscious resistance, and this defies a simple analysis.

Profound changes in the wild fishery in the decade between 1995 and 2006 have included depleted groundfish stocks, declining markets for smoked her-ring, increasing demand and rising prices for shellfish, increased lobster stocks, and new fisheries (such as sea urchin and crab). Along with this is the intense global restructuring of the food industry and the imposition of increasingly restrictive government regulations. The combined effects of these changes have led to greater income disparities and have thus contributed to fissures within

the community – fissures that have been exacerbated by the increasingly impermeable barriers to the entry of small-scale or new fishers. Complexity is related
not only to diversity but also to scale, technology, and seasonal rhythms. Herring captured in weirs close to shore require substantially less financial investment than do the mobile purse seiners, which are valued at over $1 million. Two
very different processes of capture define two very distinct levels of financial
participation. The harvest of shellfish involves expensive licences (over
$700,000 for boat and licence in 2004, according to local fishers). But there are
other significant smaller-niche activities that are important both as a way of
bridging the gap between seasons and as a means of making up for poor harvests in the larger fisheries. They include the gathering of periwinkles, clams,
and dulse (a tender intertidal seaweed that is harvested by hand during low
tide). The diversity of these traditional fishery activities has been reflected in
the cultural life of the community, which acknowledges the value of flexibility
and adaptability. As Gerald Sider's work in Newfoundland demonstrates, the
central importance of the environment as a resource base that defines the
autonomy of fishing villages is "part of the collective defence of the village
against economic and political domination" (Sider 1986, 28).

Apart from the fishery, two other crucial factors that have affected island
culture over generations need to be highlighted. The first is the powerful role of
religion, illuminated in the existence of fifteen active churches, representing
nine denominations. The second is the island situation itself, which has had a
profound impact on how the community sees the world and the extent to
which it has been affected by external forces of change, reflected in the low rates
of migration. As Agnew (1987) points out, geographical location, within which
settings structure relationships, is an important element in the construction of
"place." Both the historical links to the United States and the Loyalist heritage,
along with the isolation of the island from Canadian mainland institutions,
have had a profound impact on how the community sees the world and the
extent to which it has been affected by external forces of change. Historically,
even the small villages nestled in the protective coves and harbours along the
shore were isolated from each other. Formal communication with the mainland was not established until the inauguration of a weekly mailboat in 1838
and a post office in 1845 (Small 1997, 22). The inauguration in 1990 of a new
ferry, which almost doubled crossing capacity, was a crucial event in island history, both symbolically and practically. Practically, it facilitated the access of
islanders to the mainland and, at the same time, encouraged the growth of
tourism; symbolically, it provided islanders with a stronger sense of connection

to the rest of the world, promoting a perception of greater flexibility and connectivity. And yet, the daily reality of conscious decisions to move beyond fixed island boundaries continues to dominate islanders' lives. A conversation recorded at the ferry office illuminates this dilemma. The phone rang and was answered by the assistant, who turned to Robert:

ASSISTANT: "She wants to know if the crossing will be too rough."
ROBERT: "Who is it?"
ASSISTANT: "Could I have your name please?" [Pause.] "Mary Stewart."
ROBERT, thoughtfully: "Well, tell her it'll be more a pitch back and forth
   rather than a roll."
ASSISTANT, speaking into the phone: "Did you hear that? I guess it's not
   what you wanted to hear."

Not unlike the building of "the Road" in Newfoundland, which brought attention to the "comparative merits of tradition versus modern ways of life" (Davis 1995, 280), the new ferry represented a significant change in the community's sense of connection. And yet, there is a "love-hate" relationship with the increased connectivity, which is often seen as a threat to historic norms and practices. Islanders lobby for modern ferry services but protest when too many tourists arrive on the island. They demand increased numbers of trips but then mutter about how the new residents are buying up land. Over the longer term, there have been significant implications for the economy, as eighteen-wheeler transport trucks, cement trucks, and paving machinery could gain access to the island. Even housing was affected as it became possible to transport mobile homes or prefabricated units to the island. The greatest impact of the larger boat, however, has unquestionably been the impetus it has given to the growth of the high-stakes, high-tech development of salmon aquaculture on the part of mainland and foreign investors.

Aquaculture has become a major economic factor on Grand Manan, especially since 1995, when this study began. While it is often perceived as being part of the fishery, in fact its very different productive relations, the required scale of investment and risk, and the new property regimes of the marine Commons associated with aquaculture sites have quite new and distinctive implications for all aspects of community life. Revenues of about $11 million in 1994[10] increased to over $25 million in 1997[11] and an estimated $100 million in 2003. This growth had an immediate expansionary effect upon existing enterprises such as marine engines, boat building, truck maintenance, and trucking itself.

For example, in the five years from 1994 to 1999, the numbers of tractor-trailers using the ferry increased by more than two and one-half times (FGA Consultants Ltd. 1999). The impact upon the island cannot be overstated. Almost unanimously, if ambivalently, islanders say, "If it weren't for salmon, I don't know where we'd be." Indeed, it was not many years before that this comment seemed prescient. Major restructuring within the industry in 2005, falling prices, and the devastation wrought by a salmon infection combined to cause significant changes for the island economy and the social milieu that was intimately tied to it. In 2006, the numbers of homes for sale on Grand Manan had risen by about 50 percent from 2005, and the level of apprehension about the future surpassed any that had been discussed since I began my work in 1995. But I get ahead of the story.

An important characteristic of the demography of Grand Manan is its extraordinarily low rates of mobility. The percentage of "movers" (20 percent) in the five-year period between 1986 and 1991 on Grand Manan was less than half the national rate (47 percent) and about two-thirds the average of New Brunswick. It was even lower, at 18 percent, between 1991 and 1996. Of the people on Grand Manan in 1991, fewer than 12 percent had moved there (or returned after an absence) since 1986, and the net flow was positive, increasing by 125 people, or 5 percent. Taking births into account, at an estimated 15/1000, this indicates that net migration was approximately ninety people over the five-year period, or eighteen people per year. Even moves within the district are few. In 2006, 87 percent of the population on Grand Manan had lived at the same address for at least five years, compared to less than 25 percent of the Canadian population. Historically, most in-migration has consisted of older couples who come as retirees or women who marry on to the island (Marshall 1999). There has also been a history of low levels of out-migration. Until 2003, in the graduating high school classes fewer than 50 percent of the young people would leave for further studies on the mainland. Of those who did leave, about half would return within two years (interview with Edith Cook, former principal of Grand Manan High School, June 1997). There is no question that a sensibility described as "insular" is linked to the island identity, which is defined in terms of family roots and historical continuity and is directly associated with low rates of mobility and migration. Nevertheless, by 2004, there were indications that this pattern was changing.

One of the most obvious results of these low mobility rates and sense of geographic isolation has been the divide between "insiders" and "outsiders." This fundamental characteristic of social networks, institutional cleavages, and participation affects all social relations on the island, albeit somewhat differently for different categories of migrants. Even among spouses who have married onto the island, after decades there is still a sense of being "Other." One problem for people who migrate to the island is that they have not lived the experiences that create the "nebulous threads" of island culture. They cannot grasp the "subterranean level of meaning" that allows them to truly belong (Cohen 1982, 11). Among the three main groups of incomers who have defined migration patterns in the past (single women, retired couples, and spouses who marry onto the island), there is a notable difference between the lived experience of the first two groups and the third group.

The older people who choose to move to Grand Manan have an uneasy alliance with native islanders, but it is an alliance nonetheless. This older group tends to stay together and not to mix with "true" islanders. As one said, "Somehow we just gravitate to each other." They share understandings across a broad spectrum of social and cultural activities, such as book clubs, bird watching, organic gardening, and music, that has not been part of the island way of life. Many of these "new" residents also feel that they have to be careful about interfering when they accept executive positions on various committees, such as the museum board, the historical society, or the Chamber of Commerce. Nevertheless, for the most part these newer residents seem to have accepted that there are certain positions that will never be ceded to a non-islander. Similarly, some status jobs are restricted through unwritten rules and community norms. Spouses who marry an islander have arrived through two routes: (1) marrying an islander who has sojourned to the mainland or (2) marrying an islander after having moved to the island for a one- or two-year job (e.g., nursing or teaching). Because of their youth and the fact that they are raising families, their experience is quite different from that of the older retirees who migrate to Grand Manan (see chapter 9).

As the wild fishery declined through the decade while salmon aquaculture grew, other important changes were beginning to affect community relationships on Grand Manan. The restructuring within the global agro-food industry affected Connor's sardine industry along the eastern seaboard, resulting in the closure of the Seal Cove fish plant in December 2004. The effects of the immediate displacement of two hundred direct jobs rippled through the community. While labour shortages had been a problem on the island for the five years preceding the closure, especially because of the growth of aquaculture, suddenly both industries were being downsized at the same time. The addition of double shifts at the plant in 1999 had resulted in active recruiting in New-foundland, encouraging the migration of over one hundred Newfoundlanders to the island. The plant closure in 2004 reversed the employment situation overnight, and the Newfoundlanders found themselves without work. While some had made their homes on the island, others had returned each winter to Newfoundland. Some might "try it for the winter," but they often decided to return only for summers, while still others sent their children home for school and stayed on until Christmas. The variable patterns of uncertainty and in-betweenness that characterized the interactions of Newfoundlanders with Grand Manan was an important dimension of change that defined community rela-tionships through the period from 1995 to 2006.

Global forces of change had broad impacts upon the island, both economi-cally and socially. As well, the government's role in contributing to the changes increased in all sectors, including tourism, aquaculture, Aboriginal claims to fishing rights, education, and municipal structures. There was a growing sense of unease among islanders that change was both unavoidable and beyond the control of the community. Beginning with the paving of the road in 1947 and the centralization of the high school in Grand Harbour, the gradual erosion of village-based identities was an important change that gained momentum in the 1970s and that reached its apex in the late 1990s. For most islanders, the new connections to mainland institutions were neither welcomed nor understood. As the economy moved from a traditional fishery to salmon farming, as the internet encouraged the growth of small local companies through new export opportunities, as women asserted their educational backgrounds and attained administrative positions, as federal government initiatives encouraged an influx of new Aboriginal residents, and as Newfoundlanders in search of work migrat-ed to Grand Manan, new socio-cultural realities embedded themselves into the lives of individual islanders.

In 1995, there was a spirit of optimism that was tempered by a growing real-ization that governments were imposing greater regulatory authority – authority that was vigorously resisted by islanders whose sense of individual entrepreneur-ship had served them well for generations. The lobster fishery was booming; herring caught by seiners and weirs was marketed as both smoked and canned sardines; and, while groundfish species were becoming more scarce, the devas-tation of stocks experienced by Newfoundlanders did not seem imminent for Grand Manan. Meantime, aquaculture was being established, strongly encour-aged by the provincial government with investments from mainland and multinational corporations. While many islanders expressed reservations and some actively opposed the setting up of salmon farms, the prevailing spirit was one of optimism. Despite a hiccup in the growth of salmon aquaculture when stocks were devastated by infectious salmon anaemia in 1998, new government policies and subsidies in 2000 led to a renewed surge in growth. In 2001, the industry seemed invincible. But the crash of worldwide prices that led to mergers and bankruptcies in 2005 created another major barrier to growth. This time, confronted by the sardine plant closures and declining fish stocks, islanders were not very hopeful about the future. These economic realities were only part of the community dynamics. Between 1999 and 2001, five drug-related deaths created a sense of crisis that even five years later had not dissipated among families with young children.

In twelve years, the island had experienced every symptom and problem that the new globalized economy represented. From optimism to resilient pragmatism to worried apprehension, in the years from 1995 to 2006 Grand Mananers were challenged by social, political, and economic challenges that changed their lives forever. It was a veritable roller coaster of a decade. There were winners and there were losers; but most important was the fundamental way in which the culture of the island was transformed. Even the sense of place that permeates all conversations and attitudes about the island can be seen to be evolving, from an unwavering belief in the values of church and hard work within the wild fishery to more ambivalent understandings of the need for education, the importance of global connections, and the necessity of enhancing options. As Lucy Lippard (1997) explained in her sensitive exploration of "the lure of the local," it is inevitable that we conflate place and community because, although they are not the same, they are bound together so intricately: "Like the places they inhabit, communities are bumpily layered and mixed, exposing hybrid stories that cannot be seen in a linear fashion, aside from those 'preserved' examples which usually stereotype and oversimplify the past" (24). *Tides of Change on Grand Manan Island* is about the changing nature of place and how personal and collective identities are bound up in perceptions of place. The emphasis on community relationships and the discussion of the many implications of globalization attest to the centrality of local-global interactions that affect our sense of being and becoming. The changes that have affected Grand Manan Island may appear to be sudden in the time frame of ninety-year-olds, but they appear to be inevitable in the time frame of twenty-year-olds making choices about their future. All of the complex interactions and new relationships have forged a dynamic that is creating new kinds of boundaries and new meanings regarding what it means to belong to Grand Manan. Personal and collective identities have been irretrievably altered. The borders and boundaries that defined ways of being in the world are being renegotiated.

Organization of the Book

What follows is an attempt to untangle the complex webs of change on Grand Manan Island in ways that illuminate evolving relationships, thus providing some explanations that may eventually contribute to our broader understanding of social and cultural change. While intensive interviews have provided most of the rich and complex information upon which *Tides of Change* is based, reflecting the ambiguity, contradictions, and experiences of the people of Grand Manan in writing has been a challenge. Telling the stories of these people's lives seems to be one way of effectively conveying the complex realities of individual lives. Thus, many of the chapters that describe economic characteristics or social patterns are punctuated by personal stories that become the glue that gives life to more abstract trends and changes. The value in this outsider telling stories may be disputed by islanders who have lived the experiences being described and interpreted; however, ultimately, I believe that it is valuable to preserve experiences that, a decade later, will be transformed into different memories. Stories

> constitute our past, though the stories of our present may be completely revised before being submitted to memory and becoming our past. Not only do stories normally structure human memory, ensuring that the jumble of experience is later re-presented as if it had naturally come to us in narrative form, but memories are not structured this way, taking perhaps the form of typologies or lists which are harder for the human memory to hold. (Game and Metcalfe 1996, 76).

Not only is there value in recording stories and narratives of lived experience with regard to building historical archives, but for me, as an author, there is significant value in affirming the importance of individual lives and taking pleasure not only in trying to discover "truth" but also in trying to discover how we think about our evolving world and local communities. For me, *Tides of Change* has been a passionate engagement, both with regard to the months and years of fieldwork and social encounters and in the photographing and reflecting that have led to my having written it.

The organization of this book reflects both (1) the transition from dependence upon the traditional/wild fishery to modern aquaculture and globalized structures and (2) the ideas that help describe the many changes involved in this transition. The challenge has been to present an effective balance between the daily minutiae of island life and the overarching meanings that provide a framework within which to understand their significance. Two objectives underpin the organization and themes of this book: the first involves illuminating the complex and often contradictory patterns of change over a decade in a small east coast fishing village; the second involves exploring the links between local and global spheres of interaction, between individual stories and universal experience, and between historic ties and current trends. The links between these different levels of being reflect the challenges of ambiguity and the problems a community has in adapting to change. It was difficult, but essential, to see the problems of family dysfunction in relation to wider problems of corporate restructuring and globalization. Throughout the narrative, these two levels of reality are in constant tension, reflecting the lived experiences of Grand Mananers as part of broader global changes.

In order to establish the basic raison d'être for the island community and its long history of settlement, chapter 2 begins with the wild fishery. The diversity and primary importance of the wild fishery underpins all relationships on Grand Manan. Beginning with the earliest fishery – herring – this chapter describes the rhythms of island life that are directly linked to the seasonal patterns of fishing and to the different levels of capitalization, according to equipment and licensing requirements. As a result of historical and current conflicts and negotiated settlements regarding how the Commons of sea and land should be shared, the notion of boundaries is extremely important. The setting and resetting of boundaries occurs with a fluidity that is intricately tied to systems of governance, regime changes, formal and informal community rules, and a complex web of social, economic, and political factors. These boundaries are intrinsic to the fishery but also extend into social and spatial boundaries, for example, between islanders and newcomers to Grand Manan, between men and women, and between parents and teenage children. All of these boundaries evolve and change within the context of community life.

In order to have a framework for understanding the complexity of socio-economic and spatial changes, in chapter 3 I introduce the notion of boundaries, borders, and edges. This is followed by chapters 4 and 5, which explore the tensions and conflicts related to sharing the Commons – something that has become a growing problem for islanders. The introduction of new activities such as the harvesting of rockweed and the rapid growth of salmon aquaculture has created an escalation of community conflicts as traditional activities are seen to be threatened by new ones. Moreover, the fact that many of the new initiatives have been actively promoted by the government rather than initiated by Grand Mananers themselves has exacerbated the problems. Chapter 6 looks in greater detail at the forces of globalization and the impact that economic restructuring is having on the island. I describe the effects, for companies and for the community, of the expanding connectivity of islanders as individuals. The impacts of globalization and the related changes in company structures have had incalculable effects on the meaning of work and community identities.

The next section, chapters 7 through 10, focuses on the social impacts of all these economic changes, with particular attention to identity. The long history of the island and its relative isolation have created strong personal and collective identities that define a sense of belonging. In chapter 7, the role of belonging as both signifier and divisive marker between insiders and outsiders, or between true islanders and "from-aways," is explored in terms of its many manifestations. I allude to the relevance of this notion for my later analysis of resilience, while acknowledging that identity and sense of place are complex attributes that often resonate in contradictory ways. In chapter 8 I explore the fundamental nature of identity in the context of place and community. Historically, the villages located along the eastern shore of Grand Manan have provided the locus of family relationships, contributing to identity formation. I describe the declining importance of the villages and the role of schools, sheds, garages, and camps in creating sense of place. A central aspect of Grand Manan identity, one that has defined people's lives for almost two hundred years and permeates all aspects of community relationships, is religion. In this chapter I also examine how religion informs values, norms, and the choices that people make. It is at the centre of home and community for most Grand Mananers, even as its cultural role is being challenged and increasingly questioned.

Chapter 9 brings together many of the changing relationships experienced in the wild fishery, in new activities, and in the villages, and, in doing so, examines the concept of habitus as a way of understanding the rapidity of cultural change. In particular, the in-migration of Newfoundlanders as a result of labour recruitment by Grand Manan company owners has had a profound impact upon island culture. In chapter 10 I use the notion of resilience, paying particular attention to gender relations and youth culture, to attempt to understand the ways in which the island may or may not be able to adapt to the changes that have been described. I summarize key events that have been crucial markers in the recent history of the island, not only transforming perceptions and relationships but also underpinning the possibility of new directions and an evolving cultural habitus. Finally, chapter 11 "reimagines" the Grand Manan that was introduced in the opening chapter, suggesting some possible futures for the island and expressing the hope that the traumatic struggles of the past decade can be mitigated and used to provide a strong basis for an invigorated community of belonging.

The history of Grand Manan has been defined by its island situation and by the sea. But other environments – the shore and the land – also play significant parts in the interactions and activities that have defined the community's history for over two hundred years. These environments provide the setting for the formation of personal and collective identities. They are zones of dynamic interplay whose blurred edges contain the richness and complexity of peoples' lives. I explore each of these environments with reference to particular recurring themes and issues, such as belonging, the Commons, habitus, and globalization. Central to all of them is religion, which infuses personal identities with both meaning and tension. While these environments are crucial in the evolution of the community of Grand Manan, there is also a series of extremely important events that has precipitated particular responses on the island. These events are incorporated into the discussion and analysis throughout the book, cross-cutting and interweaving with various narrative strands. The most significant of these events include (in no particular order of importance): the arrival of Newfoundlanders in 1996 as both permanent and transient residents; the initiative of the federal government to buy island fishing licences on behalf of several Maliseet reserves in New Brunswick (especially after 2000); the harvesting of rockweed by an outside corporation, beginning in 1996; the restructuring and eventual sale of Connors Brothers (1999-2004); the introduction of aquaculture, especially after 1996 and in 2001; the initiative of the federal government and the Grand Manan Fishermen's Association to open up the Grey Zone south of Grand Manan to a summer lobster fishery in 2002; and five separate drug-related traffic deaths within just two years, between 1999 and 2001, and an additional four in 2004, all linked and culminating in arson, which resulted in criminal charges against several island men in 2006. Potentially important, but having a less direct impact as of 2006, was the 2002 sale of approximately one-quarter the area of the island to a mainland logging company.

All of these seemingly separate and unrelated events are connected to broad trends in the globalization of the economy and to market forces that are driving the changing structure of the agro-food marketing system. These events and the environments within which they are played out are described and explored through the stories and experiences of individual islanders. Ultimately, it is their struggles, successes, histories, and daily lives that define the community of Grand Manan. The stories do not form a linear narrative. They are rich, complex, and challengingly contradictory. There is no simple story. Paradox and contradiction weave together the "goods" and "bads," the stories of young and old, and the meanings for native islanders and for people from-away. The stories of virtually all families are characterized by determination, hardship, and resilience. Ultimately, however, it is the cheerful, self-deprecating humour of one elderly woman that always comes back to me as I try to unravel various meanings. In her search for work in the 1970s, she had approached a local store owner. He asked her what she was good at. She replied: "Well, I'm blind as a bat and deaf in one ear, so I guess not much!" He hired her, and there she worked for over ten years.

2

The Wild Fishery

When our pioneer forefathers had taken care of the immediate needs of shelter and sustenance, they looked about for a means of wresting a livelihood from this bleak rock. ~ (Allaby 1984, 16)

The story of the wild fishery on Grand Manan is both romantic and heroic, but it is also sad and, ultimately, a cautionary tale for governments and communities alike. Incorporating a rich variety of groundfish, pelagics, shellfish, and seaweeds, the wild fishery has defined the culture and identities of Grand Manan for generations. Its deeply embedded socio-economic structures are today being eroded through a combination of misguided policies and unassailable beliefs in the merits of efficient technologies. The shift to an industrial fishery began in the 1950s, with the encouragement and regulatory structures imposed by the government, which provided financial incentives to increased-scale technologies, thereby "marginalizing the traditional seasonality and spatially dispersed location of production" (Newell and Ommer 1999, 8). In this chapter, I describe the traditional and new wild fisheries of Grand Manan during the 1990s and early 2000s.

Much has been written about Canada's resource frontier and the notorious "boom-and-bust" cycles that characterize the towns that grew up across the country due to mineral and forestry production. The forestry sector, in particular, would seem to offer comparable situations with the Grand Manan fisheries because of its renewable character, which might appear to suggest the basis for sustainable development (Clapp 1998). Indeed, Clapp has argued that sustainability in both fishery and forestry sectors depends upon "the volume of the growing stock, its age-class distribution, and the age at which the species reaches maturity" (133). He goes on to consider the many impediments to effective management of both, coming to the conclusion that "free-market solutions are likely to be counter-productive in managing biological resources" (139). The complex interweaving of government management regimes and local strategies, especially as they are implicated in ever more intensive harvesting technologies, are part of the story to be narrated here. Fishing communities, more than forestry communities, are characterized by (1) long histories and (2) dependence upon the individualism and multiple skills of fishing families, both of which have created a culture typified by deep and complex socio-economic relations. As analyses of the Newfoundland situation over the past three decades have made clear, the political and corporate involvement in new technologies and in the restructuring of the global agro-food industry have made comparisons to the boom-and-bust cycles of the Canadian resource frontier increasingly relevant.

Beginning with the two most important fisheries, herring and lobster, in this chapter I try to give a sense of their socio-cultural significance for the island as well as to unravel the changes that are undermining the foundations of personal and collective identities. The narratives are also tied to specific places, boats, and conversations in ways that, I hope, give colour and life to the evolving community of Grand Manan.

Herring: Weirs and Seiners

Although the earliest fishing was by hook and line for mackerel, hake, cod, and pollock, it was herring that, during the nineteenth century, became the basis for a vigorous trade to the United States and the Caribbean. Of all the wild fisheries on Grand Manan, unquestionably the most romantic remains the herring weir fishery. No one is sure exactly when the weir fishery began, although there is a reference to it as early as 1797 and again in an 1849 report by Moses Perley, who noted twenty-seven weirs around Grand Manan (Doucet and Wilbur 2000, 5). In 1827, a New Brunswick law established a box size of eighteen pounds for the sale of herring (Allaby 1984, 16). They were caught just off shore in weirs built of saplings that acted as traps into which the herring swam. Later, the fishers in their dories would "torch" herring, attracting them to the surface by lighting flares over the bows of their boats and then scooping the fish aboard.

Two capture technologies and two distinct markets have defined the herring industry for Grand Manan. The earliest market in the early 1800s was for herring that was salted then smoked and sold as bloaters – a market that continued until the mid-1990s. The second market, which emerged at the end of the nineteenth century, was the sardine industry, which spawned packing plants all along the Maine coastline and included New Brunswick and Nova Scotia. For both markets there have been two methods of fish capture: the weir and, after the Second World War, the purse seiner. In 1804, Wilfred Fisher built the first Grand Manan smoke sheds on High Duck Island, just west of the main island, later moving into Woodwards Cove and Grand Harbour (Gilman 2003). As the mainstay of the island economy, smoked herring were linked not only to the weirs and boats but also to wharves, smokehouses, processing plants, and general stores. The families at the centre of the early businesses – the McLaughlins, Dakins, Gaskills, and Russells – were all associated with particular villages.

On White Head, where families such as the Franklands and Morses had settled and continue to live today, production in the late nineteenth century was so high that it was necessary to import migrant labour during the stringing and boning seasons.[1] In 1884, for all of Grand Manan, over thirty-two thousand hogsheads[2] were harvested, producing about a million boxes of herring that, in turn, were shipped off the island. By this time, Seal Cove had become the centre of the industry, accounting for about 20 percent of the total production. About 175 different smoked herring businesses existed at this time, most being relatively small and selling to the larger enterprises headed by people such as the McLaughlins and Gaskills. Interestingly, some of them joined together to form a cooperative for the marketing of their fish, which "allowed them more freedom in their personal, political and religious lives" (Gilman 2003, 117).

The weir fishery is essentially a "passive" fishery insofar as fishers depend upon the herring coming into shore where the weirs have been built to trap them. For the men there is both the mystery and the frustration of trying to guess where the fish will strike, hoping they have designed a weir to maximize the possibility of good harvests. While a pattern may seem apparent for a few years, inexplicably it will change, disproving all hypotheses discussed in the garages around the island: "Fishermen closely watch the movements of herring related to tides, winds, weather and hours of darkness; they monitor water temperatures and predators. Rowing stealthily in quiet coves on a pitch black night, they sniff the air for the unmistakable sweetish smell of playing herring" (Allaby 1984, 36). The men all agreed on one thing: "There's nothing like a herring to make a liar out of a man" (36). Regardless, Allaby, who himself was involved in the weir fishery, suggests that there seems to be a year/class relationship that results in imprinted behaviour among herring of the same spawning year, with the result that there seems to be "a rough six-year cycle of successes and failures" (ibid.).

The weirs are of simple yet effective design, and their building and location close to shore requires multiple skills. Built in sheltered areas within a few hundred yards of shore, the heart-shaped enclosures are linked to the land by a "fence" (a stretch of twine anchored on shore) that acts as a barrier to the herring, forcing/guiding them into the weir. Once within the weir the fish swim in circles, unable to find the exit. The weir itself is constructed of two layers of poles and netting, known as weir stakes and top poles, respectively, covered by bottom and top twine. The weir stakes – between forty and seventy feet of sharpened logs of ash, hickory, or fir – are bought from mainland sources at a cost, in 2006, of about $300 apiece. During the "building of the weir," the stakes are driven into the ocean bottom by a driver, the width between each stake depending upon the tidal and storm conditions of the particular site. Weirs typically have about one hundred stakes. The created circle is covered by the bottom twine ("dressed") and tied to the bottom of the stakes by divers. The top poles are slender birch poles gathered in the woods during the late spring to replace any that may have been lost to the winter gales. They, too, are dressed or surrounded by top twine, which may be lowered during the fishing season during particularly strong storms.

The history of Grand Manan weirs has responded to new technologies, such as the introduction of nylon twine (mid-1950s) that was both resistant to rot and immeasurably lighter than cotton. The use of scuba gear (1962) and the building of deeper-water weirs (1973) were developments that allowed for an expansion of the industry and an increase in the numbers of weirs around Grand Manan. One of the restrictions on the weirs' distance from shore was the length of available stakes, which were increasingly difficult to find. One solution, which appeared in 2007, involved a machine that would produce what was essentially a sleeve to connect two stakes, thus allowing them to become a long, deeper-water stake.[3]

The names of the weirs themselves indicate the attachment that the fishing community has had to them: Iron Lady, Bread and Butter, Challenge, and North Air all reflect experiences and meanings that have contributed so much to Grand Manan personal and collective identities. Following on the smoked herring industry, and beginning at the end of the nineteenth century, sardines had become an important market for the herring. The suspicion in the spring of 2005 that this fishery might be threatened was met by a mixture of disbelief

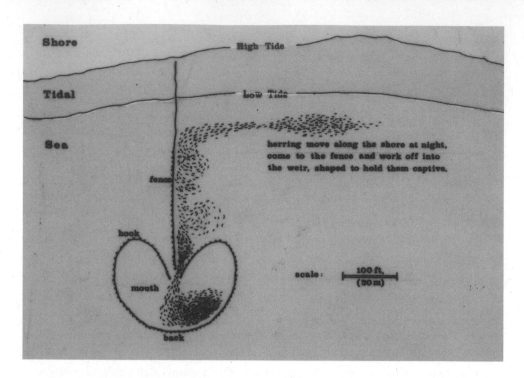

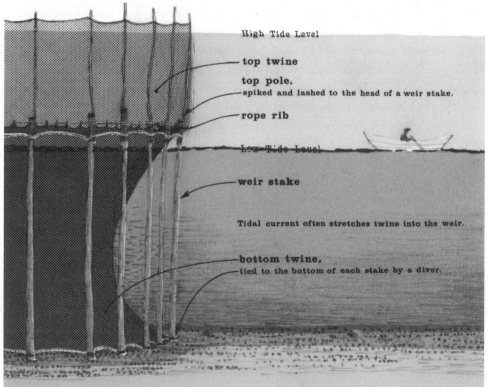

Top: Design of weirs (Allaby 1984)
Bottom: Section of a weir (Allaby 1984)

and real angst. Even as some men were considering building new weirs,[4] and others were driving stakes to repair winter damage to their existing weirs, the concerns for the long-term survival of the weir fishery were quietly being discussed in the garages and on the wharves. For many, there was a realization that old times were slipping away – the camaraderie, shared-family crews, stories of catches, and loaded carriers sailing away low in the water as the gulls swooped in for the remains. The 1990s marked another turning point, associated not so much with technology as with changing consumer tastes and the restructuring of the entire agro-food industry.

The second technology involved in the capture of herring is the mobile purse seiner – large boats that set out for fishing grounds late at night, when the herring rise to the surface to feed. The seine is a large net having acres of twine to enclose the large schools of herring. Relying upon sonar equipment, the captain seeks out the fish on his scanner and releases a small tow boat off the stern, which takes one end of the seine to encircle the school of fish while the larger boat circles from the other direction until they connect. Lead weights take the seine to the bottom, which is then pursed, drawing together the fish until they can be pumped into the hold. The clock says it is 3:00 AM, but no one is sleepy. The excitement on board as the team of seven or eight men work, each with his own task, is palpable not through shouted orders (which are unnecessary) but in the sudden burst of energy as lights go on, pulleys whirr into action, gulls soar and wheel in anticipation of a banquet, and the streaks of silver boil to the surface. When I first came to Grand Manan there was a large fleet of purse seiners, eleven in all. In 2006, with the exception of the *Fundy Mistress* owned by MG Fisheries to service its own feed plant for the aquaculture industry, there were no privately owned seiners on the island. From the peak in 1978, when Vincette Doucette reportedly caught a record amount of herring valued at $1 million (Allaby 1984, 39), the purse seiners suffered through years of declining volumes, and the industry was increasingly controlled by Connors, who built and owned a large new boat in 2005. There was no room for the smaller island-based seiners.

Of the two sources of herring, the purse seiners and the weirs, in 2004 it was the latter that accounted for the major portion of the herring catch (about two-thirds) in the Bay of Fundy[5] and that represented the most historic and embedded of all the Grand Manan fisheries. It is unquestionably a very romantic fishery. Whereas in 1986 there were still seventy active weirs around the island, by 2001 there were only twenty-six, and five years later there were only twenty. One observer described the weir fishery as being "in a state of crisis; it may cease to exist." Nonetheless, in 2005 and 2006, weirs were being rebuilt, including two behind the seawall in Dark Harbour and two that were realigned in Whale Cove. In 2007, the weirs took catches that, according to most sources, exceeded any of the previous decade. There was clearly still a great deal of interest and passion for the industry. For the approximately fifty men and women who have shares in the twenty weirs around Grand Manan, there is always much suspense around the arrival of the first herring of the season. In the decade since 1995, the herring have tended to strike at the north end of the island more consistently than they have at the south end, and in later years they have arrived on the back-of-the-island before the east side. In 1998, one of the earliest weirs to be seined was the Indian Beach weir (on 28 June), whereas Whale Cove had still seen nothing by that date. Some years fishers have waited until late July before herring find their way into the weirs. Harvests, too, vary. Brown's weir (the Mystery) captured only nine hogsheads in 2002 until one memorable series of three consecutive nights, when two thousand hogsheads entered the enclosure. According to Greg Glennie, the manager at Connors' main office in Blacks Harbour, 2002 saw the worst harvest in sixteen years, and 2003 was not much better: "But the herring are out there; they're just not coming in." Nonetheless, despite a meagre 324 hogsheads in 2003, Brown's weir took in about 2,600 a year later. When the families go south for holidays in February the community is usually correct in guessing that the weir harvests were lucrative. As I explain later, the problem for weir families involves balancing the costs associated with preparing the weir each spring against the risk that those costs may not be covered by the unpredictable herring runs.

In 1989, when I first visited the island, there were about a dozen smoke sheds still operating at Woodwards Cove and Seal Cove. But by 1995 there were only two, and they finally closed in 1997. In describing her days in the boning sheds, one eighty-six-year-old woman said: "I loved the boning. I started when I was eleven or twelve years old, standing on a step. Every night we'd make 'cots' [like sewing thimbles] for our thumbs so that the bones wouldn't pierce us. The stringing was hard work. We made twenty-five cents for 100 sticks." The women talked, sharing their joys, problems, and struggles. The proximity of home and shed, and the lack of automobile transportation, meant that mobility was constrained and that tight circles of neighbours became "family." The sheds were an extension of home; livelihoods in herring represented both the productive and reproductive as well as the community-building aspects of daily lives. Long before the purse seiners, the weirs created and sustained island life.

Smoke sheds that, until the 1970s, were nestled in virtually every cove and bay around the Grand Manan archipelago, illuminated a relationship between strong herring harvests, good markets, and a social structure defined by the demands of the industry. The 1997 closure of sheds owned by John Ingersoll in Woodwards Cove and Ron Benson in Seal Cove was caused by a shrinking market for smoked herring. For the women especially, the closure of the smoke sheds was of particular importance since they lost not only a steady income but also significant flexibility and social relationships that nurtured individual strengths and community solidarity. In 1989, the Breakwater at Seal Cove had bustled with the activity of men and women working at their respective roles. These began with unloading the herring from boats into brine, salting the fish for four to five days in large tubs that held three and one-half hogsheads (about 4,200 pounds) of herring. Then the fish were moved to the "stringing sheds," where women threaded fifteen to eighteen fish through their gills onto thirty-eight inch spruce sticks, which the men than laid onto "horses" (trolleys or

Smoke Sheds, Seal Cove, 1989

Stringing Herring, Seal Cove, 1993

wheelbarrows) and pushed to the smoke sheds. The women were paid by bundles of one hundred sticks, at a rate (in the 1990s) of eight dollars per bundle. While many women were able to fill their bundle of one hundred sticks within an hour, others took one and a half hours to earn their eight dollars. Many women preferred to work at stringing herring rather than packing fish at the Connors sardine plant because it would give them the flexibility to adjust to their children's school schedules.

From the horses, the men passed up the sticks, weighing about twelve pounds each, to the ceiling of the smoke sheds (this was called "baying up"), making several layers of hanging fish that were to be smoked by smouldering fires below. Fuelled by spruce, salted wood gathered from the beaches, and sawdust, the fires were carefully tended by one of the men to ensure that they dried the fish sufficiently in the crucial first few days so that the gills did not rot and allow the fish to drop off. Certain men were known to be especially good at keeping the fires at exactly the right temperature and thus giving off just enough smoke. The entire drying process took from three to six weeks, depending upon the weather. For the women, the easier job of filleting, or "boning," and packing the fish into wooden boxes was the final step in their season's work with herring, usually beginning in mid-November. As one woman said: "We liked the boning, but we weren't allowed to do it unless we'd put in our time with stringing first." The large herring were called "bloaters" and were packed in eighteen-pound boxes, while the smaller fish were boned and packed in ten-pound boxes. Most Grand Mananers over forty years of age had worked "in the herring" at some point in their lives. Recalling their weeks of boning, the women invariably smiled and chuckled as they told stories of the frivolity involved in this shared activity.

While the demise of the smoke sheds was mainly linked to declining markets for smoked herring, in 1996 islander complaints also pointed to the increasing intrusion of government regulations, which Grand Mananers viewed as being slightly ridiculous. In the interest of food safety, new regulations required them to have steel handles on the scoopers and to change from wooden barrows to steel barrows; and yet, as they pointed out, there seemed to be no problem with wooden sticks, wooden braces in the smokehouses, and wooden boxes for packing. One woman was unequivocal in her belief that it was these regulations that had caused the decline of the industry: "It was the government saying we had to have everything steel. It was too bad." In 1996, I stood with fifteen workers (both men and women) and tried the stringing for only two hours, discovering how difficult it is to hold a thirty-six inch stick at one end bearing about twelve pounds along its length. My total over the two hours was fewer than thirty sticks, while the tiny seventy-five-year-old woman who stood next to me chatting happily as she worked (from 7:00 AM to 3:30 PM) was not much better. I calculated that she would be earning about three dollars per hour.

Today, the cultural family-shed-village relationships continue only in memories and, with the closing of the last smoke stands in 1997, are gradually ebbing away. As recently as 1996 at Woodwards Cove the sign on one of the sheds proclaimed the ownership of RUSSELL & SONS. Gesturing around the wharf area, Hovey Russell, who died in 2002, said nostalgically, "I used to own all this." One of his sons, Bernard, emerged from his bait shed in high rubber boots, reeking of the rotted fish that he was using as lobster bait. Agreeing to stand for a photo, he commented that, because of the loss of the herring stands, the Woodwards Cove wharf was no longer as busy as it used to be. In 1994, a feed plant for the aquaculture industry had opened under the ownership of a mainland businessman. It was sold in 1998 to MG Fisheries, the largest fish broker on the island. The plant was modernized and integrated into MG Fisheries, which already had its own supply of herring provided by its purse seiner, the *Fundy Mistress*.

Woodwards Cove, 1989

The other sheds in the Woodwards Cove wharf area were also changing their functions. In 2002, a mainlander established a business whose purpose was to capture and pack eels to be shipped to Korea, where their skins would be used for leather goods. A year later it was an American entrepreneur who had taken over the premises for a new company, Fish-on-Bait, that shipped individually packed herring to the recreational fishing industry on the American gulf coast. With the demise of smoked herring, new businesses came and went, but these had hardly any connections to historical island norms and practices. Although these businesses provided a few jobs (mainly for women), in fact most Grand Mananers did not even know of their existence. The complex social relationships that defined the smoked herring industry over generations were gone, except in lingering memories and, to some extent, in the continuing status of certain families.

Beginning in the late nineteenth century, as well as smoked herring, sardines were an important market for Grand Manan's herring. In the early days of the sardine industry, Canada was mainly a supplier for US-based packing plants. Relying mainly upon Canadian supplies of herring, the Americans wanted to import the raw material (i.e., herring), and they did not want competition from canning plants in Canada. The result was a protective tariff on tinned sardines, which "effectively closed their border to Canadian products" (Gilman 2003, 1). Canneries grew up in thirty-three coastal communities along the northeast seaboard, employing about six thousand workers at the end of the nineteenth century (*Telegraph Journal*, 10 May 2005, C3). As a result, it was not until the completion of the railroad to central Canada in 1876, with a shorter American route through northern Maine (called the "Short Line") completed in 1889, that new markets were opened up and that sardine factories could open up on Canada's east coast. Canadian workers who had been employed in the American plants returned to New Brunswick to work in new factories that sprang up along the coast through Charlotte County. Started in 1893 in Blacks Harbour, Connors Brothers grew to become the largest producer of sardines in the world. Ensuring a steady supply of herring was crucial to the success of the factories. Because many fishers could not afford the cost of building a weir to capture the herring, their expenses were underwritten by the company, which then had exclusive contracts for all of their fish. As Gilman (2003, 5) points out in his history of the industry, "The lyrics of the old song, 'I owe my soul to the company store,' were true, and many fishermen spent their entire life in debt to the owners."

While most weirs were owned by islanders[6] by the middle of the twentieth century, the company continued to exert powerful constraints on the fishers. In order to prevent competing US companies from buying weir fish, Connors insisted that their agreement to buy fish was contingent on weir owners not selling to anyone else. Thus, a bidding war was always avoided. Indeed, as many fishers reported, they did not even know the price they would be paid until after the fish had been delivered. Price was dependent upon the size of fish (smaller paid better) and upon whether or not the fish were "feedy." Fish that had not cleared out their feed had two problems: first, they deteriorated much more quickly, and, second, they oozed acidic liquid that irritated the skin of the packers, leaving them with open sores on their wrists and lower arms. Connors Brothers (owned by Toronto-based Weston Corporation) determined whether they would buy the fish, how much they would pay, and when they would accept delivery.

The impact of the company on the community of Blacks Harbour was also reflected in the physical layout of the town itself. A drive through Blacks Harbour, the location of the head office of Connors Brothers and where the sole remaining Canadian Connors Brothers plant continues to pack sardines in 2008, provides reminders of the enormous influence of the company that built houses, stores, and recreational facilities to accommodate its workers. The company town has not only been part of the mining and forestry sectors but it has also had an important impact upon the lives of fishing families and plant workers. The relevance of the company town syndrome became apparent with the closing, in December 2004, of the Grand Manan Seal Cove plant and the uncertainty, in 2005, about the long-term viability of the Blacks Harbour plant. Major changes in the sardine industry, brought about through restructuring in the global agro-food industry, were having a devastating impact upon the economy and, ultimately, the culture of Grand Manan. With the loss of demand for smoked herring, the sardine industry had become the main market for the weir fish. In the spring of 2005, it seemed that even the sardine industry was threatened. One of the Grand Manan fishers remarked: "It's quite a thing when the hope for the island is weirs, and it's herring for bait!" And yet just two years later, some islanders continued to be optimistic regarding new markets for herring as salmon feed and lobster bait, both of which were growing.

The weirs create a social milieu based on the historic reticence of fishers to share news of their good fortunes and on the characteristics of the weirs themselves. There are, for example, specific site characteristics known by the island men, with respect to the nature of the bottom (e.g., rocky, muddy, sandy), exposure to storms, prevailing winds and strong tidal currents, and historical flows of herring in relation to the shape and direction of the rocky shoreline. As a result, from time to time the direction or angle of the "fence," or "wing," of a weir will be modified in the hope that better herring catches will follow. When a weir at Money Cove was to be rebuilt in 2002 for the first time in several years, the man Connors charged with the precise siting and design (Jeff Foster) consulted with eighty-five-year-old Bill Bass, who had fished at Money Cove throughout his younger years. Bill was viewed as the expert on how the herring behaved along that stretch of shoreline. Catches over the next two years seemed to confirm his ideas. In 1999, another weir, with a much less favourable history, was rebuilt after many years east of Ashburton Head. Known simply as "Albion's weir," it would cause men to shake their heads in bemused wonder that anyone would attempt to use that site. Wide open to the hazards of storms and strong tides and with a very sandy bottom, the stakes required special "doughnut" reinforcements for support and had to be narrowly spaced. As one of the experienced herring weir owners explained, "It's hard to drive, and they have to be blown out at the same time as they are driving in the stakes." The plan to put stakes eighteen feet apart was "too far," according to one opinion:. "They should be only eight to ten feet apart if they're to withstand the easterlies." As one experienced fisher explained, if there is even a small amount of "grassing up" on the twine any strong winds will topple the stakes, quickly causing a "dominoe effect." It was expensive to build and very precarious, and in two years of operation it never captured any herring. People's reputations on the island become linked to such ventures. Likewise, Buddy McLaughlin's experimentation with various enterprises, including a plastic ring weir in Dark Harbour, was the subject of many chuckles and shakes of the head. Nonetheless, by 2003, well-respected fishers had placed three additional rings around the island, all of which began to show success at capturing the elusive herring. As one of the carrier captains explained, the rings have their twine hanging from them instead of being anchored by individual stakes, and therefore it is difficult to ensure that the enclosure is properly anchored in the deep water. The real acceptance of their value seemed to be marked by their being given names, such as "Eagle Rock." The plastic rings had finally acquired some legitimacy, and gradually the idea of non-traditional weirs was no longer the butt of jokes in the garages.

Eel Brook weir

The crews who work together in the weirs tend to be brought together through family ties and often include a range of ages, from about seventeen years to almost eighty. Relationships within families and between generations create important social solidarities that are missing in the crews who work on salmon sites. My notes for one weir harvest in 1997 illustrate the kinds of working relationships and social networks that have been important to shared understandings in the community.

The three boats arrived at the Indian Beach weir, while the carrier waited just outside. As one of the skiffs circled the weir in a clockwise direction, the seine boat let out the seine net. The excitement started as the seine was being pursed. Bubbles moving around the circle tracked the diver who was keeping the net clear of jagged bottom rocks, while the men on the surface pulled their main (seine) boat slowly around the weir. Then suddenly, Jeff Naves (at seventeen years the junior man in this crew) sped around in tight circles to keep the fish away from the left wall. The real tension rose with shouted commands to "let off some rope," "pull in over there." Then Stephen called out rapid commands as the seine was pursed and brother Nathan (the diver) surfaced under the billowing net. The tide was strongly moving northward away from the carrier, and there was a danger that if it strengthened they wouldn't be able to gather in the pursed seine. Then the tackle started as they began to roll in the net. After a few minutes they decided the tide was too strong, and they would have to wait for it to turn. Everyone settled down; one man in a small boat batted at horseflies. Jeff stood stoically in the yellow dory watching as occasional small schools of herring wriggled to the top, flashing and jumping. Wayne Tucker in the main boat soaked his long-sleeved shirt in the water and put it on as a salve against the searing mid-day sun. David and Stephen (Bass brothers), in their full-length oil-skins, admitted they felt they were in a mobile sauna. When Stephen felt one of the ropes and sensed a slight change in the tidal drag, Nathan was dispatched down to check. A tug of the rope told the men that indeed the tide was turning. Nathan resurfaced and climbed into the boat. The net was pulled in, the circle narrowed, until there was a frothy sea of tiny flashing figures, still swimming in circles, occasionally interrupted by a single dogfish. Just before the captain of the carrier plunged in the black corrugated vacuum hose, he leant over and scooped up some herring, which he dumped into a bucket. He measured them, and then slit them open, checking for feed. Seconds later the box chutes

The Intruder weir

are swooshing with the tumbling silver fish being dumped into one of the
two twenty-nine-hogshead holds of the carrier.

During another seining, the work of the skilled divers became apparent.
While all weirs require the help of divers during the building and the tying on
of the twine, some weirs require their help throughout the season due to the
strong tides that can snare the seine nets on rocks. Occasionally, there are other
challenges, such as unexpected stakes that have sunk or trapped dolphins or
sharks. In 1998 and 1999, the Bradford Cove weir "caught" several unwanted
creatures. When the diver went down during a routine seining in 1998, he discovered an unwelcome presence. "Something huge; probably a shark," he reported. "Where's the herring?" he was asked. "Over there. Drop the net there,"
he advised. Meantime, his job was to chase the shark out of the weir, which, of
course, was shaped so as to ensure that fish stayed inside it. With great admiration, the men describe how Lennie Hinsdale has been able to get rid of mud-
sharks, basking sharks, and even, on one spectacular occasion, to help with the
escape of two right whales. Divers are an important part of the weir team, who
must not only have expertise but must also be able to adapt to a variety of environmental conditions not faced by the aquaculture industry.

Unexpected hazards often challenge the weir fishers to find creative solutions, demanding patience and perseverance of the entire team of seven or eight. The techniques vary, and whereas some captains ease their way into the weir, gradually establishing the location of the herring and figuring out the strategy for drawing the net around them, others will charge in, hoping to thus ensure that the herring collect in the opposite side of the weir. However, the same captain who was described as a real cowboy with regard to his brash approach showed remarkable patience when confronted by a problem that meant doubling his work time in the weir. As he began to pump the herring into the hold of the Connors carrier, it became apparent that the vacuum hose was becoming plugged, with the result that the motor had to be constantly reversed to blow out the contents and then be restarted. The problem was that the hose was small – six inches in diameter – and in among the herring were huge numbers of lumpfish, shining a beautiful blue but round rather than thin and sleek like the herring. They clogged the hose, with the result that it took a long time to seine the weir. For weir fishers, time can be crucial since the seining has to occur within specific tidal periods, usually just before low tide (when the water is most shallow) and before it starts to turn, so that there is a slack period that allows for greater control of the seine nets.

At the beginning of the herring season, the men usually hire divers to scour the bottom, checking for sunken weir stakes or other obstructions that might jeopardize the season's seining. Any broken stakes, the ten-fathom (sixty-feet) poles that are the main structure of the weir, are replaced at a cost of $200 to $400 each, depending on the type of log and where it is from. Light softwoods cannot be used because they tend to float, whereas dense hardwood hickory has become increasingly difficult to find. The stakes may come from as far away as South Carolina or British Columbia, although most commonly they have been purchased from Ontario loggers. After the stakes have been trimmed, measured, and pointed, they are rafted and floated to the weir site. The stakes arrive during April, to be arranged along the large work area at North Head where the men work on them. Meantime, the top poles have been gathered in the woods, usually young birch and aspen that are twelve to eighteen feet high, ready for being nailed and tied into place on the top layer of the weir. By mid-June the two layers of nets (twine) will have been repaired (if not the autumn before) and readied for tying onto the stakes and poles once they have been built into the weir. The entire process of preparing the stakes and top poles, driving the stakes, and tying on the twine costs the men between $10,000 and $30,000 each year, although the costs can be much higher if severe storms have wreaked havoc with the fragile structures during the previous autumn. In 2005, one of the men was away during heavy northeast winds, which are among the worst for Whale Cove weirs. While other fishers were able to get the top twine down from their weirs, allowing for freer flow of the wind and water, he was not able to do so and, as a result, suffered enormous damage to his weir over one October weekend. The loss of a significant number of stakes and top poles can mean a major outlay the following year, as was the case with this man. With the average cost of building the weir each year at about $15,000 to $20,000, it is said that, including the cost of labour for the actual seining, they must harvest at least three hundred hogsheads "to break even."

As well as having to build the weir, fishers must be concerned with potential problems with changes to the bottom of the heart-shaped structures. At the beginning of the season, at the same time the divers are hired to tie on the bottom twine to the stakes in the first step of "building the weir" they also check the bottom for broken stakes or other obstructions that might cause problems later in the season. It is not unusual, during the first seining, for men to bring up unexpected and unwelcome surprises, such as sunken stakes. When this happens, there is a flurry of careful activity aimed at releasing the stake from the nets without tearing them or losing any herring that might be trapped in them. It is a delicate procedure fraught with tension and specific shouted instructions. The risk is losing perhaps $4,000 to $5,000 worth of herring (thirty to forty hogsheads) and having to repair seine nets that may be needed again the next day. A lot is at stake, and skilled teams are crucial to the working success of any weir fisher.

The skills are also apparent when unexpectedly large numbers of herring fill the weirs, making sales, or finding "market," more difficult. In some cases there may be several weirs all with herring and only limited packing capacity at the fish plants. In this case, rather than buy all its product from one fisher, Connors will spread the market over several weirs, with the result that a weir may not be able to sell all of its herring at the same time. This means that it is important to be able to divide the herring effectively so that none is killed in the process and that the amount arrived at is just what Connors has agreed to buy. It is a delicate and very skilled operation that takes time and patience. There is no one better at this than Neil Morse and Bud Brown, whose two thousand hogsheads in a single harvest meant that the harvest had to be divided into three.

For most island herring fishers, in both the weir and seiner sectors, the herring harvests represent only one part of the annual income. Rhythms defined by the seasons involve participation in lobster, sea urchin, and groundfish fish-

The Mystery, dividing the harvest

eries as well as the gathering of dulse and clams and periwinkles. The unpredictable harvest is an overwhelming reality that, every few years, threatens to drive men out of the fishery altogether. In speculating on the long-term viability of herring catches, some of the men focused on the incursion of salmon farms, but others pointed to overfishing, especially in spawning areas. "The spawning beds along the Nova Scotia shore have been ruined," said one man, and "there's a real danger we'll lose the Upper Fundy areas too. Seiners go in there, but the men know it's not a good idea. DFO [Department of Fisheries and Oceans] has been told, but there's no regulation." Another factor cited in the lower herring catches was the loss of cod, which "used to chase the herring in." Some years the challenge is in the unexpected interruption of the seasonal rhythm, as in the winter of 2006–07, when herring were spotted around the island and even became trapped in some of the untied weirs. Both the Jubilee and Winner weirs took so much herring that the men scrambled to take advantage of the New Year's gift. The team with shares in the Winner in Whale Cove was able to tie on enough twine to seine the weir, capturing, according to reports, over two hundred hogsheads. Their harvest was sold into the American market for bait for the lobster industry since the fish plants were not open. But for most weirs, despite the appearance of the herring that winter, the problem was that men were already committed to their other fisheries – lobster, sea urchin, and scallops. As one man said, "You've gotta be ready to go 24/7 when they come in like that," and usually it is too difficult to find an available crew of seven men. Moreover, the winter gales from the north and northeast can make harvesting almost impossible.

The transitions between the various activities within the traditional fishery are determined by a variety of factors related to both government regulations (seasons and quotas) and degrees of family proprietorship with regard to capital and licences. For example, one lobster fisher had decided not to build his weir in 2004 because of low catches over several years; instead, he rigged his boat for halibut trawls, a fairly recent fishery on Grand Manan. His frustration at the very low government quotas for halibut (early season was 500 pounds per week and later season 250 pounds per week) was palpable as he and his son worked to clean some nets and fix equipment one early July day. With other groundfish, such as pollock and haddock, increasingly difficult to find, he wondered what other fishery might allow him to keep working during the summer months.

As I discuss in greater detail in the next two chapters, government regulations not only determine quotas and seasons as a way to control harvests but they also define fishing boundaries at many scales. Weir sites cannot be less than one thousand feet from each other (always a contentious and difficult measurement to agree upon), and the territories within which seiners may operate are also strictly regulated. With regard to the weirs, which are totally dependent on the fish swimming to shore at night and coming to the surface to feed, there is a mystifying and often frustrating unpredictability that challenges even the most successful of the fishers. Many of the weirs have been in the family for generations, with some changes in partnerships (distribution of shares) evolving over the years. The Mystery, for example, originally built in 1936 by Floyd Brown, is now "owned" in part by one of his sons, Bud, who has two of the seven shares. The other five shares are held by Neil Morse. Many of the formal regulations that govern today's fishery, based on decades of understandings among family groups, were formulated in the 1970s. At that time, the Department of Fisheries and Oceans (DFO) established rules whereby the families might retain claims to the weir sites as long as the weir is built each year. They allow three consecutive years of non-operation, after which the family retains right of first refusal, although anyone can then obtain rights to the site. There is an annual fee of $100, collected by the federal government, for the "privilege" to the site. With the rapid growth of salmon aquaculture, a factor that has emerged as particularly important in the negotiation for rights to weir sites, the sites themselves have been subject to rising market value. In effect, the introduction of aquaculture and government regulatory structures have combined to create a new market and to affect the value of the sites. Since the family rights of access are guaranteed only for as long as they build and operate the sites, the marketability (and monetary value) of the sites has a limited life for a specific family. For weir owners who decide not to build their weir for three years, the only value that remains is in the twine that has been measured and fitted precisely to the size and shape of each weir.

One interesting impact of technology upon the weir sector has involved the process of negotiating markets for the herring. The reliance upon cell phones has changed the nature of communications, which have become more private, allowing for speedier negotiation with Connors. This has occurred since the summer of 2003, when digital technology (as compared to analog technology, which was vulnerable to personal scanners) was inaugurated on the island. Traditionally, the early morning ritual of "feeling in" for herring was followed by a hurried call (usually from a boat phone, which could be and often was monitored) to Connors to negotiate a market for the fish. These pre-dawn phone calls conveyed the expected amount of fish, to which Connors responded with its indication of how much of the harvest it would buy. Money would be discussed only after the carrier captain who arrived to take the herring had checked them for size and feed content. Establishing a market for the fish in the early morning hours over a boat phone meant that any and all interested fishers quickly learned of the day's catches and activities for the various crews. Today, the advent of digital cell phones, which cannot be monitored with scanners, has meant frustration for many, especially those on the margins of the weir fishery whose involvement never seems to be as lucrative as they had hoped. Any information is speculative unless the weir fishers decide to give it out themselves.

The negotiation to buy depends upon several factors: (1) whether it was a Connors weir, in which case it has priority; (2) how many other weirs had captured herring and in what quantities; and (3) how much plant packing capacity was available. Occasionally, a weir might have to wait a day or two before it could be seined, although usually, with regard to modest amounts of thirty to sixty hogsheads, the total catch would be accepted. The necessity for a crew to "split-off" herring into several lots always entails both greater skill and coordination among the crew and a great deal more time than is required to simply bring in the entire haul. The danger in splitting is that the fish might be smoth-ered and thus rendered useless as sardines. The prices paid are determined only after delivery to the plant and are related to (1) the size and uniformity of size of fish, and (2) whether or not they contain feed.

Asked about the patterns of the herring runs from year-to-year, the men inevitably shrug, smile, and throw up their hands: "Who knows?" One fisher was quite philosophical: "It doesn't matter how many records you keep, you can't guess where the herring will be." On the other hand, questioned about salmon farms, the response is less casual but no more enlightening. As salmon sites began to proliferate in the mid-1990s, conversations frequently turned to possible implications for the wild herring runs. At first, people were willing to make links based on their experiences in the immediate past. The most com-mon suggestion was that herring would avoid areas "downstream" of salmon sites because they are known to avoid their own dead. As much as fishers like to speculate on the causes and patterns of herring runs, they tend to be quite cir-cumspect in attributing any direct feed-related impact of salmon sites to specif-ic herring catches. On the other hand, as I describe later, the advent of salmon aquaculture has had huge repercussions on the herring industry in other ways.

Whale Cove

While the wharves and the smoke sheds have been the working centres for the herring industry, one place in particular represents the coming together of fish harvests, economic realities, and the stories and myths regarding social rela-tions on the island. Whale Cove, located at the northwest end of the island and nestled between bustling North Head and the cliffs along Seven Days Work, is a cove of quiet beauty and artistic elegance, rich in history and stories, and the centre for early morning detectives trying to elicit information about the weir harvests. Knowing who had caught herring, how many, and whether or not there was a market for the fish were all important topics for speculation, ques-tions, and listening. On the stoop of the twine shed in Whale Cove, called the "office," one man would sit whittling away at a stick. He would be joined later by someone else, who would sit beside him looking out at the fog and listening for the sounds of gulls around the weirs. Then another man would wander down, stand, nod to the two seated men, and comment on the winds or the baseball game the night before. Silence. Then the speculation might begin.

ARTHUR: "I wonder what they got in the Jubilee."

BILL: "I hear they got a bunch."

ARTHUR: "Oh yeah?"

BILL: "Well, according to Herbie, who heard from Peter, Blain, and Stacey, they say there's a bunch."

ARTHUR: "Enough to take out?"

BILL: "I guess."

Ten minutes later, Don, who has shares in the weir with Blain and Stacey, drives in. Arthur and Bill speculate as to whether or not he will admit to anything. After a few minutes, Don, in an ironed shirt and tie and grey pants, ambles over to Arthur and Bill swinging his cane. Greetings all round.

ARTHUR: "Good take over at the Jubilee?"

DON: "Seems like it."

BILL: "What d'ya think's there?"

DON: "Well, I'll tell ya. And it was at high tide, mind you. Seems like there's a good bunch."

ARTHUR: "Well, that told us a lot!"

BILL: "You were there with them, weren't you Don?"

DON: "Yup. Well, and I dunno what it might say at low tide, but there might be fifty [hogsheads]. Yep. And I guess maybe more."

Meantime it has been established that the "drop" is up, which means that they don't want to lose what is there. They will be seining the next day.

On another day in 1998 in the month of June, which is considered early for the herring in Whale Cove, news of herring in several of the weirs generated enormous excitement among the men. Even an outsider could feel the thrill and apprehension as the men from the Winner weir raced to get their bottom twine on, knowing that other weirs had caught huge amounts the night before. The previous year they had had the best year since 1986, and there were high hopes for a repeat. One of the shareholders, who had been able to clear over $7,000 in the three-month season, was overheard saying to one of his partners that they were lucky that they were "actually paid to do this!" While one of the wives was less enthusiastic because the herring had arrived on graduation day at the high school, the men were all smiles as they pushed the trailer load of twine down the ramp. Grumbling about pesky kayakers who were taking up valuable space, and mumbling loudly about tourists' cars that were parked too close to the launching area, the men kept moving. The crew included men whose ages ranged from mid-forties to late seventies, plus a diver who was hurriedly called to check for stakes and other obstructions along the rocky bottom. Apart from the crew with shares in the weir, other islanders gathered to watch the action. One roared off in his truck to collect his eighty-year-old buddy so that, by the time the carriers arrived in Whale Cove, there were more than ten observers, all speculating on the catches according to the time it took to pump out the seine net or how far into the water the carrier sat as it pulled away from the weir. One of the men estimated that they pumped out at a rate of about two hogsheads a minute. Someone else pointed out that "the guardrail is down" (i.e., the hull in the water). One carrier seemed to have taken only twenty hogsheads on board, while the *Michael Eileen* took about fifty-seven. For everyone, the 21 June harvest was a celebration.

Another day, in mid-July 2000, Paul Brown arrived with his dory on his truck, unloading it with the help of a couple of the men who were there to check on the weirs. He had been doing some repairs and was concerned that he might not have enough paint for the inside creases. The men discussed the relative merits of Moore-Clarke and Sherwin Williams, the costs of paint, and agreed that quality wasn't what it used to be. They turned their attention to comparing the boats pulled up on the slipway, focusing on the flare on the bow of one of the boats. They noted that it should be higher in the prow but concluded, "What can you expect, it wasn't built by Robert Ingalls." According to the men, Robert had built 110 boats on the island but now was retired. Then Blain drove in and asked if Stacey had been around. He checked the motor on his boat and noted some dripping water, indicating that Stacey must have been out earlier. They all commiserate about the lack of fish. "Albion got ten hogsheads, but he's not tied down so maybe he lost 100!" Someone else suggested that there might be six hogsheads in the Eel Brook weir: "They say there's twenty-five in the North Air [at Dark Harbour]. Lennie says Wayne will put the twine on at Bradford Cove on Tuesday." Someone else drives in, leans out of his truck, and joins the speculation, complaining that "there's no gulls around." Bill has an idea about the gulls. The dumps were recently closed and maybe the gulls have gone to the mainland. Paul Brown leaves, thinking maybe he'll try to get some dulse "even though the dulse tides are dead" and "they're giving easterly for the next two days," which is not good for drying dulse. The mood is gloomy. Don arrives and, more optimistically, says that he has been checking his books and that, in 1993, the fish didn't arrive until 18 July, at which point the fishers went on to take over 1,400 hogsheads, the best year ever. Bill remains unconvinced, arguing that they need "three steady weeks" to break even and that that doesn't look very likely this year. Then the arrival of Robbie Young brings news that they got herring in the Mumps weir off White Head, which brings a measure of cheer to the small group. Occasionally, the gloomy atmosphere is lightened by a dose of wry humour, as when Herbie opined that "the fishermen are an endangered species" and suggested that they get in touch with Sheila Copps, the then heritage minister, and ask her to make the whole island a heritage site.

Compared to the casual but purposeful information-seeking gatherings in 1995, in 2006 only a couple of the older men could be found at Whale Cove on a Saturday morning. Changing popular tastes for herring, smaller markets, fewer fish, and the retirement of the older men from the fishery all conspired to make these mornings at Whale Cove a solitary time for personal observation. Gone were the gatherings of men speculating on the catches; gone were the flurry of skiffs being pushed down the slipway to get out to the weirs early to check for herring; and gone were the exchanges of non-information that characterized so many of the gatherings at "the office." These exchanges of guesswork and local gossip constituted a community of mutual understandings that provided a basis for a sense of belonging and collective identities. As the weir industry becomes threatened by corporate restructuring, so are the social and cultural meanings associated with it undermined. First it was the smoked herring; now it seems to be the sardine industry whose packing plants may be closing. As I discuss in chapter 6, the corporate restructuring of the Weston Corporation may represent a further erosion of livelihoods should the sardine plants close. This could herald the denouement of Grand Manan culture as it has been sustained for two hundred years.

In the meantime, the prospects for the weir fishery are mixed. On the one hand, as of 2007, the men seemed optimistic that prices were increasing because of the continuing healthy lobster harvests, which require large amounts of bait, and because salmon aquaculture requires herring for the wet feed that is a basic component of the feed given the smolts for their first few months in the marine environment. On the other hand, there was concern that the fish plants that have traditionally operated on the island and at Blacks Harbour would no longer be packing the herring; that, instead, the restructuring throughout the global agro-food industry would involve changing the location of sardine packing from Canadian plants to places in Africa or China, where labour costs are much lower. The herring would continue to be caught, but the processing would not be done in New Brunswick; rather, the herring would be frozen and shipped out, and Connors would control even more of the process through its ownership of both weirs and seiners. In 2008, such speculation has been been muted by the prolific 2007-08 winter harvests of seiner herring, which allowed the Blacks Harbour plant to freeze about sixty thousand pounds for packing in the early summer months, when regular harvests might be problematic.

For one optimistic entrepreneurial fisher there continues to be opportunities in traditional activities on Grand Manan. After twenty years in the salmon aquaculture industry, in which he had invested both time and money, Wayne Green felt forced to sell out to the larger companies during the restructuring of 2005-06. Looking for new opportunities, he decided to invest in the two most traditional activities on the island: gathering dulse and catching herring. He argued that the prices for herring were rising and that, because of the strong demand for salmon feed and lobster bait, there was no sign that they would come down. He bought two sheds in Woodwards Cove; he obtained several weir privileges along the back of the island, one of which was operational in 2006 at an old weir site, the North Air; and in 2007 he was searching for a drying facility for dulse somewhere "up the island," where drying was more possible than in fog-shrouded Grand Harbour and Seal Cove. Even as many islanders were bemoaning the loss of jobs at the fish plant and on aquaculture sites, he was creating new opportunities for himself with typical Grand Manan entrepreneurial optimism.

Handlining: Groundfish

The historic importance associated with herring is only slightly more significant than that associated with groundfish. While hake disappeared many years ago, pollock, haddock, and cod continued as key species in the diverse fishery around the island until the early 1990s. In the nineteenth century, hake was a major export from North Head not only as dried and salted fish but also as hake "sounds" (i.e., air bladders), which were the basis of a product used in the making of beer: "Thus the hake sound market was determined by beer consumption in the United States; in one particularly intemperate period, hake sounds sold at a North Head auction in 1878 for the unbelievable price of $1.00 per pound" (Allaby 1984, 17).

In 1989, there were still about fifteen fishers who depended on handlining for pollock, cod, and haddock, bringing in several tons every season between May and October. By 1995, while draggers and gillnetters were still catching many fish, the handliners were catching smaller and smaller amounts. By 2000, there were only three boats that continued to handline, with their sales being mainly to islanders who vied for the catch as the men arrived back at the dock before sunset. Islanders know their fish, and handlined fish have not been drowned: the meat is fresh and whole. So as the word went around that someone was due back, he would be met at the wharf by eager buyers. In 2002, there were only two men still handlining. One described his summer: "I've been out since May, and I got my first fish on August 1, my birthday." By 2005, he was the only person who tried to catch some handlined fish, but he had no luck at all. It cost him $800 in fuel, and he was not able even to feed himself. Somewhat wistfully, he said that he had sold his lobster licence about twelve years earlier, "when no one was making any money, and lobster was only $2.50 a pound." That same summer, islanders complained that the only haddock they could find was in the form of small fillets that had been brought in from Nova Scotia. Whereas in 1995 stories were told of hunting up and down the island for fresh vegetables in spring, ten years later islanders were having to search for their groundfish to secure the twenty to thirty pounds that everyone wanted to put in the freezer for the winter. The availability of a staple diet of pollock or haddock and potatoes, squash, and turnip was no longer a foregone conclusion. As in Newfoundland, the degradation of the fish stocks was affecting more than economic realities; it also affected food choices on a daily basis (Neis et al. 2000).

The disappearance of the groundfish stocks was not the only factor contributing to changing diets on Grand Manan. Many mixed farms that provided milk, eggs, meat, and vegetables were an important part of the island economy until the 1970s when, coinciding with the introduction of a modern supermarket, they were abandoned. Nonetheless, virtually everyone still has at least a small garden plot, providing a significant amount of vegetables such as green beans, potatoes, squash, turnips, beets, chard, peas, lettuce, and cucumber. Strategies to foil the voracious deer can be seen from the road, varying from elaborate fences made of stringing sticks (for the smoked herring) covered by nets to swaying aluminium pans and plastic bags. Regardless of their tactics, gardeners usually have to concede a quantity of produce to the deer, even as they rush to gather in ripened crops to be pickled or frozen for the winter. The dependence upon locally grown produce is considerable, and even in the middle of winter it is not unusual to share a family meal that consists entirely of island produce.

Nonetheless, the fish component of the meal has become a challenge. Despite the increasing reliance by fishers on a wider range of species, such as halibut, crab, and sea urchin, for the most part these are not species that are eaten or enjoyed by islanders. Even salmon is usually only available in the supermarket during the summer season when tourists are on the island. When I told one of the fishers that the price of salmon in our Montreal markets was five dollars a pound, less by two dollars than the price of cod, his response was: "Well, you know which I would buy!" The loss of the groundfish species represents not only a significant degradation of the marine ecosystem and an economic loss for the island but also an important change in local diets. As less and less fish is available, more meat and semi-prepared foods such as fish sticks and coated chicken are being eaten. In part, at least, this dietary change is reflected in higher average weights among islanders.[7]

The plight of communities affected by the declining groundfish stocks has been well documented, from Newfoundland to the northeastern seaboard of the United States. In his study of Digby Neck in Nova Scotia, Davis (1991) suggests that the decline of the small boat sector is affecting the ability of fishing communities to muster political support for their interests and that the highly capitalized sector is jeopardizing the sustainability of fish stocks generally. One of the most moving and well documented accounts is that by Richard Adams Carey, whose descriptions of personal stories interwoven with documented scientific reports provide a depressing testament to people's stupidity and greed. As he shows, communities invariably understand the trade-offs between the profits of bottom-destroying draggers and hook-fishing technologies. Faced with total collapse of the eastern seaboard groundfish fishery, community meetings debated the various alternatives to existing practices, aware that sustainability was going to be a difficult challenge. In one scenario, the fishers were offered limits on the numbers of hooks per boat (3,500) and on the numbers of days at sea (less than 100 days/annum). But the dominance of draggermen on the governing board ensured that any significant curtailment of dragging operations on Georges Bank would be vetoed (Carey 1999, 183).

In a case study of northwest Newfoundland in the early 1990s, Palmer and Sinclair (1997) explored the fate of "domestic commodity production" (DCP). Intricately linked to social structures, DCP was shown to have persisted despite the government's attempts to "rationalize" the fishing industry through individual transferable quotas and the gradual concentration of capital. Resistance to government efforts to introduce market-oriented measures that would improve "efficiency" inevitably created tensions within many small communities throughout Atlantic Canada, especially with the increasing divisions among categories of fishers based on levels of capitalization (McCay 1999). For communities, the stakes in these controversial gear conflicts are too high to be negotiated without rancour. Even the mythological sharing for which small communities are noted cannot assuage the independent competitive sensibilities of fishers who have invested in expensive technologies and new equipment. On Grand Manan similar versions of the fate of the groundfish are told. One fisher summarized the problem as follows: "There's not a whole lot going now for the boys in summer." For many fishers there is a sense of impending collapse not only of the wild fish stocks but also of the entire social order as they have known it for generations. Yet, the residents of small fishing communities are "incredibly inventive in creating adaptive strategies that allow space for local action within the structured environment of international commodity production" (Palmer and Sinclair 1997, 9).

In 1996, an interview with Randy Parker, who had not grown up in a fishing family, reflected many of the frustrations and challenges that typified lives on Grand Manan. Having depended on dulse for many years, a broken toe the previous year had affected his balance on the slippery rocks and now limited his ability to work the same hours. Therefore, he was trying to buy into various sectors of the fishery, including groundfish, lobster, and sea urchin. While the government allocation of a limited numbers of licences made sense to him, he was frustrated by the high costs of licences due to the bidding process on the island. For groundfish the cost of a licence had risen from thirty dollars ten years earlier to about $20,000 in 1996. Added to this was the problem of individual transferable quotas (ITQs), which had been inaugurated in the mid-1980s. There were two types of quotas: one was a group quota (thirty licences) for boats over forty-five feet; the second was the individual quota for boats that were less than forty-five feet. But this second option was now closed as all available licences were already taken by long-time fishers. In fact, there were only four of these, of which only one was being used in 1996. Knowing this, Randy had tried to "borrow," or have transferred to him for the summer, one of the three other licences. In each case someone had got there before him. Moreover, the group quota had been filled. So he was effectively shut out of the groundfish sector. Some of the men who owned herring seiners had groundfish licences attached to their million-dollar boats and were not using them because the quotas were too low to cover the expense of running these large boats.

Therefore, in their case, they usually opted to sell their eighty-ton quotas to the companies for two-thirds the price they could get for fish. In other words, for them, there would be no expenses: only profit for doing no work. As well, the companies, aided and abetted by government subsidies, now had control of most of the fish quotas and were in the process of buying up the rest. For a young, hard-working fisher, the rules set by the government all worked in favour of the large companies and already-invested fishers.

One of the policies many of the men complained about was the "use-it-or-lose-it" policy. Despite DFO denials that such a policy existed, in fact many fishers told of people who had lost their handlining licences because they were not being used. Islanders did not blame each other for holding onto or trading licences; their critique was focused on the rules that the government had in place and that seemed to favour the "big guy." Again, we see the issue of boundaries emerging out of government policies. One fisher argued that the government had caused increasing divisions among islanders because of rules such as those associated with the ITQs – divisions that were intensified by the lowering of quotas as stocks were depleted. Moreover, the increase in holders of multiple licences, he said, had been necessary because of the smaller catches, meaning that fishers needed to maintain incomes by diversifying their activities. Nevertheless, the complaints about ITQs and company ownership, which were such issues in 1996, had ceased to be part of any discussions in 2005. The issue was irrelevant because there were no more groundfish. This tragic irony was replicated in another part of Atlantic Canada, in northwest Newfoundland, where the small fisher had survived, in part, because of poor government planning and lack of enforcement and, in part, because of the collapse of the very stocks they thought they could preserve (Palmer and Sinclair 1997, 13).

While 1999 had reportedly been a "good year," and a handlining licence was still valuable at a price of $20,000 in 2000, few licences were being bought by 2004 as harvests continued to decline. In 2001, there were about seventy-five licences still on the island, although, according to the Grand Manan Fishermen's Association (GMFA), only about forty-five people were actually involved in the fishery. Whereas in 1996, the eighty-ton quota for cod and seventy-ton quota for pollock could easily be caught within the first couple of weeks of each month, by 2005 Grand Manan fishers were unable to find any fish. Even the large gillnetters were not able to guarantee supplies to the local brokers. Even so, when one of the fishers sold his two gillnetting licences in 2004, one of the older men shook his head in dismay, wondering why a young man would want to sell out. For the last handlining fisher, Melvin Ingalls, the sad reality was that several fuel-consuming trips in 2005 produced no groundfish at all. For Melvin Ingalls, who had no other licenses, it appeared that 2005 would mark his final year as a fisher. After fifty years fishing, he seemed resigned to the prospect of retirement. Even the draggers were being given up; while there were three in 1996, there was none in 2005. And only one gillnetter continued to fish in 2005.

A couple of fishers, ever on the search for alternative fisheries, tried halibut trawling. While it did provide activity for limited periods of time, the quotas were so low that just a few large fish would quickly reach the limit, which was reduced in 2004 from 500 pounds per week per boat to just 250 pounds. With the entire Bay of Fundy only allowed to harvest 239 tons, each boat had a seasonal quota of twenty-seven tons, which did not provide enough income to make up for the loss of other groundfish.

The pressure on stocks coincided with government efforts to "rationalize" the industry by declaring a "core fishery" that would be defined according to sectors and income. A major departure from existing policy, the move to exclude a high percentage of active fishers from their status as "core" was meant to eventually consolidate the fishery within fewer enterprises that, it was charged, would make it easier for the government to control. Regardless of the motivation, the impact was significant. In March 1996, the DFO announced its intention to define the core according to several stringent criteria, one of which involved a prohibition against the transfer of licences to children. There would be a low-income cutoff of $25,000, and handlining (the traditional method of catching groundfish) would be excluded. The announcement was met by vigorous protests across the Atlantic provinces, including on Grand Manan, where those not to be included as core participated in a sit-in at the DFO office in North Head. While the outrage ensured that these criteria were never enacted, the government did exclude weir fishing and all niche activities, such as dulse harvesting and clamming, from being part of the core. This had implications for the membership of the Grand Manan Fishermen's Association (GMFA), which controlled who could vote at meetings and who would be eligible for draws to enter new fisheries (such as sea urchin and crab). Even in 2004, it seemed that new regulations that would "professionalize" the industry were being considered. Upset by what he saw as an effort "to destroy all us little guys," one fisher exclaimed that the DFO was pushing through new rules without listening to the communities: "At first when you hear it, it sounds OK. But, reading the details, you realize they're out to get rid of us all."

Lobster

As Bill sat in his kitchen talking about his early days lobster fishing, it was obvious that his nostalgia was not related to wanting to go back to it. Compared to the forty-five-foot-plus boats the men use today, his small dory seemed a vulnerable craft upon which to have to depend in the heavy tides and winds that characterize the Bay of Fundy. In the late 1930s, he and his father and Hant Stanley would walk about an hour and a half from their homes in North Head out to Money Cove where they kept their dory and would spend the week setting lobster traps, about ninety in all, and hoping for even one lobster in each. Working together, the three men might haul three hundred pounds on a good day, compared to the 1,500 to 3,000 pounds that today's Grand Manan fishers depend upon during the first early weeks of each season in November: "At 15 cents a pound, divided by the three of us, it wasn't a lot of money." Sometimes he would work alone, in which case he had to stand in the dory, pushing the oars (i.e., skiffing), then bracing himself as, with one hand, he pulled the trap up: "And they were hemp ropes and wooden traps … it was hard work!" Asked whether he ever felt in danger, he said: "Oh yes, but that's the way it was." When they worked near shore they would time their advances to every third swell, when there was a longer space between waves. But occasionally they would get caught by a "comber" – a wave over eight feet high. Once when a comber caught them the dory was almost vertical: "It was a miracle it didn't flip right over." Bill figured he had almost been drowned over a hundred times, and this would not be unusual for most fishers on the island.

In the decade between 1995 and 2005 significant changes in the lobster fishery were related both to new technologies and to government policies related to the Maliseet First Nation of New Brunswick. Also, the loss of diversity within the traditional fishery meant that lobster became increasingly important as the mainstay of the traditional fishery. As one journalist asked, "Who would have predicted the day when the once diverse and prosperous Bay of Fundy would be almost totally dependent on bottom-dwelling creatures?" (D. Wilbur, *Courier Weekend*, 13 May 2005, A4). The collapse of the groundfish species affected all sectors of the wild harvest, including handlining, gillnetting, and longlining. But, as fishers observed as early as 2001, and the science community posited in 2005 (Frank et al., 2005), the collapse of groundfish had even broader implications than did the loss of an economic species. It had significant impacts on the ecosystem itself as well as on the increased stocks of crustaceans.

One multiple licence holder described the diminished finfish stocks as non-economic, while saying that the lobster stocks had "gone crazy." He also commented that the once controversial ITQ's were no longer an issue because there were no longer any fish. One of the island women described the impact upon her children: "The kids don't even know why it's called a fishery. They've never seen their father bring home a fish!" Within the traditional fishery, the one bright spot in 2005 was lobster. In the continuing strong harvests, high prices, value caught per fisher, and certainly with respect to status within the community, by any measure lobster was the most lucrative and most desirable of the fisheries. It represented a value of over $15 million in sales to Grand Manan. By 2006, however, ominous signs were being reported by some of the fishers. The desirability of entry into this fishery created dual problems that I explore later. One problem was the escalating costs for lobster licences, which was inevitably associated with their profitability. The second was much more controversial as it was directly caused by federal policy, which bought up licences that were then given to Maliseet reserves on the mainland, thereby introducing a level of demand that resulted in the precipitously rising values of the licences.

Setting Day

The second Tuesday of November is a special day on Grand Manan. Children are excused from school; older sons away on the mainland arrange to be home for a few weeks to work on the boats; mothers prepare special dinners at night; the stores are well stocked with goods; the wharves are alive with teams of working men and families with cameras; people move briskly. It is the start of the lobster season. For weeks the men have been repairing traps, checking tags and trawl lines, painting buoys, and finally stuffing bait bags with foul-smelling herring and fish heads. In the final week leading up to setting day, traps have been moved onto the wharves, piled and stacked by cranes and men, crowding the lane between to barely the width of a half-ton truck. Two days before, everyone is talking weather. In 2002, the Saturday conversations referred to three different forecasts: wind from the southwest, from the south, and from the east, and some rain, lots of wind, light wind. Some people recorded the weather at the equinox on 21 September and believed that whatever the winds were then would predict the weather for the entire quarter. One fisher summed it up: "The jet stream is a straight line across the country, so the systems usually move through quite quickly. As long as the wind and rain coming in tomor-

Painting Lobster buoys: Liscombe Green and great-granddaughter Shelby Calder

row [Sunday] clear out and there is twenty-four to thirty-six hours for the swell to die down, it'll be okay." That year the 12 November setting day dawned warm, with a bit of overcast and light breezes, just fine for the 7:00 AM start.

By 6:30 AM the wharves, still shrouded in darkness but lit intermittently by boat lights, are alive with activity and excitement as men clamber down ladders and crawl over boats to reach their own and carry out last minute checks on the ropes and buoy lines. Years earlier, the start had been indicated by a gun shot. Now it is indicated by the inching forward of boats, like yachts at the start line of a race, until one engine roars or someone yells that they have heard on the radio that another wharf has left; then they all rev into action. The watching DFO officer commented that maybe being ten minutes early was tolerable. But later we heard that the patrol boat had been out in the channel to slow down those who had beaten the clock. The eighteen boats at North Head had left at 6:46 AM, while at Ingalls Head the forty-five boats were gone four minutes later. The officer told the story of a fisher a year before who was feeling frustrated by all the new larger, faster boats, and figured that he deserved a bit of a head start in his older, slower boat. And so he had left twenty minutes early. The DFO did not chase him, as they have been known to do; rather, it was his fellow fishers who warned him that he had better not do that again.

Over thirty years, delays caused by weather have occurred only a few times. One of those was in 2005. Talking with the men as they worked on their equipment days before the 2005 season opened, it was apparent that everyone was apprehensive. There had been indications the previous year that some areas were not as bountiful as in the past, and for those fishing in the mid-channel, the lobsters seemed to be getting smaller. Then, as weather reports indicated strong winds and heavy seas for setting day, the DFO officers began to make the rounds of the wharves, checking on how the men felt about postponing the opening by one day. The dilemma was that, while the large boats, captained by experienced men, might survive heavy seas, the risk to the smaller boats or to crews who had less experience would be too high. And yet, as the officer explained, if there was not an official edict postponing setting day, the large boats would go, and, undoubtedly, the smaller ones would as well since a delay would mean the loss of significant revenue potential: "If even a few decide to go, they'll all go, and the small boats won't be able to handle the seas." The southerly, thirty-knot winds were expected to change to westerly gales of fifty to sixty knots. "It'll be a real mess!"

With boats loaded with all the traps they could carry, not only was there little space for the men to work but there was also real danger that the top-heavy piles of traps would capsize the boats in the heavy seas. As decisions were being made, stories of past delays began to be told. The last time North Head did not go was in 1991, said one fisher, although the Seal Covers had gone because their inshore ledges were protected by the outer islands. In 1994 it was the turn of Seal Cove fishers to stay home, while North Headers were able to set their traps on the scheduled setting day. This time, in 2005, with areas overlapping and larger boats in all the harbours, everyone felt there had to be one rule for the entire island. At a lunchtime meeting on the day before, the fishers voted to delay until Wednesday, 9 November. As news spread of the delay, the entire rhythm of the island seemed to change. Suddenly the intense activity on all the wharves slowed down, the men were ambling, casually chatting, stopping in their trucks to check up on news. The relief was palpable. The next day, sunny and clear, belied the strong westerly winds that shook trucks down at Southern Head and that would have endangered boats and crews. To delay had been a wise decision.

With the unexpected postponement, the fishers' attention was temporarily diverted away from their own urgent jobs to the activities of others. One topic that several men were discussing was the very large new boat that had arrived at Ingalls Head two days earlier. Looming over all the other boats at the wharf, the freshly painted maroon boat attracted attention not only for its colour and size but also for its ownership – a First Nations reserve on the mainland. The boat had been built in Nova Scotia and was being crewed by Nova Scotia fishers. Never had an outside crew been allowed to fish in Area 38.[8] Asked about it, the DFO officer explained that the licence was an island licence that had been recently sold and that there were no rules against off-island crews running the boat. One fisher threw up his hands: "It's the Indians; what can you expect. They have their own rules." Some of the men suggested that they should have been more proactive with regard to showing their antagonism to the Native presence right from the beginning. This important topic is dealt with in detail in chapter 4.

Loaded lobster boats before setting day, Seal Cove

Setting traps is a strategic operation, requiring planning and the systematic stacking of the traps in boats according to measured lengths of traplines that correspond to the depths at which traps will be set. Leaving the wharf, a boat heads out with carefully planned trips, the traps on trawls of eighteen, twenty-five, or forty fathoms, set precisely so that buoys are not submerged. Occasionally, a misjudgment means that a trap has to be retrieved, markers reapplied, and the sorting in the stacks realigned. Each of the two or three trips will target different areas, with different wind and tidal conditions. Even for experienced fishers the anticipation and apprehension of setting day – ensuring that ropes are not tangled, that routes are accurately planned with corresponding trap-lines, and that crews are well-prepared – can be extremely stressful. One fisher acknowledged that his "knees had been knocking together" with the tension of the early morning start. Catches during the first week of lobster season can account for 40 percent of the year's harvest. "You might as well not go if you miss the first week," is a common refrain.

Setting day has its share of stories as well. One of the distinctive boats on the island, formerly owned by Don Hatt, is named *My Dear Boy*. In describing the events leading up to the naming of the boat, Bradley chuckled as he recalled his early learning days. He had been a crew member for Don Hatt and Kenny Brown, who had told him to release a trap whenever Don said "yup!" As Don and Kenny were chatting, Don kept nodding and saying, "yup," "yup," "yup." Bradley, tensed and nervous, grabbed a trap and threw it overboard each time he heard a "yup." Suddenly, Don realized that far too many traps had gone in and exclaimed: "My dear boy! My dear boy!" Which is how the small red boat came to be named.

Until the 1990s it was rare for women to be included as crew members on lobster boats. Today it is not unusual to see even teenage girls going out with their fathers on setting day, although their work comes only in later trips when they do the banding. As one strong young woman described the work, "I can bait bags for a few days, and then when they start hauling I do the banding for a couple of weeks. But my back can't handle setting." There's scarcely an able-bodied man left on the island for the first few days of lobster season. Even men who normally do not work at any other time in the fishery are hired on as crew. People arrive from the mainland to stay for three weeks or more; the New-foundlanders delay their return home in order to earn extra money during the first intense weeks of the season. Hauling traps can begin as early as mid-after-noon on the day they're set, although usually this only happens due to the

Top: Lobster setting
Bottom: Three generations setting traps

eagerness to see what might be in a few. But on Wednesday morning the boats are out before daylight. The dilemma is always the weather. One year Thursday was extremely rough, with gale-force winds blowing from the southeast. Many men stayed home. Others went out and returned early. But, said one fisher, "Once you're out you can't come back too soon. There's your reputation to think of."

Island attention has increasingly focused on lobster in recent years because of the precipitous decline in groundfish catches. When new homes are built, new vehicles are bought, and new boats are ordered, the community feels the success of the lobster fishers. In 1999, the catches and prices were the best in many years, according to even the most dour of the fishers. For Grand Manan Area 38, the catch had increased from 528 tons in 1998 to 851 tons in 1999, a 61 percent rise. Similar harvest increases occurred in 2000 in the adjacent area known as "top of the bay." There, the increase from 1998 to 2000 was almost double, from 590 tons to 1,040 tons. Two years later, the news of large catches was spreading through the community only two days into the season. With reports of a one-day harvest of 2,200 pounds and 3,600 pounds for two of the younger men, everyone was excited. But some of the men were more circumspect.

Trying to explain the growing numbers of lobster, fishers speculated that the decline in the groundfish might offer a reason in that it was suspected that they feed on lobster spawn, which now would have higher survival rates. Someone else wondered if there had been a slight rise in water temperatures and that this was encouraging better survival and growth rates; or, perhaps it was related to the feed around the salmon cages. Other fishers, however, had a more gloomy explanation. Talking about the large boats, the influx of Aboriginal fishers, and the shrinking diversity of choice that was characterizing the fishery in 2005, two of the men felt that the increasing harvests were directly related to increased effort, especially with the size of traps. As one of the Ingalls Head men explained, the average size of the trap had gone from the small old wooden traps, (about thirty inches long and domed rather than square) common until the early 1990s, to thirty-six-inch squared wire traps and, more recently, to forty-two inch and even four-foot traps. He said that, in moving to the larger forty-two inch traps in 2005, he now had to make three setting trips rather than the two that had been sufficient in previous years. But he also pointed out that the larger traps were certainly catching more and larger lobster. He was convinced that the harvests have peaked and expressed concern that they cannot be sustained at the current level of effort. His views were corroborated by a fisher in North Head, who said that his harvests had stabilized two years earlier and that it was only the rising prices that brought them slightly higher incomes. Of great concern to all islanders was the fact that early reports of the catches in the winter of 2007-08 suggested that even the deep-water trawls had not taken the quantities of past years. These pessimistic reports were made even gloomier by low prices due to the high Canadian dollar and the impending US recession. Indeed, the concerns being expressed reflected a reality not far away, in Northumberland Strait on the northern shore of New Brunswick. Throughout 2005, reports indicated a decreased harvest, but it was the independent report from Prince Edward Island in June 2006 that confirmed the imminent collapse of the lobster fishery there (Saint John, CBC News, 7 June 2006). Recommendations that the DFO should buy back 15 percent of the licences were being met with mixed feelings in that region. For some Grand Mananer lobster fishers, the situation in Northumberland suggested that caution for the future might be in order.

Even though fewer than eighty families own lobster licences, the effects of good stocks and high prices are felt throughout the community. Home renovations brought business to the local hardware store, and winter holidays took children out of school. Said one non-lobster fisher: "There'll be lots of travel this week before setting day, because of the plastic" – that is, credit cards. People would be shopping on the mainland in anticipation of lobster season opening. Again in 2001, there were eight new boats arriving on the island, sparking discussions about the size of the engines and how fast they could travel. One comparison between a company-owned boat with an $80,000 engine that was "slower than Brian's boat with a smaller engine" provoked interest in the modifications being made at a local boat shop that allowed the vessel to plane better. Competition is intense. But 2001 was the year of the Trade Center disaster in New York. Prices for lobster and scallops plummeted as air traffic was curtailed and markets crashed. Starting at $4.50 per pound, prices for lobster did not rise until after the Christmas season, with the result that local fishers netted less even with higher catches. With exactly the same catch, one fisher reportedly had $15,000 less income than he did a year earlier. In 2007, the problem was the currency markets, which had devalued the US dollar, causing a more than 30 percent rise in the Canadian dollar in six months. This directly affected prices, which opened at $4.50 per pound in November that year. The apprehension about possible stock declines was being abetted by the sales prices. Nevertheless, not everyone agreed with the dire predictions. At the same time as one complained that he had 2,100 fewer pounds throughout the spring season, and that he had also been hurt by the threefold increase in the cost of bait and the 35 percent increase in the cost of fuel, another man said he had the best spring ever. The guessing as to where the lobsters were moving would continue.

Sea Urchin

For many of the marginal fishers on Grand Manan, the possibility of entering non-traditional wild fisheries offered some hope of new income. There had been a couple of men involved in sea urchin during the 1970s and 1980s, but the markets were unreliable and eventually they gave it up. Then, in 1988, greater interest from the Americans led to an exploratory fishery that was formally initiated with the granting of six permits in 1992. At first few people were interested. But opinions quickly changed as it became apparent that, for the six who did apply and were granted permits (not licences), the excellent market and high prices paid by the Japanese had opened up a very lucrative new fishery. In 1996, it was decided to allow seven more licences (seven permits were converted to licences), six for Grand Manan and one for mainland Maliseet who were beginning to demand entry into the fishery. How the six new island licences would be allocated was a problem decided by the GMFA and the DFO. Under a little known provision, entitled Special Eligibility Criteria (SEC), it was decided to exclude any non-core fishers from entering the draw, thereby precluding entry by dulsers and weir fishers who did not have other "core" licences, such as lobster or scallop. When questioned about the decision, the GMFA contended that the DFO had insisted that only core fishers should be eligible for the draw and that "we have to pick our battles." According to a GMFA spokesperson, the DFO "did not want new people in the fishery." Said one weir fisher in 1998, "those that 'have' make the rules that keep out those who 'have not.'" There was resentment among many of those excluded, and this resentment continued to ferment ten years later.

In 2005, the sea urchin fishery, conducted by dragging large chain pocket nets along the seabed, seemed to offer an ongoing lucrative source of income, despite lower catches compared to the first few years. Regulated by a total quota of 430 tons, each of the thirteen boats is allowed to harvest up to thirty-three tons. While the price has varied from a high of $2.20 per pound in 1999 to a low of $1.20 in 2001, at an average price of $1.70, it continues to be a profitable fishery. As a boat unloaded its harvest under the watchful recording eye of the

Sea urchin net

dockside observer, Marilyn Locke, twenty-three containers of sea urchins with about eighty pounds in each were moved into a waiting truck. The estimated value was almost $3,000. A measure of how lucrative sea urchins can be may be seen in the value of the licences. In 2002, one sea urchin licence was reportedly sold for $375,000, up from $265,000 a year earlier. Moreover, by this time the government's policy to give mainland Maliseet reserves access to the marine fishery was affecting all of the lucrative fisheries on the island. The result, as with lobster licences, was inflated values that Grand Manan fishers could not afford. By 2004, eight of the thirteen sea urchin licences were held by off-island Aboriginals. For those who had multiple licences, it became a question of whether to fish for lobster or sea urchin during the winter months. Occasionally, a licence would be "rented" for the season in order to retain ownership while allowing someone else to use it.

Nevertheless, for the five Grand Mananers still involved in the sea urchin fishery, strong indications of declining stocks in the winter of 2006 forced them to reassess their quotas and seasons. In mid-February, after a twenty-nine-day season, there was a meeting of all licence holders, the eight First Nations licences represented by David Bollivar, a non-Aboriginal consultant for the three reserves involved. It was decided, without any pressure from the

DFO, that the 2007 season should be shortened in order to encourage greater regeneration of stocks. All the men had commented on the decreasing size of the urchins and the greater percentage of below-minimum measures. While, based on numbers of rings, they knew that some of the urchins were as much as twenty-five years old, there seemed to be no clear scientific understanding of the reproductive cycle and what could be considered a sustainable harvest. The men were setting their own limits without specific data that could support their efforts at maintaining the stocks. However, as Tony Lyon pointed out, even with a cutback to a fifteen-day season in 2007, anyone could double their working time and effectively harvest the same amounts in a shorter period. The problems of establishing effective and acceptable fishing regulations were well understood by the fishers. In the spring of 2007, the quota was indeed cut to 30,000 pounds, which the men were able to catch in twelve days. Compared to the two-month season only five years earlier, when quotas were at 166,000 pounds (successively dropped to 132,000 pounds and then 100,000 pounds) the sea urchin was now merely one more threatened fishery.

In various ways, the sectoral changes on the island have been directly related to the growth of salmon aquaculture (the socio-economic impacts of which are discussed in chapter 6). Yet another fishery it has affected is that of the sea urchin. At first, those involved in the sea urchin fishery were concerned about the extent to which they had "lost bottom" due to privatized areas that were essentially out-of-bounds to urchin fishers. However, they began to wonder about the possibility of the salmon feed actually attracting sea urchins. In 2004, when one of the largest salmon sites had been totally harvested and was fallow, waiting for new smolts, several of the boats were invited to try harvesting sea urchins across the bottom of the site, in among all the anchors and stanchions. Their success encouraged them to try again in 2006, when the salmon cycle again had emptied the site. This time there were no urchins. As one of the fishers said, "For whatever reason the feed doesn't seem to attract them." He added that, contrary to perceived wisdom and to his surprise, the bottom was "really clean."

One innovative experiment that involved collaboration between a local entrepreneur and a postdoctoral student involved a sea urchin facility established in 2000 to explore the possibility of roe enhancement. Having done his doctoral work on kelp growth rates and nutrient values, Chris Pearce was interested in the possibility of sea urchin aquaculture. In partnership with islander Kenny Brown they established an experimental facility in Woodwards Cove, complete with five different grow-out designs that included raceways, tubs, outdoor trenches, and small cages enclosed within a lobster pound. With two graduate student assistants, Pearce was able to establish a program of feeding and enrichment that would boost the roe production. His aim was to increase the percentage of roe from the wild average of 4 to 5 percent to 10 percent. As he said, the sea urchin stocks were already in decline and aquaculture offered one way in which the lucrative markets could be sustained. The sea urchin fishers had requested a decreased quota for the first time in 2001, from 166,000 pounds to 132,000 pounds, indicating their concern for the long-term viability of the harvest. One of only ten research facilities in the world, the Grand Manan unit represented an important research initiative with global implications.

In his plans for commercial production, Pearce calculated that a 25,000-square-foot facility would be needed, at a cost of over $1 million, including interior infrastructure. He felt that land-based systems ultimately would be the most efficient, economic, and environmentally safe choice. He also argued that, rather than trying to grow urchins from eggs, it would be more viable to build an enhancement facility using hand-harvested urchins that would then be "bulked up" to 10 percent or 15 percent of their weight. That the facility was finally closed when Pearce left the island two years later was unfortunate. Pearce had been positive about the long-term prospects and had been very impressed by Brown's willingness to take the financial risk required for this innovative initiative. Unfortunately, it could not be sustained long enough to realize benefits for the island. That most Grand Mananers did not know of its existence is testament to the low profile that most of them keep when exploring new ideas.

Crab, Quahog, and Scallop

Another experiment on Grand Manan involved a crab fishery that was introduced in 1995–96, with the allocation of three permits. Then, in 1999, a draw among the core fishers allowed an additional three boats to obtain permits (not licences) to participate in the pilot project. While the DFO continues to support the experimental nature of the crab fishery, the men who are participating enjoy a significant benefit. As well, because of its relatively unregulated nature, other fishers are able to capture some of the harvest as by-catch in the lobster fishery. Indeed, in large part, the Grey Zone fishery has continued because of this crab by-catch. The profitability of the crab by-catch has affected local efforts to certify it as a full fishery. Neither the lobster fishers nor the GMFA was active in encouraging that the fishery become a full-licence regulatory structure as too many lobster fishers were able to benefit from the unregulated nature of the crab by-catch. There are no official figures on the harvest, and everyone seems content to maintain the existing status quo.

Always on the lookout for new possibilities, a couple of Grand Manan men invested in a specially designed rake that could pick up quahogs from the sea bottom. After a couple of seasons they gave up because of uncertain markets and low harvests that barely paid for their fuel costs. The self-designed rakes lay beside the wharf at Seal Cove for several years as testament to the perpetual efforts to diversify and to maintain marine livelihoods.

One of the most lucrative fisheries on Grand Manan has been the scallop, beginning in the early 1980s. Again, the establishment of boundaries has been a crucial issue for this fishery, especially with respect to international juris- dictions. The Canadian territorial marine boundary had been extended from twelve miles to two hundred miles in 1976 and was ratified by the United Nations Convention on the Law of the Sea (UNCLOS) in 1982. This major change in the definition of national access to the fishery resource and the pos- sibility of excluding foreign fishing fleets had vast implications for the evolu- tion of the Canadian fishery, affecting all sectors. For Canadian scallop fishers a major decision of the International Court of Justice (ICJ) at the Hague in 1984, by which a significant portion of rich scallop grounds were awarded to Canada, meant that an assured new territory for scallops would allow for the growth of that sector on Grand Manan. The result can be seen in the migration figures for the island, with many families returning to take advantage of this lucrative new fishery. Compared to 1971–81 when almost 9 percent of the pop- ulation had left the island (net out-migration), by 1991 the returning islanders had compensated to the extent that fewer than 1 percent represented out migra- tion (Statistics Canada 1981, 1991).

Licensing can be for either a mid-bay or full-bay area, with provisions to avoid areas in seasons that overlap with the lobster season. One of the most fre- quent problems in the traditional fishery has been gear conflicts. The impacts of scallop draggers on lobster beds are obvious, and generally the men respect regulations that seek to avoid such overlapping jurisdictions. Nevertheless, constant monitoring by the DFO can engender tensions that occasionally esca- late into violent conflicts. In 1997, the DFO had ruled that every boat had to carry official observers (this was later changed to one in five boats). The prob- lem for island scallopers was that no explanation was offered and their naturally suspicious tendencies were not alleviated by any discussions. According to local informants, even the observers on the boats had to rely upon cell phones when awaiting instructions. As tensions mounted, and the DFO made no concessions with regard to communicating what they were actually trying to observe, finally one family turned hostile upon returning to the wharf. Court action ensued. Unfortunately, while the DFO monitoring and regulatory system is meant to provide for a sustainable fishery, fishers often feel left out of the process through which regulations are made and enforced. In a lucrative fishery such as scallop, the intense and competitive desire to make money can push captains to extend territories or hours outside those permitted under the regulations. In 1997, the shortening of the allowed harvest time from all day to what one fisher described as "Banker's Hours" – 6:00 AM to 6:00 PM, meant that there would be smaller harvests with the same fuel consumption.

Provincial boundaries in the fishery occasionally become issues in the granting of rights to the Bay of Fundy fishing areas. Whereas lobster areas are clearly defined, the fishing zones for scallopers have been more changeable, partly due to the nature of the dragging technology and partly due to overlapping seasons. The extension of allowable areas onto the Georges Bank in 1984 was a major addition to the harvestable zone for Grand Mananers. Then, in 2001, the DFO bought up a significant quota from a Nova Scotia family in order to offer it to the New Brunswick Maliseet. This, of course, as with the lobster licences, ensured that the price for quotas soared, putting them out of the reach of the average fisher. While the DFO action did bring more of the Fundy quota to Grand Manan, it also provoked provincial discussions about the transfer between provinces of the fishing quotas – a practice that was subsequently terminated.

Dulse and Other Niches

If the sea is at the heart of individual and collective identities on Grand Manan, the shore reflects the nexus of relationships between land and sea, men and women, children and parents, tourists and Grand Mananers, and new technologies and old. It is the boundary between land and sea, the edge of harvesting activities, and space of meaning within a context of dynamic social relations and shifting identities. Upon the shoreline is written the history of Grand Manan, and this is where we see the playing out of the many complex relationships that define the island. The shore incorporates both the beaches and the intertidal zone, where rockweed, kelp, sea lettuce, dulse, and various crustaceans such as periwinkles and clams can be harvested. It is along the shore that one most appreciates the mutability of the environment, with daily, weekly, and annual changes in tidal detritus, storm erosion, and stream deposits. Entire beaches can suddenly be transformed from narrow sloping stretches of sand to wide steep banks of sand and rocks strewn with rockweed and kelp thrown up by powerful northeasterlies. Women check the beaches every day at low tide for the sought-after beach glass. One cannot ignore receding coastlines that threaten houses, beaches, and roads, even if the clearcutting of forests remains hidden from view. Marine environments and land ecologies, wild fisheries and community activities all intersect and mingle in these extraordinary zones of transition. With the growing intrusions of government commercialism, which has encouraged the rockweed harvest, for example, and the increasing numbers of summer residents buying up large tracts of beachfront properties, these zones of transition have become areas of both cultural significance and conflict.

The activity that is most associated with the shoreline is undoubtedly the gathering of dulse. After the weir fishery, there is probably no more characteristic and historically important activity on Grand Manan than the harvesting of dulse. It contributes to the economy of the island, sustaining two export companies, as well as providing a substantial income for many families and individuals. But apart from its economic value, there is an intrinsic cultural value attached to dulse. It is found in paper bags beside the men in their trucks, taken out with the seiner crews for early morning snacks, offered in elegant living rooms as snack food, and sent to families who have moved to cities across the country. Dulse is part of the daily diet, as much a cultural icon as a source of employment. When the first dulse appears in late March down at Deep Cove, islanders are there to scout it out. Small amounts are harvested for personal consumption, dried on back porches, and treated like a delicacy.

For Grand Mananers, the shore has been the gathering place for the dulsers for generations, those intrepid islanders whose work day, even at 3:00 AM, begins four hours after high tide when the ebb is almost at its lowest. They arrive in their trucks and waste no time transferring the burlap bags and plastic baskets to their dories. Within minutes they will be pushing out the dories or hauling them across pebble beaches or up and over the steep rocky ridges of natural sea walls to set out for their favourite dulse areas. From late April until November, dulse grows at the most extreme lower edge of the intertidal zone, giving the harvesters about two hours of gathering time between the tides. Back-breaking work, gathering the dulse requires agility and steady footing to preclude accidents. The harvesters must dry it themselves, laying it out in single layers on stone beds (the "drying grounds") that have been specially prepared at, in some cases, a cost of several thousand dollars. In ideal drying weather (warm [not hot], breezy, low humidity, and sunny), the dulse can be dried in six to eight hours, including a complete turning. The drying process is vulnerable to the vagaries of east coast weather, especially since dulse is very tender and spoils easily if rained upon or if it dries too quickly in a harsh sun. Days that are foggy, wet, and cool pose particular problems because the freshly picked dulse cannot be kept in the sacks for long duration. In these conditions they are "tied down" in burlap bags for up to six days at the edge of the tidal zone so that they continue to be washed by the tides and protected from light, which damages the tender seaweed. Even these holding spots are carefully selected places where "the water is colder and runs well."

Women picking dulse at Dark Harbour

Dulse held over on foggy days

There are about thirty-five families for whom dulse is the major source of income; but another fifty to sixty people also pick it on a regular basis, using it to supplement other income from the fishery or fish plant work. For many islanders, it is a niche activity between fishery seasons and/or fills gaps when a fishery is down. As one lobster crew described it in 2006, "The lobsters are late this spring at this end of the island [North Head] so we're going out only twice a week. I'm getting dulse over in the Passage on in-between days." Although people such as the intrepid Carolyn Flagg have been known to dulse even at seventy-five years of age, because of the physical demands most harvesters are much younger. Carrying sixty-pound sacks of wet seaweed over slippery rocks while wearing rubber boots that have no ankle supports is a challenge for anyone. The value of dulse to the local economy has been estimated at about $1 million.

The first commercial dulse can be picked in late April in the "Passage," between Ross Island and Cheney Island. Considered to be lesser grade because, due to light and tide conditions, it is tough and tends to be dirty, buyers pay about a dollar less per pound (dried) than they do for the Dark Harbour dulse picked along the "back-of-the-island." The ratio of wet to dry is slightly differ-

Drying dulse at Flagg's drying grounds

ent for the two areas, about 8:1 for Passage dulse, and 6:1 for Dark Harbour dulse. Not only must harvesters take into account drying conditions but they must also respect the tides. Dulse can only be harvested during the most extreme tides. What they call the "drain" tides offer access to the lowest areas of the intertidal zone, where the best dulse can be found. When the tides are less extreme, depending upon the alignment of the earth with the moon and sun, the possibility for large harvests is much reduced.

Even islanders not actively involved in harvesting dulse understand the importance of the intersection of good weather and strong tides. During the drain tides of mid-June, early in the season when the dulse is at its most tender, islanders buy their supply, often about ten pounds (dried) that will be stored in their freezers. Commenting on the poor weather in 1998, Helen said: "There's good money in it, but in this weather [foggy, damp, and cool] it'll never dry. It's too bad because they've had such good tides." The precariousness of the inter-section of various environmental conditions determines whether or not families will secure an income. Always exploring different options that might mitigate unpredictable weather, one dulser considered whether freezing it might be a way to save the dulse. Paul Brown had tried it once for two weeks, but it took a day to thaw and then it was very cold spreading. Altogether, he concluded, it was not worth the trouble. Another possibility had been to use salt water tanks. Again, it had been tried, but soon the men realized that the costs involved in keeping the water circulating were not worth it.

Once dried, it is baled in large rolls and taken to a buyer, either to Atlantic Mariculture or at Roland Flagg's company, now run by his son. Flagg, who has the largest export market for a variety of intertidal seaweeds and crustaceans, buys and sells dulse, kelp, sea lettuce, and nori as well as clams and periwinkles. In general, he buys only top-quality Dark Harbour dulse, while Atlantic Mariculture, which grinds much of it up for various uses (such as cattle feed and health food products), sells it in bulk to buyers in the United States and buys mainly Passage dulse. Atlantic Mariculture was established in the early 1970s by two mainlanders, one of them an American. In 2000 it was selling, mainly for export, about 100,000 pounds of dried dulse a year. Dulsers would receive $2.50 per pound for Passage dulse and $4.00 per pound for Dark Harbour

Rolling dried dulse

dulse. By 2003, those prices had risen slightly to $2.75 per pound for Passage dulse and $4.50 per pound for Dark Harbour dulse. By 2008, prices had increased significantly during a season when harvests were the best in years, one bright spot in an otherwise morose spring. For Atlantic Mariculture the two sources of dulse (Passage and Dark Harbour) accounted, respectively, for two-thirds and one-third of its total volumes. The Passage dulse is ground into three grades, from flakes to powder, while the Dark Harbour dulse is sorted and shipped unprocessed in bags. Whereas most of the Atlantic Mariculture product is in powder or flake form, Flagg sells almost exclusively whole dried dulse in large bags.

Flagg started his business in 1970, with the sale of periwinkles to Ontario, and later ventured into dulse and other seaweeds. In 2001, he estimated that his gross revenue might be around $350,000 for whole dulse, with only five hundred pounds of dulse a year sold as powder. In 2000, he had sent out 240 T4 employment insurance (EI) receipts, which he said represented about thirty families who would be working full time and about another seventy individuals who would be working part time. It is significant to note, however, that while there might be about one hundred people involved in dulsing in any one season, in fact most families on the island have participated in the harvest at some stage of their lives. Islanders who today are in their fifties and sixties describe their days dulsing when they were first married, trying to make ends meet. While the activity is dominated by male harvesters, many women have dulsed and continue to do so. Flagg's business expanded around the world, and in 2000 he had shipped a full container to France, consisting of, by volume, 40 percent dulse, 40 percent nori, and 20 percent sea lettuce. Already planning to retire in a couple of years, leaving his business to his son, he described the fun of making global contacts and expanding his markets through the internet.

Dulsers have particular intertidal zone areas that they regard as "their own," while informal rules of access that have evolved over many years help to monitor harvesting intrusions. But, as one former dulser described, "People used to have their dulse grounds in front of their camps, and no one would encroach. No longer." The boundaries between the informal territories have blurred, and the rules have been tested in recent years by the influx of migrant labour to the island. One dulser was blunt in talking about the newly arrived Newfoundlanders: "They don't understand; they come too close to where we always pick." Another islander, complaining that the Flaggs had actually "imported" people from the mainland to do the harvesting, was dismissive of their impact. He said that he "saw two fellows left off at the Whistle Point without a boat. They were picking for about four hours and had only a bucket! They hadn't a clue what they were doing." Some years there are changes in the quality of the dulse, apparently associated with unexplained water conditions. One year the dulsers complained that the dulse seemed to have "almost disappeared." One harvester described his usual gathering area, a stretch of two hundred yards of rocky shore, as "looking as though someone had waved a Javex bottle across it." Other stretches had small roots, with only four to five inches of dulse. He had never seen it like that. Shrugging his shoulders, he said, "Maybe I'll have to get a woodlot." Grand Mananers invariably consider alternatives, even as they continue their habitual work. Despite his pessimistic survey, dories could be seen up and down the shoreline, and on the rockwall at Dark Harbour the glistening purple seaweed was spread out for drying.

Speculation about the low dulse harvests through 1999 and 2000 focused on a variety of explanations, including: over-harvesting due to the sudden influx of Newfoundlanders; climate warming, with more sun bleaching out the dulse; inappropriate picking methods (associated with newcomers) that destroyed the possibility of regrowth; longer tidal cycles that allowed for higher rates of harvests before the tides retreated; and an increased periwinkle population that might be eating the dulse. People seemed unanimous about the lower amounts, but there was no consensus on the reasons for it. In March 2007, one of the dulsers said that they had agreed not to start harvesting until 1 May to "give it a chance to spore and spread."

Probably the most interesting expert harvester of sea vegetables was Paul Brown, who not only gathered dulse and experimented with various options for salvaging it during the long interludes of damp foggy days but who also picked nori, kelp, and sea lettuce. Largely ignored by other islanders, these other seaweeds also had important markets and sold at a premium if treated properly. They offered islanders particular advantages. For example, they are available earlier in the season, before the dulse is ready in the spring and, especially in the case of nori, are often available at higher intertidal zones so that harvesters are not dependent upon extremely strong tides for their gathering. The procedures for cleaning and preparing the seaweeds are often more specific than they are for dulse, which may explain why only two islanders participate in their harvest. Sea lettuce is dried, then washed with vinegar and again with sea water, and then it is dried flat. For this "they pay a lot," said Paul, "$12.00 per pound." He wasn't sure how the product was eventually used, although he speculated that it might be used like nori (i.e., in sushi). Sea moss, another desirable product used as a stabilizer in many food products, also requires special drying conditions that involve a solid surface and careful handling. One July day he had returned from collecting several different seaweeds, including nori, kelp, and sea lettuce. The twenty pounds of nori he carried in a basket would take longer to dry than would the equivalent amount of dulse, but it would dry down in a ratio of about 3:1 (rather than 6:1) and, at five dollars per pound, would be sold to Flagg at a higher price than dulse. He estimated that he would net about thirty dollars when it was dried. The three large sacks of sea lettuce would be worth, he guessed, between thirty-five and fifty dollars each, depending upon the market price, which had been dropping since French buyers located a local supplier. The basket of kelp he had was a new harvest for him, and, at the time, he was not sure that Flagg would be interested in buying it. Assuming he had indeed collected the right species, Flagg would pay him six dollars per pound dry, about 50 percent more than the current price for Dark Harbour dulse. He was hopeful that his three hours' work might bring him about $200. He explained that price varies according to local preferences, which for Japanese buyers is black nori, which has grown with minimal sun exposure (when not protected from sun by the shade of cliffs, nori is lighter in colour and less desirable in the Japanese market). He admitted that it is hard work but that "you're on your own, and you can come and go when you want to."

One of the frustrations for those who enjoy the flexibility of dulsing is the bureaucratic system that establishes different income levels for claiming Employment Insurance (EI) benefits. In other words, with regard to qualifying for EI, working on salmon sites counts as "labour" and requires a minimum of $2,500 income, whereas dulsing counts as "fishing" and requires $5,000 income. The two categories cannot be combined. Yet again, government regulations intrude in ways that establish boundaries within the community, provoking resentment and frustration. As an islander complained, her son could not work for a few weeks on a salmon site and then move into dulsing because the incomes would be in separate categories, neither being sufficient to claim EI. The changes to EI in 2000 caused much grief among the most marginalized fishers on the island – people who struggled to harvest dulse on wet slippery rocks or to dig clams in the slimy mud of tidal flats. As one seventy-seven-year-old woman, who lived in a tiny clapboard house with peeling paint and broken steps, explained, the new system required a total of 520 hours over twenty-one weeks rather than 420 hours over twenty-three weeks. That would make it extremely difficult for many islanders to qualify. "They'll have to go on welfare," she said, "so it'll still cost the government money!"

The issue of licensing for dulsing would be raised from time to time because it was virtually the only activity not regulated by the DFO. Some people felt that licensing would be a good idea as it would keep out new entrants, especially during the years of the sudden in-migration of Newfoundlanders. During the rockweed standoff in 1996, it was anticipated that the dulsers would be among the most vociferous opponents. In fact, during the early protests they were never heard from, and it was only later, when it seemed that rockweed would continue to be harvested, that people involved in the niche activities raised their concerns. One observer speculated that their silence may have been related to not wanting to raise the issue of licensing because then they might be included under the same umbrella of new regulations. As with all the marine resources, regulations were seen as both (1) useful for ensuring a measure of conservation and equity and (2) the cause of increasing disparities and unfair limits on incomes. Even the possibility of licensing provoked emotional responses on both sides of the question.

Periwinkles and Clams

As well as the seaweeds, periwinkles and clams are important parts of the niche economy because of their flexibility and easy accessibility, involving no capital investments and few regulations. While fewer people rely upon periwinkles and clams than upon dulse, they nevertheless represent another key safety valve for many islanders. "Wrinkles" (as periwinkles are called locally), for example, attract many islanders for short intervals when other work is scarce (e.g., during June, when the herring have not moved in or at the end of the lobster season into July). Even in the storms of winter it is not unusual to see a couple of men in their high boots, carrying pails across the rocks at low tide. While it is difficult to know their value to the island economy in monetary terms ($160,000 has been mentioned), for individuals they provide a significant niche for extra income. Because there is no licensing required, there are no precise records of how many people are involved, either regularly or sporadically, and all payments are in cash. In 1999, it was reported that Flagg had bought 28,000 pounds of periwinkles over a period of only five days, which, at one dollar per pound, would have brought the pickers $28,000. According to most informants, about seventy to eighty dollars worth of periwinkles can be collected on a tide, which many people consider to be "good money."

It was suggested that the numbers of periwinkles seemed to be increasing and that possibly this increase was linked to fewer sea ducks, which feed on them. Islanders are very sensitive to the changing environments upon which they depend, and they are constantly trying to figure out patterns and relationships. Regardless of whether the problem is changing amounts of dulse or periwinkles or unexpected herring arrivals or lobster harvests, fishers keep records, watch each other, and speculate about the future.

Unlike harvesting periwinkles and dulse, clamming does require a licence, the number of which is limited. As a result, there is a changing market value for these licences on the island. In 2000, a clam licence was valued at about $1,500, according to one informant, although, according to another, it might cost up to $6,000. Occasionally, the relatively safe clam beds of Grand Manan attract off-islanders such as Gordie, who came over from the mainland in 1997 in order to take advantage of the open clam beds. Because of the lethal red tide algae that had spread along the eastern seaboard, many of the mainland clam flats had been closed. In 1999, Gordie was able to dig enough clams on a tide – about eighty-five pounds – to ensure that he would have twenty pounds of meat to sell at $7.50 per pound, thus making $150 per day. Working six days a week, he said, he could make $1,000 per week – good money for him. The back-breaking work occupied him all through the winter and summer, when he wore rubber boots and used a specially designed shovel while clamming along the Thoroughfare (the channel between the main island and Ross Island) and at the Passage (the channel between Ross Island and Cheney Island).

For some islanders, such as ninety-four-year-old Smiles Green, the delicious clam itself is sufficient enticement to dig enough for a meal. One mid-June day he set out to dig a bucketful, which took him about an hour. Then he phoned to invite me for supper. In the meantime, he had them all shucked and cleaned, and his daughter-in-law fried them in a light batter of egg and cracker crumbs: a feast by any definition.

Other traditional activities that have sustained island food requirements over generations are gradually becoming less significant. Collecting gulls' eggs in early June was one of the activities that most older islanders recalled with fondness, both for the day's outings for entire families and for the reward of the large brown speckled eggs. The key was to ensure that they had not been fertilized, which was easily tested. The eggs would be put into a pail of water, and if they bobbed up at all it meant that there was a blood vein in them causing an air pocket and, thus, the flotation. These were to be rejected. An egg had to stay horizontal on the bottom of the bucket. Duck hunting was another of the traditional shoreline activities that helped provide for the larder. While even the fishy-tasting mergansers were historically hunted, in 2000 only black ducks were occasionally hunted by a few of the men during February. The neighbours of the men who enjoyed the hunt provided them with gun shells and, in return, would receive "a brace of ducks [four], beautifully cleaned, plucked and then singed. They would be washed with Ivory to remove the taste of burned oil and feathers, then rinsed. They were very greasy, with not much meat."

Also along the shore people would collect rockweed for their gardens. They would gather truckloads at a time to enrich the soil and to ensure healthy vegetable crops. Some islanders continue to harvest rockweed for their gardens, although gradually people are beginning to use chemical fertilizers, especially for the growing numbers of flower gardens.

Those who depend upon the niche fisheries (dulse, periwinkles, and clams) have a variety of reasons for their involvement. Randy had been raised in a non-fishing family and had neither the capital nor the expertise to run his own boat. He began dulsing for income; however, following an accident that resulted in the loss of his toes, he was unable to sustain the long hours of balancing on the slippery rocks and knew he would have to take up a more lucrative activity. He moved into fishing, working as a crew member before being able to buy his own boat and licences. Stephen enjoyed the independence of dulse harvesting and the fact that it entailed no individual financial outlay. Working the tides at all times of the day during the season (April to October or November) he was able to earn a respectable income (about $275 on a tide, for one hundred pounds of dried dulse), which he supplemented by crewing for a lobster fisher as well as, in some years, tending weir. Depending on the tides, people usually pick for about three hours, although occasionally it is possible to pick for up to four hours, gathering perhaps twelve baskets that would weigh about seven hundred pounds wet. The flexibility of these fisheries was a mixed blessing insofar as the hard work and long hours came at a price to Stephen's home life, which included young children. For his part, David moved into dulsing due to a terrible accident on a scallop boat that resulted in the drowning of several men. He escaped but would never again engage in scalloping. He began to gather dulse and periwinkles, which he sold to Roland Flagg, and to do odd jobs such as mowing lawns. With periwinkles at a dollar per pound for large ones and eighty-five cents per pound for smaller ones, he could gather a five-gallon bucket on a tide that would sell for between sixty and seventy dollars. On a tide he could harvest over 150 pounds of dulse, which would be equivalent to twenty-five pounds dry, depending on where it was gathered. For Passage dulse, he might earn about $100 for his three hours work of gathering, laying, turning, and baling. This would not account for the cost associated with fuel and boat maintenance.

The niche activities described here provide good incomes for some families but not without a cost. The literally back-breaking work and hazards of balancing on slippery rocks in the dim light of dawn or dusk are serious physical constraints on harvesters as they age. For other families, these activities have been important ways of supplementing other seasonal incomes, such as those provided by the scallop or lobster fisheries. For an outsider, the flexibility and adaptability that are characteristic of the niche harvests also reflect the work culture of the island more generally. As islanders move from one sector of the fishery to another, they are moving with the rhythms of the resource and the environmental conditions, responding to the natural cycles of change. In many ways, these work behaviours are carried into their capacity for finding solutions when the environment does not offer reliable harvests. Creativity and entrepreneurship have defined the adaptive strategies of islanders for generations, and the niche activities seem to best capture those admirable traits that are so central to the identities of the Grand Mananer.

In the next three chapters I explore the many ways in which sharing the marine Commons and the land Commons has evolved over several decades, but especially in the twelve years of my study. People involved in the wild fisheries allude to various government strategies to grant access to the fishery resources while trying to ensure sustainability – strategies that often have the unexpected impact of creating community divisions and social tensions. On Grand Manan, which has depended upon the fisheries for over two hundred years, there is a wariness towards government intrusions that is the product of decades of increasingly onerous regulatory regimes. This wariness permeates all discussions, largely because of perceived disparities that have been caused by such new regulations as ITQs and the establishment of a "core" classification. Social networks are intricately implicated in the regulatory regimes that create boundaries. The collision of traditional norms, values, and traditions of voluntary sharing with new realities of government controls, technologies, and markets is transforming Grand Manan's collective identity and its relationship to the world. Together with these economic and regulatory factors are the new migrants and the increasing contacts with the mainland through summer homes and tourism, which also contributes to the formation of new cultural meanings. As I discuss in the next chapters, new boundaries and edges of transition define future possibilities and hopes.

3

Boundaries, Edges, and Spaces of Meaning

Islands, because of their isolation, are revelatory, places where boundaries
are wafer-thin. ~ (Nicolson 2001, 140)

The wild fishery that has defined community relations and daily lives on Grand
Manan over generations has necessarily been subjected to an evolving series of
regime changes – governance structures established initially by the community and increasingly by federal and provincial levels of government. These
rules and regulations, some of which have been mediated by market forces, are
complicated by informal community-negotiated norms. Within the community there are also complex norms of behaviour related to spatial characteristics
of land use and social relations that have evolved and that may be subjected to
conflicts that can impair the community's ability to adapt to new realities. All
of these spatial, social, and economic dimensions interact to create a complex
web of boundaries and edges of transition. They equally affect the conduct of
fishery and community relationships, governance structures, and personal identities and notions of belonging.

In physical landscapes and biological domains, boundaries and edges are
crucially important, affecting wildlife management, biodiversity, erosion control, and microclimatic zones, along with many other areas of ecosystem relationships:

In questions of biodiversity and land fragmentation, the ratio of edge to
interior habitat is critical. In aesthetics, views are often dominated by
edges. In forestry, edges are commonly characterized by crooked trunks
and blow downs. In agriculture, edges are often the source of pests, as well

as predators controlling pests. In soil erosion control, boundaries cause wind speed changes and turbulence. (Forman 1995, 81)

Using physical metaphors to understand human relationships and interactions can be fraught with the danger of producing misleading ideas and images. However, the physical characteristics that describe and define boundaries and edges provide some interesting and helpful frameworks within which to characterize levels of complexity and the multifaceted nature of social, cultural, and economic changes. In the social world, boundaries can be seen as setting limits and defining separate areas of activities and interaction. They may be established through formal institutional processes of negotiation, as with borders between nation-states and provinces. Or they may be informally agreed upon separations of hunting territories – for example, those determined between Aboriginal hunters whose family histories have determined territorial grounds according to animal populations and seasonal rhythms. Between these formal and informal definitions of boundaries, there are infinite varieties of meanings and processes that mediate social relations within communities.

Each landscape element contains an *edge*, the outer area exhibiting the edge effect, i.e. dominated by species found only or predominantly near the border. The inner area of a landscape element is considered the interior or core, and is dominated by species that only or predominantly live away from the border. A *border* is the line separating the edges of adjacent landscape elements. The two edges combined compose the *boundary* or boundary zone. (Forman 1995, 85)

One way in which I consider boundaries is in terms of deterritorialized abstractions, defined with respect to fluid relationships. Sunil Gupta is an artist who has explored the ways in which modern relations and globalized structures are affecting our socio-cultural milieux. A project entitled "Disrupted Borders" illuminates the complex and myriad ways in which political, economic, and cultural dimensions have been transformed, leaving indelible imprints upon populations, landscapes, and relationships (Gupta 1993). An internationalism that appeared to offer hope of dissolving boundaries of conflict seems to have been stalled, she suggests, with borders becoming "ever more fortified." Calling upon the work of photographers and visual artists, she explores the notion of fluid boundaries that redefine the ways in which people are able to situate their lives. Concrete, specific local examples are numerous on Grand Manan, providing insights into the tensions of rapid community change. Property boundaries assumed through generations, for example, may come into dispute as land is subdivided and families sell tracts for which documentation is incomplete. Rights-of-way that have been used without question for decades may be challenged as pressures from tourism businesses transform the importance of beach access. Boundaries are not immutable, and in times of rapid change they are especially vulnerable. They become the battlegrounds in the struggle for establishing future rights of access and potential incomes. Legal, customary, or informal, all boundaries are interwoven into the social fabric of community relations and culture. They contribute to the meanings that individuals describe in their attachments to place, and, ultimately, they are inextricably linked to belonging.

David Sibley has explored the idea of borders in human relationships and the evolution of personal and collective identities that inform cultural change. He points out that "the boundary question [is] a traditional but very much under-theorized concern in human geography" and that groups and individuals who erect boundaries of exclusion can suffer from lives that become constrained, ultimately precluding resilience (Sibley 1995, 32). While the "need to make sense of the world by categorizing things on the basis of crisp sets" can provide security and comfort, there is also the danger of closed and non-adaptive responses to new possibilities.

Similarly, Edward Said describes the importance of how boundaries affect the impact of polar opposites. On the one hand, he says, they reinforce identity, and, on the other, they undo identity. He argues that there is a "perilous territory of not-belonging" in which people can become refugees with no sense of a community of belonging to provide them with a sense of rooted being (Said 1984, 51). In the final chapter, where I look at how Grand Manan might be "reimagined," I consider the possibility that, for some islanders, the changes will create a new world in which they can no longer feel that they belong. Bauman (2001) discusses the transformation of community identities that accompanies major economic shifts in another way. He describes the "cutting away from the web of communal bonds" that occurs when industrial activities separate livelihoods from households. Referring to Marx and Engels, he employs the image of "melting all solids" to describe the process of recreating social and productive relations within community settings: "But the melting job was not an end in itself; solids were liquefied so that new solids, more solid than the melted ones, could be cast" (30).

Polanyi (1944) describes the disembedding from social structures of the economy that accompanied modernity. Regardless of the metaphor used, the idea that the transformation of economic relations through new technologies, fishery activities, and global structures would directly affect social worlds that had been isolated in time and space by the island situation is crucial to understanding the profound nature of change on Grand Manan. Embedded in these changes has been the increasing role of government, technology, and growing levels of capitalization that are contributing to growing disparities between families and groups of fishers on the island. These disparities (or boundaries) create divisions that affect social cohesion and create problems related to youth cultures.

On Grand Manan the ultimate boundary is between insiders and outsiders, between those who are acknowledged by the community to be native islanders and those who will always be described as "from-away." As Lucy Lippard (1997, 292) has suggested, local life "is all about communicating across boundaries." She argues that the homogeneity of North American life threatens our senses of place and that we need to be "looking around and listening to each other" (ibid.). In the decade upon which I focus, even the boundary between insiders and outsiders is being dissolved through the addition of new and distinct identities. The migration of Newfoundlanders to the island beginning in 1991, but especially after 1995, established a new way of being in island culture – one that was neither "from away" nor islander.

As described in chapter 6, the historical backgrounds of the Newfoundlanders provided a completely different context for their move to Grand Manan, with implications both for their own self-identities and for the ways in which Grand Mananers accepted them into the community. The boundaries between Newfoundlanders and islanders have characteristics that differ from those historically applied to those "from away." The significance of new groups of immigrants was highlighted in an interview with a Newfoundland woman who had lived permanently on Grand Manan for seven years. Describing some of the difficulties, she commented that the recent (2001) moves of Aboriginal people to the island might take some of the attention away from the Newfoundlanders, so "they'll leave us alone." The boundaries between "us" and "them" might become less visible, less important, she thought. Perhaps just as important will be the more recent in-migration of retired couples who seem to be initiating a new "critical mass" of cultural transformation. The announcement in early 2007 of a new medical couple (married doctors) who had been enticed to the island by offers of a house specially purchased by the village council was good news for the community. But again, it suggested a changing notion of island belonging that would have to be more open to other cultures, in this case a family of Bangladeshi origins.

Within the context of changes on Grand Manan, the concept of habitus (Bourdieu 1984), explained in chapter 9, helps to give substance to the idea of dynamic boundaries and zones that define norms, values, and beliefs within social spaces for men and women, youth and parents, visitors and fishers. Again, as the following narratives makes clear, these boundaries are gradually changing, being transformed through complex networks of new values and new relationships. The early economy of Grand Manan was equally based in the fishery and farming, making islanders largely self-sufficient and outside the mainland money economy. This independence remains an important characteristic of islander identities, contributing to a pervasive entrepreneurial spirit and culture of adaptation within their specific island spaces.

In her exploration of the idea of boundaries, Barbara Morehouse (Pavlakovich-Kochi, Morehouse, and Wastl-Walter 2004, 20–1) suggests that "much of the power of 'boundary' as a concept lies in a certain kind of faith, faith that, through drawing lines, one will cease to be lost; that one will have form, substance, identity, protection, and shelter. Equally powerful is belief in boundaries as limits to be exceeded, lines to be crossed, barriers to be crashed." She argues that the relevance of both spatial scales and time scales to the construction of identities informs how we should understand the ways in which meanings are both affirmed and challenged. While spatial and time scales cannot be explicitly linked, taken together they are a useful means for "understanding continuity and change in the location, configuration, functionality, and importance of individual boundaries and portions of boundaries" (20). In a sense, the powerful geographic boundary of their island situation has been an overwhelming presence for Grand Mananers, for whom other boundaries of difference – between villages, churches, genders, and various degrees of fishing success, for example – all contribute to layers of personal identities that are both celebrated and resisted.

For men and women, the experience of gendered space has been a crucial formative reality that has affirmed and reflected gender relations over generations. For women, the male territories of the garages and sheds have involved boundaries that, while informal, nevertheless clearly defined exclusionary social milieux. Doreen Massey has explored the relationships between spatial relations and power structures, arguing that the uses of space and people's move-

ments through space reflect their sense of empowerment and their social interactions in the community. Historically, the woods have been spaces dominated by men for hunting and making their small camps. As well, the camps and
hunting platforms asserted claims over particular hunting areas. For the fishers,
perhaps the most overt claims to land spaces have been those allocated to work
activities, including net repair, preparing weir stakes and top poles, and launching boats. With growing incursions by outsiders, such as tourists, handwritten
signs at the Whistle Beach launch area and a NO PARKING sign at Whale Cove
have testified that local fishers feel that visitors have been transgressing important spaces of meaning and work. Sharing these common workspaces has been
fraught with tensions and negotiated solutions.

Newman and Paasi (1998) have also examined the relevance of boundaries
for understanding social patterns and the formulation of social theory. They
argue for greater awareness of the multidimensional nature of boundaries, the
importance of geographical scale, and the increased incorporation of environmental perspectives and multicultural approaches into boundary studies. They
categorize the ways in which boundaries can define fields of interaction within
communities, including boundaries as barriers, natural features, points of conflict or cooperation, and contexts for action.

An edge is a distinctive kind of boundary. In biology, edges are intrinsically rich zones of transition, having a biodiversity representative of at least two
biological zones, whether they are forest edges or the intertidal zones along the
shore. For birdwatchers, the edges of meadows or marshes are special environments that harbour a rich diversity of species. According to Professor Herman
at Acadia University, the biodiversity of non-human species is most threatened in the ·middle ground. Arguing against the conventional wisdom, he
points out that it is the populations in the edges that are most able to survive.
It is the marginal groups that are strongest (CBC interview, *Maritime Noon*, 23
May 2003). Applied to human populations, the idea of survival and inherent
strengths has merit. While these edges exist in nature, they may also evolve
within the context of social boundaries that are in the process of transformation. Edges incorporate constant ebbs and flows of tension and resolution,
and eventually they may become new and distinctive zones of social interaction. Edges can be both existential and creative, their characteristics determining social relations on the island. According to Turner, Davidson-Hunt, and
O'Flaherty (2003, 439), "Cultural transitional areas – zones where two or
more cultures converge and interact – are similarly rich and diverse in cultural
traits, exhibiting cultural and linguistic features of each of the contributing

peoples." While they go oh to argue that resilience and flexibility may be a significant benefit of these cultural edges, or transition zones, my work on Grand Manan suggests that the dynamic flux of relationships in these areas is more contradictory and less benign than they propose, certainly in the short term.

New spaces of meaning and personal and collective identities emerge out of the flux that occurs at the edges of social relationships. As Massey (1997) and McHugh and Mings (1996) have pointed out, a focus on place specificity provides an excellent basis for understanding diversity and difference and the inequalities generated by the changes wrought by social and economic restructuring. Massey's concept of place as a progressive entity with open boundaries emphasizes the social and cultural heterogeneity of places that are constantly being reconstructed.

In the preceding chapter I described the importance of the wild fishery as the fundamental defining characteristic of Grand Manan, alluding to how, throughout history, spatial relations are related to governance structures. The following two chapters focus specifically on how sharing the Commons of both land and sea is implicated in social relations, a phenomenon that is key to explaining tensions and conflicts within community culture. The period from 1995 to 2006 was one in which many new zones of transformation were being created and new spaces of meaning linked to identities and belonging were being redefined. Large areas of the Commons, on the sea, shore, and land, were being challenged, and it became necessary to share them.

4

Sharing the Commons:
Marine Spaces

Some of the most complex and contentious issues in the fishery pertain to marine-space management, encompassing the concepts of the Commons, property, and accessibility. According to Davis and Bailey (1996, 253), property is "as much an expression of social and political, as it is economic, relationships." The problem is in defining who has right of access, and by whose authority or mandate. Historically acquired rights, government regulatory regimes, and the changing dynamics of technologies and markets have all had roles in redefining the nature of access to marine spaces. Resource regimes regulate actions that govern both appropriation of the stock and the provision of its benefits. Scholars have explored the legal and moral justification for buying and selling something that has been defined as public property (Ommer 2000; Eythorsson 2000; Ostrom 1987, 1990; Liebcap 1989). Ommer has looked at the changing patterns of sequential occupancy in the east coast fisheries, describing the difficulties of defining ownership that arise partly because they are intrinsically related to the impossibility of settlement that allows a claim (as on land). She argues that the histories of different technologies, visions, and agendas have been underpinned by changing motivations and ethical frameworks in the context of differing ideologies: "However, what is fundamental is that, *in terms of access and control,* any one group or person who can claim and maintain exclusive access to a fishing ground holds *de facto* effective ownership of that fishery" (Ommer 2000, 119, emphasis in original). Marine systems are what Ostrom (1987) terms a common-pool resource, involving a dual challenge: (1)

to organize in a way that allows withdrawal of benefits within a system of multiple and competing uses and (2) to allocate production units among them. The concern here lies with the transformation of fishing access into capital that supports the increasing dominance of corporate interests and the alienation of local communities from their historically embedded relationships with marine resources.

Decisions as to the nature of property rights and who has access are crucial to the functioning of local economies and the social relationships within them. This is because the property rights institutions structure the incentives within society and also determine who will be the main actors (Liebcap 1989). The role of governments in establishing the nature of regulatory regimes is a crucial element in how access to the Commons is defined. The formal rules that they develop provide the framework within which communities assert claims on various resources and evolve their working and social relationships. Informal rules and mutually acceptable rules of compliance often overlay, or are embedded within, the formal structures. The lobster fishery provides an excellent case study of these mutually reinforcing formal and informal networks of rules. The proposal in early 2007 to fundamentally alter the nature of licensing provisions for fishing in Canada generated confusion, suspicion, and concerns that have not been allayed at the time of writing. The regime that the government has established is a mix of common property and privatized commodification of fish that varies between fisheries and that is fluid over time. As I describe for the lobster fishery, local rules and informal agreements can be an important part of the regulatory institutions.

Critics of the idea of common pool resource management argue that the Hardin "tragedy" will continue to dominate outcomes and that there is a need in Canada to change to a stronger economic, or market-oriented, system (Hardin 1968). In a column for the *Globe and Mail*, for example, Jeffrey Simpson decried the Canadian "mixture of some economics and various forms of welfare" that he said defines the Canadian fishery management system (4 April 2007, A15). Pointing to the relative success of new programs in New Zealand, Australia, and Iceland, Simpson suggested that Canada should "ditch common property resource regimes" and adopt a "share catch fishery, also called a transferable quota system." Citing a report by the non-governmental organization (NGO) Environmental Defense, he pointed out that quotas can be "given" to individuals, cooperatives, or communities that would be defined by scientific calculations. The report released early in 2007 was an analysis of "incentive-based management." It studied the impacts of these management tools on fisheries stocks in several countries and argued that those known as "catch shares," or "limited access privilege programs" (LAPPs), seemed to offer the best hope for aligning the interests of fishers and ecological concerns.[1] Ultimately, market disciplines would make it in everyone's interest to preserve the stock, a case that seems to be borne out in the experience of the three countries that have adopted it. What is not considered is the continuing impact of governments' subsidizing increasingly large boats and technologies that encourage "more efficient" harvest of marine resources. Efficiency and welfare do not necessarily complement each other. These issues are further addressed when I look at the lobster fishery (below).

For the most part, governments have been reluctant to assign property rights to marine areas large enough to cover roving species "viewed by politicians as interfering with the politically popular guarantees of the right to fish" (Liebcap 1989, 76). But aquaculture has not posed the same dilemma. Marine spaces of forty to seventy acres have been withdrawn and essentially privatized by the aquaculture companies, who have been emboldened and subsidized by the provincial government (Marshall 2001a). Whereas boundary permeability in the lobster fishery has been negotiated within historical and informal community arrangements (Acheson 1987), and residence in a local community in Newfoundland provided the *de jure* right to compete for local fish (Andersen 1979), in aquaculture there can be no such informal, community agreements.

Aquaculture and Herring

Herring weirs and salmon aquaculture farms represent two extremes in time, technology, corporate structure, and social relations. They intersect in the negotiations for suitable sites. The competition for good aquaculture sites has focused on weir sites because they provide protection from prevailing winds, storms, and strong tides; good water circulation; and access to a harbour. This has meant that, beginning in the early 1990s, weir sites became increasingly valuable, marketable commodities. For fishers who no longer wanted to participate in the weir fishery, there was an opportunity to lease their sites to aquaculture companies. Suddenly, a site that might have sold to a local family for $12,000 in 1990 became a financial investment that grew to about $1 million in 2004.[2] In 2001, Connors' purchase of a site for a reported $300,000 concerned local weir fishers because of the possible conflict of interest. Moreover, a weir site located in an area vulnerable to storms, or a site in shallow water, would be valued less than those sites with good protection and harbour access. By 2001, virtually all of the viable sites had been allocated.

The basic requirement of aquaculture for access to and control over marine sites created a new "resource" in that it resulted in redefining marine sites as private property and attributing market values to them. This move to essentially privatize marine space represented, according to one scholar, a "distinctive form of neoliberal practice that uniquely combine(s) private industry and government regulation" (Mansfield, 2004 565). It is a major "geographical transformation in the political economy of the oceans." The specific sites that have been given value are described as "weir privileges." For generations, families on Grand Manan have built and tended their heart-shaped weirs, located near the coastline in water usually about six to eight fathoms deep at low tide. Characteristics of shoreline conformation that might direct the feeding herring inward – protective shoals, islands that offer shelter from destructive storms, and strong tidal currents that wash through the area twice daily – all describe ideal conditions for both herring weirs and aquaculture cages. The historic weir privileges, originally acquired through family lineage and, more recently, controlled by the federal government through licence fees, became highly desirable locations during the early 1990s and especially after the new policy of 2000, when it seemed the industry was on an unlimited growth trajectory. However, that the marketable value of weir privileges as aquaculture sites had limits became apparent in the spring of 2005, when a major restructuring of the industry occurred, mainly as a result of declining profit margins related to low market prices for salmon. As islanders scrambled to protect their investments in the industry, sites that had been valued at about $1 million in 2004 were estimated to be worth less than half that one year later. The optimism that characterized the rapid growth of aquaculture through the 1990s until 2003 was quickly eroded as market prices for farmed salmon plummeted in 2004. As the industry restructured, there were calls for government aid, and eventually the federal and provincial governments offered support of $30 million in July 2005, which did not preclude bankruptcies and the sale of assets that spring.

In the early years of negotiations some fishers were able both to retain their weir privilege and to obtain rights to a salmon site. In the later period of development of the industry on Grand Manan (after 1999), islanders bought up former weir sites with the intention to "build" them for a year and then convert them to a salmon farm, earning rent or investing as a part owner. The old weir sites around Wood Island, for example, suddenly became attractive to local

entrepreneurs who purchased the privileges, built a weir for one year, then negotiated a salmon farm contract with one of the large aquaculture companies.

Typically, the leases were contracts for three to four years and were held by
an islander for about twenty years. For example, in 1997-98, one site was the
subject of a bidding competition between companies whose offers varied from
$20,000 per year for three years to $60,000 annually for four years. Another,
more favourable, site had purportedly garnered $75,000 annually for three
years for a local weir fisher. In another case, three men who had never fished
herring weirs before bought a weir privilege, and the next year Stolt Sea Farms
prepared the application for a salmon site (at an estimated cost of $50,000) and
eventually took over management of it. According to one industry critic, the
three islanders reportedly received between $120,000 and $200,000 annually
for "doing nothing." However, the new market for weir privileges meant that
some families were able to negotiate high selling prices. One islander bought
up four privileges, one of which cost over $100,000, with an agreement that
there was to be a commission paid out as a percentage of profits every two years.
The years during which the expansion of the industry was at its peak, 1995 to
1998 and 2000 to 2001, were years of great instability and debate, but also of
high optimism, on Grand Manan.

While the New Brunswick government clearly supported the growth of
salmon aquaculture, the DFO's regulatory jurisdiction of the weir sites and the
active hostility of many traditional fishers to the new, high stakes industry
were both factors that created a difficult social milieu for community change.
Within the community there were, in the early period prior to 1998, clear lines
of separation between the families who had been quick to take advantage of
leasing weir sites and becoming involved at various levels in the aquaculture
industry and those who resisted any growth of that industry. And yet, these
clear lines soon became blurred as more and more young people found jobs on
the salmon sites, thereby implicating more and more families who, however
begrudgingly, acknowledged the value of jobs for their children. The tension
between traditional fishers, who argued that salmon farming would degrade
the environment, especially affecting lobster stocks, and local representatives
of salmon companies, who were based on the mainland, reached a peak in the
spring of 2001.

As a direct result of a new provincial policy, published in October 2000, a
new system of site allocations took effect in January 2001 (New Brunswick
2000). Fifteen site applications around Grand Manan over the next three

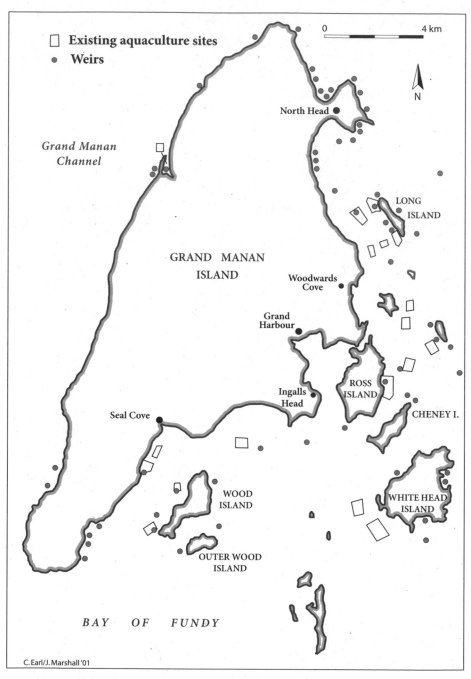

□ **Existing aquaculture sites**
● **Weirs**

0   4 km

N

Grand Manan
Channel

GRAND MANAN
ISLAND

North Head ●

LONG
ISLAND

Woodwards
Cove ●

Grand
Harbour ●

ROSS
ISLAND

CHENEY I.

Ingalls
Head ●

Seal Cove ●

WOOD
ISLAND

WHITE HEAD
ISLAND

OUTER WOOD
ISLAND

BAY   OF   FUNDY

C. Earl/J. Marshall '01

Aquaculture and weir sites in 2000

months generated an outcry from many island residents. Among the most vocal were the lobster fishers, who felt that their traditional fishing grounds were about to be alienated from them. The inshore ocean Commons was to be privatized in a way that would far surpass in effect the "privileges" of the weir sites. The salmon sites, marked by privacy signs, occupied several acres in area: they were leased from local weir fishers but controlled and managed by mainland and (until 2005) Norwegian companies. And the effluent produced by the sites threatened the natural ecosystem. Unlike weirs, the sites had no family history associated with them, and any understanding of their interaction with the natural environment was negligible. This was a high-technology, tightly controlled industry whose owners were unknown to most islanders. As government approval for various sites was announced, hostilities grew, until in May 2001 a blockade of two wharves by the lobster fishers resulted in physical altercations and a counter-blockade of the island airport by aquaculture workers. Tension was high. Nonetheless, when the media requested interviews, typical island reactions were muted since no one wanted to be seen to take sides because everyone had relatives on both sides of the question. One lobster fisher was quoted in the newspaper: "The fight for the waters of the Bay of Fundy has changed the way of life on the island" (*Telegraph Journal*, 25 May 2001).

In 2001, the fifteen applications were eventually screened to allow six additional sites, bringing to twenty-four the number of aquaculture sites around Grand Manan. The year 2001 was pivotal. In many ways it serves as a boundary between the ongoing dominance of the traditional fishery within island culture and collective identities, and the transition to increasing reliance upon aquaculture, with concurrent social implications related to jobs for youth, money and drugs, and the in-migration of new permanent residents both as seasonal site workers and as personnel to fill highly skilled jobs (such as manager of the new hatchery). Owned by the only independent salmon company with direct island ties, (and described in more detail in the next section) Fundy Hatchery was built in 2002 and had its first full production year in 2003-04.

The social impact of aquaculture has been, and continues to be, the single most important change experienced by Grand Manan in its history. Despite a local description of postwar changes on the island that depict a comparable level of growth in the late 1940s and early 1950s, this recent period seems to involve a much deeper and more complex network of changes. The earlier description hardly touches upon the depth of meanings associated with current changes: "Progress with a capital P had come to Grand Manan. The high school was open, and the little hospital was proving its worth. The fishing boats

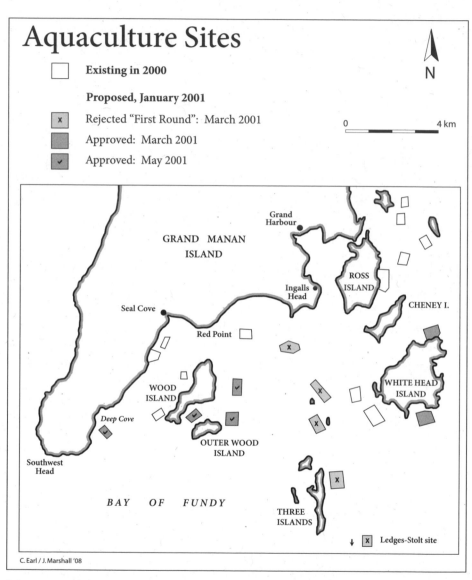

Aquaculture Sites

N

☐ Existing in 2000

Proposed, January 2001

☒ Rejected "First Round": March 2001

▨ Approved: March 2001

☑ Approved: May 2001

0   4 km

Grand Harbour

GRAND MANAN ISLAND

Ingalls Head

ROSS ISLAND

CHENEY I.

Seal Cove

Red Point

WHITE HEAD ISLAND

WOOD ISLAND

Deep Cove

OUTER WOOD ISLAND

Southwest Head

BAY OF FUNDY

THREE ISLANDS

☒ Ledges-Stolt site

C. Earl / J. Marshall '08

Salmon aquaculture sites in 2006

were bigger and better, equipped with ship-to-shore radios. The main road was paved, and lo and behold, there was an airstrip at North Head" (Small 1989, 81). Despite concerns about the wild fishery and the tensions that began to surface between the two sectors, the growth of aquaculture had created a spirit of optimism in the period leading up to 2001, and this was reflected in the growth of the population, with an increase of almost three hundred people between 1996 and 2001. For young islanders in particular the future seemed positive: they could earn steady dollars immediately upon leaving high school, and there were new businesses opening while others were expanding (notably trucking, two new fish hatcheries, and the machine shops). Overwhelmingly, however, it was the change to a wage-based economy, over which local residents had almost no control, that fundamentally altered both productive relations and personal and collective identities. One local observer suggested that aquaculture has essentially "commercialized" the island.

Regardless of the increased presence of salmon farming, and its enormous economic impact, for many island fishers the weirs continue to provide both substantial incomes and rewarding work for teams of seven or eight men. There has been much speculation about the impact of salmon sites upon the weir fishery but very little scientific information. Because of the overlapping site requirements, and the close proximity in certain areas of salmon farms to weir sites, the fishers carefully record the yearly fluctuations in total herring harvests and in the specific weirs that successfully capture significant amounts. Whereas the weirs south of Bancroft Point and around Wood Island have not taken much herring for at least a decade, the weirs on the back-of-the-island (Bradford Cove, Money Cove, the Gully) and those in Whale Cove (notably Brown's Weir and Neil Morse's Fish Head) and around Swallowtail (the Intruder and the Iron Lady) had remarkably consistent harvests over the period from 1995 to 2004. None of these three areas has any salmon farms nearby. In 2004, the Gully and Money Cove weirs, owned by Connors and managed by islander Jeff Foster, together brought in an astounding seven thousand hogsheads. At an average price of $125 per hogshead, those two weirs alone would have earned almost $1 million in a four-month period. While reliable figures are notoriously difficult to obtain, many men keep careful records on a daily basis, collecting the information from conversations throughout the day.

The most common suggestion was that herring would avoid areas "downstream" of salmon sites because they are known to avoid the dead of their own species. One fisher, in a 1997 interview, said that initially he had fought against salmon farms because he believed fish farming was "killing the herring runs."

But he acknowledged that he could not prove this and only pointed out that his Long Island weir had not fished well for seven or eight years, ever "since salmon farming took over." For the early first few months that young smolts are in the cages they are fed a mixture of dry feed and "wet" (i.e., herring) feed. It was felt that the herring would avoid areas where this wet feed was being used. Apart from the fact that in 2004 a far lower percentage of wet feed was used, and for a much shorter period of time in the grow-out cycle, than was used ten years earlier, and given the historic unpredictability of herring runs, it would take at least two decades worth of data to establish the veracity of any relationship between feed and herring runs. Today, the men are much more circumspect in attributing any such relationship.

On the other hand, in the location of one salmon site just above a weir at Pat's Cove, there was considerable debate based on the direction from which the weir traditionally captured the fish. The original proposal for the siting of the salmon farm was seen to be in a direct path with the incoming herring; therefore, the site was forced to move further away. Another issue raised by the weirman's association was the use of "pingers" in the salmon cages to scare away seals. In a 1998 letter to the association, the salmon growers denied that the pingers would have any impact upon the herring runs. Still another concern was the excrement itself and whether or not it would contaminate the sea bottom under the cages. While the weir fishers continued to express doubts about the impact upon their herring catches, lobster fishers began to speculate that the salmon cages actually supported lobster because they provided extra feed. While some fishers suggested that catches were higher close to the salmon sites, other islanders complained that some lobster tasted of salmon feed. The boundaries between weirs and salmon farms are constantly being monitored, and eventually any relationship between salmon farms and herring runs may be more clearly established. In April 2005, the widely reported results of a British Columbia study (Krkosek, Lewis, and Volpe 2005) was able to show a clear relationship between the siting of salmon farms and the spread of sea lice in the wild population. However, the challenges of establishing relationships to already unpredictable herring runs would appear to be quite daunting. When one weir fisher complained in 1998 that a new tourist establishment was creating too much light and that this prevented the herring coming inshore, islanders laughed, pointing out that they used to use flares to attract the fish. But island fishers will continue to look for patterns, relationships, and explanations to help them predict and gain an edge on their neighbours.

Weirs and salmon cages in Dark Harbour

The co-existence of the herring fishery and salmon aquaculture – one a traditional fishery, the other a modern industry – has been questioned by researchers and media commentators. Weirs have historically been owned by family groups as "privileges," registered with DFO and subject to a small annual licence fee. Increasing involvement by Connors, which bought majority interests in four weirs along the back-of-the-island form the 1990s to 2004, suggested growing corporate control of all aspects of the herring fishery, and this corresponds to the global trends towards integrated, consumer-led marketing systems for food. Weirs, which are dependent on unpredictable runs of herring and rudimentary fish traps, stand in stark contrast to salmon aquaculture, which is a high-investment enterprise dominated by large corporate interests. Similarly, the involvement of families in the weirs (beginning with the early morning visits to the boat launches), the teamwork, and the informal gatherings at the shore to watch the boats at work in the weirs reflect an attachment to the sea formed of shared experiences, hopes, frustrations, and struggles, none of which is replicated on the salmon sites. Very labour intensive, the weir fishery demands a lot of knowledge and a variety of skills.

With regard to salmon aquaculture, site managers, usually Grand Manan men between thirty-five and forty-five years old, are hired by the mainland companies and have differing levels of responsibilities, depending on the specific contractual arrangements of each site. In the early period of 1995 to 1998, many relied on learning by doing, and many mistakes were made. But gradually, as a technical school on the mainland that offered aquaculture courses began to graduate students from its one-year program, men came to their jobs with improved levels of basic knowledge. For some, such as Nathan Bass, whose natural talents and commitment have contributed to his success at on-the-job learning, the opportunity to be a site manager has been an important factor in his ability to succeed on an island where success has traditionally been defined within the traditional fishery. Whereas one company will encourage the site manager to organize and supervise the building of the site, including the design of anchoring methods, another company will undertake that responsibility with a mainlander, and, only upon completion of the site, will the manager be hired to supervise the feeding of the fish. Site managers on Grand Manan are paid a salary that varies widely, from about $35,000 to $65,000 per year, but for some there are no additional health and pension benefits.

For the vast majority of site workers, the work is unskilled and subject to seasonal vagaries. While young men frequently leave school as soon as they legally can (age sixteen) in order to work on the sites at wages beginning at nine dollars per hour and rising to about twelve dollars per hour, the average age for the strenuous work, which involves lifting fifty-five- to sixty-six-pound tubs or bags of feed, is about thirty years. One man in his mid-fifties who had lost his job moved onto a salmon site, but he was able to work for less than three months due to the back-breaking lifting that was required. Leaving the wharves about 7:00 AM, the small flotilla of refitted former lobster boats carry the groups of three to five salmon workers out to the salmon sites, where they work about twelve to fourteen hours a day in the summer months, feeding salmon three times a day. One observer of the changing employment patterns among young men just out of school expressed concern that in "the early eighties they made more shucking scallops than they do now. These young guys don't realize that the economy can change quick. Grand Manan has gone up and down so much over the years." Another astute observer decried the change in a dominant pattern of waged labour that completely altered the meaning of work.

Compared to the weir fishery, salmon aquaculture offers more steady work in season but without the variety of skills and activities. The nature of work in the weir fishery may include everything from stake preparation to mending twine to a variety of roles during the actual harvest. As well, the traditional weir fishery allows for greater flexibility in participating in other fisheries (such as dulse and periwinkle) or other jobs (such as those available on the ferry). Among the teams of men who work in the weir fishery there is a camaraderie and satisfaction that is absent from those who work on aquaculture sites. Compared to the work in the weirs, work on the salmon sites is tedious and unchanging. Apart from the site managers and the skilled divers who check for net repairs, structural problems in the cages, and fish mortalities, the variety of jobs and skills is limited. The result is frequent boredom and, for many workers, the opportunity to indulge in drug use. While the problem of drug abuse on Grand Manan is not directly caused by the advent of salmon aquaculture, two factors certainly show that there is a relationship. One is the amount of money paid to young workers who continue to live at home in a community where choices for activities are minimal. The second is the at-work boredom and relative isolation from community-enforced norms of behaviour. When a committee of concerned citizens approached the aquaculture companies requesting a "no tolerance" policy for drugs on the sites, they were refused. As one committee member told me, "They won't comply because they can't enforce it," repeating what she had been told by a company representative. Although there may be exceptions, in the traditional fishery on-the-job use of drugs and alcohol are rarely a problem because of the known risks and long histories of ocean deaths in all families.

The transformation of social relations on the island has been from one defined by family lineage and teams of men of all ages who all contribute to an unpredictable harvest to a situation of predictable daily work demanding few skills and requiring mostly young strong men. Working as unskilled labour in an industry that responds to corporate strategic initiatives and global market patterns, individuals have no control over their daily working lives. There are no shared stories. The gathering of men on the wharves or in Whale Cove to hear news of how many hogsheads were caught, and the speculation as to when the fish will come in or how loaded a carrier is as it leaves a weir, are fundamental to a sense of oneness between the environment and the community. The conversations, stories, and myths that grow out of the experiences in the weir fishery have been part of island history and collective identity for generations. Aquaculture offers no such continuity. For social relations the transformation is not across a zone, as it is between different fisheries; the change is radical, representing an entirely new dimension, across a chasm into a different world.

In New Brunswick, the early development of the industry, according to government spokespersons, was based upon a philosophy that had promoted and anticipated local participation in the ownership and development of aquaculture sites. The responsibility of the provincial government (rather than the federal government, which is responsible for weir licences as part of the traditional fishery), aquaculture policy was not systematically articulated until after the disastrous epidemic of infectious salmon anaemia (ISA) in 1997–98, which spread through the Passamaquoddy Bay sites, culminating in a moratorium in the establishment of any new sites. In 2000, a radically new philosophy was promulgated and was set out in a new policy that was to be implemented with the allocation of new sites in the spring of 2001.

Until 1999, the province had encouraged site development and set minimum standards for locations and company entry into the industry. Owners of weir privileges were able to conduct negotiations for salmon sites with interested companies, essentially renting out their sites over a specified leasing period. In the expansion that took place between 1991 and 1995, from two sites to ten sites, and then an additional eight sites between 1995 and 1998, negotiations hinged on the leasing arrangements related to term, contractual arrangements for building the cages and anchoring them, and the eventual management and "ownership" of the smolts and salmon. Controversially, the owners of the weir privileges themselves did not have to invest either labour or capital in the development of the site, allowing the company free rein over a now "privatized" part of the marine Commons. Typically, the lease arrangements would be for three or four years, at $25,000 to $45,000 per year, depending upon the timing of the negotiation and attributes of the site. The desirability of these sites created a new market for weir privileges that had, for generations, been informally traded between family groups. During the negotiations, a variety of arrangements was possible, including outright sale of the privilege for $250,000, which in fact did not occur (until the major upheaval of 2005-06), though it was theoretically possible. Especially after the announcement of the new policy in October 2000, when the government lifted the moratorium and it was apparent that the number of potential sites was rapidly diminishing, weirs that had not been built for years suddenly became marketable commodities. Unaffiliated individuals organized new business groups to build a weir, thereby establishing their rights to a site. They were then in a position the following year to lease it to a mainland or foreign company as a salmon farm.

The earliest site, established in 1980 in Dark Harbour, was built by two local entrepreneurs, Glen McLaughlin and John L'Aventure. During the early years, despite the challenges and hazards of on-the-job learning, including devastation by disease, it was possible to maintain profits because the prices for salmon remained high throughout the 1980s. In 1989, another business partnership established a second site at North Head. Two years later, a third site was established in the Long Island area near North Head. However, the original North Head site met substantial local opposition from lobster fishers who claimed that the area was a nursery for lobster. As well, the inauguration in 1990 of a large new ferry provided another argument against the site as the ferry captains complained that any malfunction of the boat would be dangerous for both the site and the ferry itself. The site was closed in 1991, just as others were being considered for the areas around Long Island. Then John l'Aventure moved his early Dark Harbour site to Seal Cove Sound.

Over the next few years, to 1997, new sites were built under a variety of contractual and leasing arrangements that involved varying degrees of local control over the hiring of labour, sources of feed and smolts, and when fish could be harvested for market. Supported by the entrepreneurial acumen of McLaughlin and islander Carmen Cook, a net company was established (SeaNets) that was eventually bought out by the main worker, who had arrived from Newfoundland in the early 1990s. As well, a company (Northern Plastics) under the ownership of l'Aventure was formed to manufacture sea cages. This presaged an important strategy that characterized the business enterprise of John l'Aventure. With the development of Northern Plastics, Fundy Aquaculture effectively controlled a significant part of the start-up costs and was also able to benefit by sales to other sites. Eventually, as described below, he added a hatchery. The original one was located in Grand Harbour and operated for only three years, from 1995 and 1997. His second attempt involved a larger and more technologically sophisticated operation, built in 2003 but only surviving until the forced sale of the entire company in 2005. The addition of a fish processing plant (2003) in order to establish a totally integrated organization was meant to allow him to control his major costs and, crucially, the timing of the entire flow-through process.

For John l'Aventure, the most entrepreneurial and forward-looking of local investors, the crucial strategy was to develop a totally integrated process that would make him independent of the control that large multinationals held over prices, supplies, and markets. As well, it was crucial to ensure timely flow-through for each stage of the grow-out process, from eggs to fry to smolts to marketable salmon and, finally, sales. This flow-through was dependent on factors such as date of acquisition of eggs, fry, or smolts; feed deliveries; and access

Top: Fundy Aquaculture: Conversion of smoke sheds
Bottom: Fundy Aquaculture: Salmon processing (2003)

to markets. Beginning with the problem of mortality rates for smolts and the high cost for each young fry, and extending through the entire system to processing, l'Aventure recognized that cost control in a highly competitive market would be key to survival.

Unfortunately, as well conceived as his strategy was, ultimately, in 2005 he seems to have suffered from basic lapses in business practice: overextended bank loans, insufficient hands-on management, and overly optimistic trust in relatively untrained employees. On 16 May 2005, he was forced to sell a company that he had struggled with and poured his life into for twenty-five years. Aquaculture is a high-risk, high-stakes industry in which, unlike the poultry industry, many of the factors of production (grow-out rates, mortalities, water temperatures, market prices) are extremely difficult to control. L'Aventure's determination to have an integrated company that would maximize the opportunities for local control certainly provided the basis for success. Its relatively small size, however, militated against having the degree of flexibility enjoyed by larger companies. On the other hand, whereas an independent owner such as l'Aventure could choose to buy feed (the single greatest cost factor) from any one of several companies, most salmon sites were owned by multinational corporations such as Stolt and Weston, which affected prices and limited their feed suppliers to specific companies. For example, whereas the New Brunswick company owned by Glen Cooke charged $1,200 per ton for feed in 2005, Heritage, owned by Weston, charged site operators $1,540 per ton. With feed representing almost 90 percent of total grow-out costs, this difference of over 20 percent was obviously a major issue for salmon site operators and, ultimately, for the profit margins of the smaller companies.

During the early years of aquaculture on Grand Manan, the complexity and diversity of the contractual arrangements between site "owners" and the large mainland companies related to length of lease, ownership of cages and equipment, ownership of the fish, and employee relationships. So, whereas some sites were built for and owned by Stolt (such as those on White Head Island), and only managed by a local islander, the equipment for others, such as Ross Island Ventures owned by Kenny Brown and the Seal Cove site owned by Glendon and George Brown, was locally owned. In other words, boats, cages, and nets were all owned by the islander, who then rented out the cages to Stolt or Cooke to grow out the fish. The variety of operating agreements extended to the fish that were being grown out. For sites such as those owned by Ron Benson and Kenny Brown, for example, while Stolt "owned" the fish in most of the cages, each site also had cages dedicated and owned by the local site operator and owner. While this offered the possibility of greater flexibility regarding when to harvest fish, thereby taking advantage of higher market prices, the logistics of transporting feed and smolts demanded that the islanders use the same corporate supply chain for their cages as was used for those owned by Stolt. In other words, control effectively remained with the foreign or mainland company, even when local entrepreneurs "owned" their cages. Moreover, in tight markets the multinational company had priority selling rights, which constrained the ability of the "independent" islanders to sell their fish at the most lucrative and opportune time. This was a serious problem because of the need to clean out sites and restock during the prime growth period provided by the summer months. The implications of these constraints permeate many aspects of socio-economic relations on the island. While islanders are clearly apprehensive about losing control over their lives as the large companies take over, they also acknowledge that these companies offer some advantages. For example, it was pointed out that the benefit packages offered by a company such as Stolt is usually better than those offered by local owners. However, one worker was unequivocal in his feelings: "What's the difference? Islanders exploit you, so the big companies do it. What's the difference?" For example, in 1997 and 1998, there was a major outbreak of ISA, which attacks the liver function of salmon. Islanders whose sole interest in aquaculture was as "landlords" were not adversely affected; rather, it was the companies based on the mainland or in Norway that suffered the losses. For independent owner John l'Aventure, on the other hand, the disease was catastrophic, almost bankrupting him with losses of over $1 million. The process of getting rid of the disease involved disinfecting everything: pens and nets had to be destroyed; chains and anchors were disinfected and reused; the wooden parts of scows were removed and burnt; clothing was destroyed. By 2000, when the province was developing its new policy for aquaculture, it had become apparent that these contractual arrangements were a source of difficulty for the mainland companies, who were unable to borrow money from the banks or ensure sale of assets because, in fact, the "assets" were intangible contracts. Moreover, there was resentment that some islanders were simply landlords, who had taken no financial risk and yet were collecting an annual rent for their weir privileges, which had become salmon sites (Marshall 2001a).[3]

The new policy enunciated in October 2000 effectively prohibited new entrants into the industry and established areas around the island that delineated "single year class sites." This was known as the Bay Management Area Framework. In other words, to avoid the epidemic spread of disease between grow-out years, the government decreed territories that could not contain different ages of fish. All fish in one site would be on the same grow-out cycle. The problem for salmon farmers was that, in planning for buying feed, hiring labour, and operating boats and scows, there was an important economic dimension that dictates the need for two sites (i.e., two different grow-out cycles) in order to allow for balanced production flows.[4] This created instant demand for second sites for all companies in the context of a two-year grow-out period and the economics of labour and capital efficiency. Following the announcement of the new policy, therefore, there was a race to acquire additional suitable sites, with a concurrent rise in the value of weir privileges. Indeed, the weir privilege itself became not only marketable but also a form of leverage for obtaining bank loans. One of the local investors talked about his "sweat equity," criticizing the weir fishers who, as landlords, leased their sites. He supported the new policy, which effectively prevented non-invested people from entering the industry: "Why should they be allowed in? We had to work hard not to be sucked up by the pressure of the mainland companies. I have no sympathy for the guys who are hollering now."

In 2001, when the moratorium on new sites was being lifted, apprehension among Grand Mananers focused on two issues. The first was the concern of traditional fishers that aquaculture was taking over large areas of the marine Commons, "taking bottom" from the lobster and sea urchin fishers. As one media commentator noted, "No, I'm not anti-aquaculture at any costs; I just don't think it should take over" (D. Wilbur, *Courier Weekend*, 2 February 2001). The second issue was the concern that the big companies were receiving preferential treatment from the government. As the mayor at the time, Philman Green was asked to solicit the support of municipal council for the expansion of the industry on the island. As he pointed out, "economic development is important, and we can't slough it off"; at the same time, he acknowledged that some fishers felt that council should not be actively supporting off-island corporate interests. As a result, council sent a letter to the New Brunswick Department of Agriculture, Fisheries and Aquaculture (DAFA) registering their concerns and requesting fair treatment for local investors. This was indeed a pivotal year for the island. Even as new sites were being approved, storm damage in March devastated one of the sites, destroying five of twelve cages. The financial risks being assumed by several Grand Manan investors were substantial and, over the next few years, would prove to be unsustainable.

In the early years, the community resisted the introduction of aquaculture. With concerns about the alienation of lobster beds, "taking the bottom," space on the wharves, and a changing social hierarchy, Grand Mananers were apprehensive about impending changes. Lawrence Cook's media "fight for the waters of the Bay of Fundy" expressed the pent-up frustrations of many islanders who feared for their livelihoods (*Telegraph Journal*, 25 May 2001). With the lifting of the provincial moratorium in October 2000 and the rush to establish additional sites under the new policy, it was obvious to the community that state and commercial interests had overtaken local control of both the economy and the marine environments. The competitive interests between local fishers and the basis of personal and collective identities were being transformed. The introduction of new productive relations, especially with the dominance of waged labour, removed the intricate embeddedness of socio-economic relationships that had defined community structures for generations. One local fisher was adamant that aquaculture threatened his livelihood: "When rats are cornered, they're going to fight. That's how we feel" (April 2001). At the beginning of 2002, several groups formed a coalition to oppose further development of aquaculture on the island. Spearheaded by Connors, the GMFA, the Fundy Weir Association, and one other partner met to discuss how they might stop further growth. In fact, nothing came of that initiative and, instead, a Connors subsidiary – Heritage Salmon – was formed. The tense relationship between weirs and salmon sites was to continue for another several years as the inexorable forces of globalization gathered momentum.

Lobster

The regulations and rules that define who has access to the lobster fishery are an interesting and complex mix of legal and informal arrangements that have evolved over generations. Number of traps, season, size and fertility status of lobster, as well as a specific territory are the critical legal constraints that the DFO established in its allocation of the resource and its attempts to ensure a sustainable fishery. In addition, however, while the DFO clearly delineated Area 38 as the area within which Grand Manan fishers may set traps, and it has established the season as being from the second Tuesday of November until 29 June, the island fishers have superimposed on this a secondary level of informal community rules to establish the territories for individual fishers. There are three main areas: "inshore," "ledges and south," and "deep water," or "mid-channel." Each of these is fished by different people, depending upon their home wharves, size of boat, and investment in new, larger traps and trawls. Over the decade after 1995, however, these traditional areas, mainly because of evolving technology, moved towards larger boats and traps. The official setting day (always a Tuesday) was originally determined based on the religious culture of the island, which prohibited working on Sunday. Because the bait bags have to be stuffed close to the actual day of setting, a Monday opening of the fishery would have required the men to fill their bait bags with the foul-smelling herring and fish heads on a Sunday. Therefore, Tuesday was accepted as the official day for opening the lobster season.

There is probably no more interesting system of informal community rules regarding the division of the marine Commons than that which exists for the lobster fishery. The work of James Acheson (1987) on the Maine lobster fishery illuminates the problem of "boundary permeability" and the role of political manoeuvring with regard to differential boundary-maintenance mechanisms within the context of community norms. His studies showed that access to the lobster fishery and territorial limits to individual fishing grounds were directly related, and defended, through a clearly defined set of normative rules that had been established over generations. In Newfoundland and elsewhere as well, "residence in the community and economic dependence on local marine resources gave one the minimal *de jure* right to share a common base of operations and compete for local fish" (Andersen 1979, 304). Andersen's work in Newfoundland also examines issues of public and private access, focusing on the regulatory strategies that evolved as a mix of formal-legal constraints and community-based rules of access and privileges. He describes how Newfoundlanders have developed a variety of techniques by which they influence, if not directly allocate, and balance their numbers and effort against fishing opportunities (329).

On Grand Manan the informal rules have historically provided effective frameworks for allocating territories for the individual fishers. The traditional areas for setting their traps have been partly related to distance from their home harbours, a restraint that has been considerably weakened over the past decade with the increasing numbers of large and powerful boats. Until recently, one could predict that boats leaving from Ingalls Head, for example, would fish to the east and southeast of the island, whereas those leaving from North Head would fish around the northern ledges. Seal Cove fishers have historically set traps southward, around Machias Seal Island and the Ledges. A study by a master's student at Dalhousie University explored the processes of negotiations and

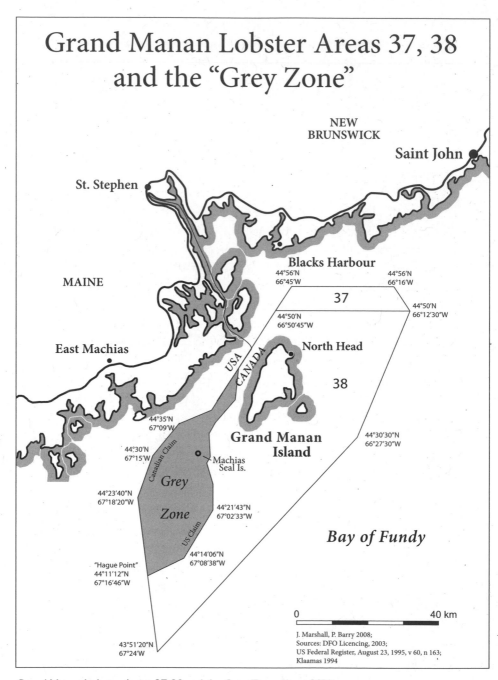

Grand Manan Lobster Areas 37, 38 and the Grey Zone (Area 38B)[i]

agreements whereby these territories were established. She discovered that there were three distinct areas, defended by unwritten codes, which, in turn, reflected the social relations on the island (Recchia 1997). The first and oldest of these areas was the inshore, divided according to generations of family harvests. The second was the area to the south known as the Ledges, developed more recently as the bigger boats and advanced LORAN technologies permitted longer trips as well as the more precise navigation and demarcation of setting spots. This area Recchia described as something akin to the "wild west," with those who were constantly pushing the limits, imbued with a strong individualistic competitiveness, being the first to assert claims in the area. In fact, within a few years of having written her thesis, the division of areas and basis upon which the men selected their favourite areas had changed dramatically.

A third area was discovered in the early 1990s and kept secret for a couple of years as the men would land their catches surreptitiously during the night, sometimes at different wharves to further deceive any curious onlookers. When other fishers became aware of the excellent harvests of very large lobsters that were being brought in, a lottery system was inaugurated through which seven boats would be allocated specific trawl lines based on GPS readings. In other words, advanced navigational technology allowed an informal system to be negotiated – a system that worked for several years. Then, in 2000, three other fishers decided to enter this lucrative zone, which was located in mid-channel in the deep water of the Bay of Fundy. The problem was they refused to cooperate with the lottery system. It became a free-for-all, in which the biggest and fastest boats were able to determine who got the best fishing grounds. While the fishing in this most distant part of Area 38 was lucrative throughout the 1990s until 2003, concern about the stocks surfaced in the 2004 season, when harvests in this area declined (reportedly, they had declined even further during the winter season of 2007–08). Moreover, the catches began to bring in soft-shelled lobster, which indicated late moulting and which fishers speculated might be related to changing water temperatures. By 2006, many fishers, using long trawls and large traps, were fishing "outside" near the shipping lanes at the far boundaries of Area 38. Others, mainly those who did not invest in large new boats, resisted the apparently good catches there and continued to fish "inshore." Overall, however, there was much less adherence to the boats' village wharves in the various areas selected.

The official management regime comprises a combination of a set number of licences, a set number of traps, seasons, size, and whether or not the lobster are "berried" (i.e., contain spawn). The DFO allocates a fixed number of lobster licences to Grand Manan (Area 38). Set at 136, some of these are "half" (or partnership) licences when they are held on one boat with a "full" licence. A full licence allows 375 traps, and a half licence, which can be added to the same boat, allows 188 traps. This means that a father and son could have a total of 563 traps. However, it also means that they could divide the licences between two boats and, effectively, add an additional 188 traps to their fishing effort. This became an issue when some men began to sell their licences to Aboriginals under the new federal policy established in the wake of the *Marshall* decision in September 1999 (see below). According to a DFO official, in 2005 there were about 46,000 traps set in the Grand Manan lobster sector.

While the lobster fishery was only one among many of the fisheries open to Grand Mananers in the 1950s and 1960s, and in fact some fishers sold their licences for a few thousand dollars, by the 1990s lobster had become the richest source of income for island fishers. Those who held one or more of the 136 licences became the ones who could afford new vehicles, occasional trips to Florida, and the most trendy basketball shoes for their kids. Licences rose in value from about $10,000 in 1990 to $25,000 in 1992 based on the increasing harvests and rising market value per pound. Sold within the community, these licences were valuable assets that passed from father to son and were associated with specific boats and equipment. However, as lobster catches became more lucrative and the cost of purchasing a licence rose beyond the reach of most families, companies became increasingly dominant factors in the structure of the lobster fishery on Grand Manan. Both through family groups and three large brokers, throughout the 1990s a few individuals began to own an increasing percentage of the lobster business. As one observer pointed out, "Lobster is not really small, independent boat enterprises anymore; it's mainly companies." While the crab fishers on Fogo Island in Newfoundland resisted the idea of concentrating licence ownership and control (McCay 1999), Grand Manan fishers seemed resigned to the inevitability of doing so as the value of licences escalated. The consolidation and increasingly exaggerated differentiation of local elites was occurring in ways that were not unlike what was happening in other parts of Atlantic Canada (Davis 1991). Not only did the structure of business interests change but also the size and power of boats increased, with concomitant impacts upon fishing efforts. The most obvious change with the advent of larger boats was their ability to reach the outer limits of Area 38 and to carry larger traps on mile-long trawls. The increasing success of some fishers did not go unnoticed. In 1997, seven new boats provoked much discussion on the island, with observers driving to the boat shop and to the wharves to engage in careful inspection. Every year thereafter, several new boats appeared. Then, in 2004, it became an open secret that one of the largest family groups had ordered a large new boat with a radical new design from Newfoundland, purportedly to be ready for the new lobster season, which was to open on 9 November that year. As particular families consolidated their power within the industry, they also became active and supportive members of the GMFA. The socio-economic implications of change in the lobster fishery were mirrored in the political arena.

Although precise figures are impossible to obtain and calculations are notoriously difficult because of the various combinations of full and half licences as well as company involvement in financing arrangements, a conservative estimate of annual gross sales for an individual licence would be $200,000.[5] One fisher provided the following summary of an average individual lobster licence: Catch 30,000 pounds; 2004 price, $5.75 per pound; gross income, $172,500. These figures do not account for family groups that may control six or seven licences or for boats that may have a single licence but support two families that have father-son crews. When business costs (such as the crew share, insurance, boat operation, maintenance, depreciation, and licence fees) are factored in, the net income provides the boat owner with a "good living" that may or may not allow for a Florida vacation. Indeed, even as they tend to underestimate gross revenues, all the lobster fishers acknowledge that catches increased steadily over the decade of the 1990s. On the other hand, because of the significant increase in the value of the Canadian dollar, prices in 2005 and 2006 did not continue to rise. From 2001 to the end of 2005, the value of the dollar had increased by almost one-third, and this was reflected in the prices that fishers were paid for the lobster, which mainly depended upon the American market. Nonetheless, it is undeniable that, since about 1990, the lobster fishery has provided the highest average family income of any of the fisheries. As a result, lobster families have high community status and are identified as those "who have." For the families who began to consolidate and increase the number of licences they controlled, the rewards were significant. Moreover, individual fishers could get into the lobster fishery through the financing provided by these family groups in exchange for a guaranteed supply of lobster. In other words, the companies, with construction of pounds and lobster cars for longer-term storage, now controlled supply and price through controlling multiple boats and licences.

But these significant gains have also been buffeted by important changes that have challenged the supremacy of the lobster fishery. The growth of aquaculture, government encouragement of an Aboriginal fishery, and government encouragement of a new summer fishery in the disputed Grey Zone opened up questions of community control and the role of leadership as well as illuminating the confluence of local and global issues of change. The two events that changed the nature of the lobster industry forever on Grand Manan are (1) the introduction of Aboriginal licensing in 1996 and the DFO's aggressive buying up of island licences after the *Marshall* decision in 1999 and (2) the opening of the international Grey Zone to a summer lobster fishery in 2002. Both of these events, which are fundamentally about sharing the Commons, triggered major upheavals on the island and created divisions that continue to this day. Both events were instigated by the federal government, and both, as a direct result of the government's actions, combined to drive up the price of licences in an artificially created market. The 2007 proposal for a fundamental rewriting of the Fisheries Act, promulgated as Bill C-45, raised the bizarre possibility that the government intended to take away the capital rights to licences, which it had essentially created, by precluding the ability of fishers to transfer licences. Before examining this incredible proposal, let us first explore the impacts that the federal government's Aboriginal policy wrought on the community of Grand Manan.

In 1999, the *Marshall* decision (see below) essentially opened up the possibility for Aboriginal claims to the marine fishery. Policies developed by the Departments of Indian Affairs and the DFO, without consultation with affected communities, were designed to encourage Aboriginal participation in all sectors of the traditional fishery in the Atlantic provinces. As the government poured millions of dollars into subsidizing the entry of Aboriginals into the fishery throughout the Maritimes, the value of individual lobster licences (not including the boat and traps) soared from $100,000 in 1997 to about $350,000 in 2004. In the autumn of 2005, two "pieces of paper" – a lobster licence and a crab permit – cost the Maliseet Woodstock reserve $900,000, which was paid in total by the Canadian government. Grand Manan fishers no longer had the possibility of buying into the lobster fishery if the family was not already participating in it. A major community fissure had been created as a direct result of the federal government's attempt to correct generations of incompetence in the treatment of Canada's Aboriginal population. For Grand Mananers, the Aboriginal policy directly impeded any possibility of their entry into the lobster fishery. This lack of accessibility not only affected the potential for improving family incomes but it was also seen as fundamentally unjust.

For many who were actively fishing, the decision of some lobster fishers to sell their licences to Aboriginal people was seen as a betrayal of the community. "People would rather sell to the Indians than protect their family heritage," said one woman. The price of a licence, which went from about $3,000 in 1977, when "we could hardly make a living," to $10,000 in 1990 and to $25,000 in 1992, seemed to increase hugely, but it was based upon lobster harvests and market prices. However, the rise in price to almost $400,000 in 2005 was of a completely different magnitude, and individuals could not absorb it. In effect, the government's Aboriginal policy had created a false market based on what the market (i.e., the DFO) would bear rather than on a realistic assessment of a repayment schedule. Assuming a conservative gross earnings of $150–200,000 for an annual lobster harvest, one fisher suggested that a more realistic value for the licences would be $150,000, which could be repaid over a period of ten years. The combined impacts of lucrative harvests and government Aboriginal policy caused a major restructuring of the industry, with the result that companies increased their control.

While many fishers recognized that the government was trying to ensure secure income for Aboriginals on reserves, they were also suspicious of its long-term agenda. Several fishers mentioned the idea that the government ultimately wanted to decrease fishing by withdrawing licences from the community. One even said that he had overheard a DFO official acknowledging that what was desired was a decreased effort in the lobster fishery. The buying up of licences that were then to be given to Aboriginals, but that were still technically under the control of the DFO, would allow the government to retire those licences should the Aboriginal fishery be abandoned. Alternatively, fishers expressed concern that those licences would be sold to the already powerful brokers on the island, further constraining access by the small family unit. This period marked a radical shift in the socio-economic and political structures of the community. New relationships were forged between fishers, large and small; between the GMFA and lobster fishers; and among everyone in the wider community.

As *Tides of Change* is being written, fishers are speculating about more government intervention and the possible imposition of quotas. While the framework of the Canadian regulatory regime has relied upon seasons, size of lobster, and number of traps to ensure sustainability, Grand Manan fishers worry that the DFO might introduce a new regulatory system. Acknowledging that allotting the number of traps to be used is susceptible to cheating, there is a consensus that quotas will work against those fishers who work harder and who test new areas for harvest, thereby introducing a disincentive. "A quota would cut me back. If I catch only my share, then I'll lose," said one hard-working lobster fisher, while at the same time acknowledging that even the trap system is subject to abuse: "There will always be someone who tries to beat the system. Enforcement is very difficult." However, stringent monitoring during the first two weeks usually catches some who try "to bend the rules." For the two boats caught with short lobsters in 2005, fines were a nuisance but were certainly not as effective a deterrent as would have been the proposal (never acted upon) that punishment should involve a ban from fishing during the first two weeks of the subsequent season. This, according to several informants, had not been implemented because of concerns that it would not withstand a court challenge based on the human right to a living. But it was acknowledged that it would have been an extremely effective deterrent. If there is one point of unanimity among lobster fishers, it is that the most important two weeks of the lobster year are those opening two weeks, during which catches usually make up about half of what is caught in the rest of the season combined.

Another problem with using number of traps as a regulatory mechanism has emerged since about 2001, when larger traps were brought in initially for the deep-water fishers but were increasingly used by everyone who could afford to change their "kit." The larger traps caught more lobster. As one worried lobster fisher told me, "Everyone else is getting them, so I have to too if I want to keep up. Now on setting day it takes me three trips rather than two because of their size. The catches are up because the effort is up." In 2005, he had changed from thirty-six-inch traps to forty-two-inch traps. Some of the men were setting forty-eight-inch traps, weighing about seventy pounds each. But this increased effort and the higher catches were causing many of the men to express concerns about sustainability, very much supporting Hardin's (1968) much maligned thesis about the "tragedy of the Commons." However, as Hardin (1974) himself acknowledged in a subsequent article, the mitigating factors of community rules and norms could change what he had described as "open access" to a common property regime. As reports of the catches began to percolate around the island by the end of the week, the news seemed to be mixed. Whereas the inshore fishers seemed satisfied, the offshore catches were not as good as they had been in past years. One fisher said that the "highliners" per day in their first week were getting five thousand (and even up to ten thousand) pounds but that the average was about 1,600 pounds: "That's good, but not like it has been in the past ten years." One fisher who had pulled one hundred traps one day in the first week brought in 1,300 pounds, which was considered good, but at thirteen pounds per trap not as good as he had hoped for. It seemed that in 2005-06 there were increasing disparities among fishers' catches, varying from sixty to seventy thousand pounds down to about five thousand pounds, according to one informant. One of the journals reported an "unusual but generally good season," which was made successful, in part, "by the warmer water temperatures that were still holding at 38–40 degrees in mid-January" (*Navigator* 9, 2 [2006]: 15). For some, the concerns were that low catches would seriously affect their ability to maintain mortgage and loan payments, especially if they had invested in large new boats and equipment. In the spring of 2006 there was a prevailing mood of apprehension. As often noted, "When one fishery is down, there's always another one that does well." This time people wondered if that legendary truism would be apt.

With the lucrative returns from lobster, it is hardly surprising that the government saw an opportunity for the Aboriginal reserves, even though they were inland communities that had never before been involved in the marine fishery. For the community of Grand Manan several significant issues have been raised by the withdrawal of licences from Grand Manan fishers. Not only have the artificially high values driven local buyers out of the market, but now the revenues are leaving the island. Money earned is no longer spent in the community. The eighteen Aboriginal licences (in 2006) represent 13 percent of all licences and 13 percent of potential income for the island. Furthermore, as the licences have been split on local boats, instead of the previous 1.5 licences, now a father-son team can sell one part, retaining a full licence (375 traps) and then converting the second part to a full licence to be sold to an Aboriginal person for $400,000. This means that now the fishing effort in the sector has also increased by 33 percent with regard to that licence. According to a representative of the GMFA, the increased pressure on stocks creates a serious potential hazard to sustainability. The decline in the other fisheries, especially groundfish but also in the lucrative sea urchin fishery, has put even more pressure on the lobster fishers. For an island that has always relied upon multiple licences and the flexibility and adaptability that a diverse fishery allows, the imminent collapse of any one fishery creates concern for the sustainability of the others. When combined with the increasing weight of regulatory authority and the DFO's determination to discourage the entry of new individuals into the fishery, the fishing community feels constantly under threat. Even as the increasing lobster harvests in 1999 and 2000 brought a renewed sense of hope for some, those without licences were fearful for the future of the diverse fishery that had supported the island for generations.

Meantime, for father-son teams, the splitting of licences and purchase of new boats continued through 2001 and 2002. While financial commitments stretched the resources of some families, the possibility of selling to the Aboriginals in a few years made the investment worthwhile. Nevertheless, the muted criticism of those who were selling their licences was tested further in 2005 when a large new boat arrived from Cape Breton. Not only was the boat under an Aboriginal licence and from Nova Scotia, but the entire crew was also from that province. The restructuring within the lobster sector, largely caused by the federal government's Aboriginal initiative, has affected not only ownership of licences and boats but also trucking and distribution networks. The high prices of lobster licences have effectively precluded new individuals from entering the fishery, except under the financial arrangements provided by the companies. Any available licences that were not sold to the Aboriginals were being bought by one of several large companies, including Special K, MG Fisheries, and Helshiron. These groups own about twenty-six licences, or 20 percent of all licences on the island, added to the 13 percent owned by the Aboriginals. But at least two of these groups also control another forty to fifty licences through their financing arrangements, whereby fishers agree to sell exclusively to a particular company in exchange for generous loans. With the supply guaranteed, the companies can then invest in large trailer trucks. As well, four lobster pounds owned by several of these same interests, and two brokers, served as distribution companies, through which all of the smaller lobster fishers sold their catches. Increasingly, in the decade between 1995 and 2005, the lobster industry had become concentrated and was controlled by relatively few families. Speculating about the future, one observer predicted that, when the "present generation of lobster fishers is gone," everything will be company-owned.

The Aboriginal Fishery

The only certainty about the *Marshall* decision is that it has created
enormous uncertainty for the Maritimes. ~ (Coates 2000, 187)

In September 1999 the Supreme Court of Canada handed down a major decision related to Aboriginal rights of access to the fishery resource. Called the *Marshall* decision, it forever transformed the relationships of Aboriginal peoples in the Atlantic region of Canada. The decision was based on a judicial interpretation of the Peace and Friendship Treaty, 1760, with particular focus on the "truck houses" and Indian trading agents of that time, which were interpreted as the equivalent of today's commercial network. In his survey of the initial impacts of the decision, Ken Coates pointed out that the media coverage had focused on the decision as a "problem" for small communities in the Maritimes without acknowledging the counter-argument, which is that the decision would provide "hope to a people who have faced generations of discrimination" (Coates 2000, xii).

As a legal text, situated within the growing body of Canadian jurisprudence on Aboriginal rights, the *Marshall* ruling speaks to the contemporary authority of pre-Second World War treaties, the growing power of First Nations groups, the need to share the diminishing natural resources of the country with the original inhabitants, and the determination of vulnerable non-Aboriginal resource communities to survive in the face of conflicting pressures. (20)

At the time of the decision, editorials tended to support the Aboriginal position with regard to their aspirations and needs. And perhaps it was this apparent media support that the government responded to in its subsequent actions as, almost immediately, it provided major DFO funding for buying licences, boats, and the infrastructure to facilitate the entry of the Aboriginals into the fishery. Nonetheless, from the perspective of Grand Manan, the government's intention to rectify generations of colonialist policies through new DFO programs was misguided, poorly considered, and outright unjust. By late 2000, three Maliseet reserves in New Brunswick had begun to make incursions onto Grand Manan: these were Tobique (located in the Upper Saint John River Valley at Perth-Andover) and Kingsclear and Woodstock (both of which are closer to Fredericton). All these communities had populations of between eight hundred and 1,300 people. For fishers who had struggled for over two hundred years to maintain livelihoods within the context of a sustainable fishery, the sight of new trucks, large new boats, and multi-million-dollar facilities being brought to the island by Aboriginals who had no tradition within the marine fishery was anathema. They were outsiders who had been given millions of dollars to participate in a fishery that many Grand Mananers could not afford to enter. While Grand Mananers recognized the need to change the prospects for the Maliseet in New Brunswick, the government's strategies certainly raised questions about its awareness of how they would affect the community and the key institutions that had sustained

Grand Manan for generations. As Coates (2000, 188) points out, "While the *Marshall* decision gives the Mi'kmaq and Maliseet new opportunities, it lacks the authority, clarity and all-party support necessary to truly represent a foundation for a completely new relationship."

Emanating directly from the Supreme Court decision and bolstered by decades of growing unease with current policy solutions to the "problem" of First Nations in Canada, the DFO suddenly announced a new policy to encourage Aboriginal entry into the marine fishery. Apparently with little input from the Department of Indian Affairs and Northern Development (DIAND), the policy seems to have been formulated by the Prime Minister's Office without any communication with participating communities. Essentially, the federal government agreed to buy up licences from non-Aboriginal fishers in coastal Atlantic communities and then give them to First Nations reserves (not to individuals) to be distributed as band leaders felt was most appropriate. As well as these licences, the agreements (signed in 2000 and again in 2003) provided tens of millions of dollars to the reserves, again giving the band councils almost complete freedom to allocate this as they saw fit. One reserve, for example, decided to invest in a fish hatchery to serve the aquaculture industry. When it ultimately failed three years later, as reported in the media, the reserve requested compensation from the federal government. The financial awards given to the three Maliseet reserves of Woodstock, Kingsclear, and Tobique directly affected Grand Manan. The reserves themselves are all located in the interior of New Brunswick, with the result that, historically, their livelihoods have depended upon forestry and some river fishing. None had ever participated in a marine fishery before coming to Grand Manan. Each of the reserves used its money in different ways and developed different relationships with the island.

Historical Contacts

Many islanders claim Passamaquoddy Aboriginal heritage. A Grand Manan history teacher described how she had tried to encourage her students to learn more about their family histories as well as about the history of Grand Manan. She asked them to bring in their family trees, with as much information as they could discover. With amusement, she said that almost half the class seemed to have been descended from the same nineteenth-century Aboriginal woman. While there are wide discrepancies in estimates of Aboriginal ties (up to one thousand, according to one interested islander), there seems to be some consensus that there are about seventy-five islanders who have papers that establish their Aboriginal status (*Courier Weekend*, 2 November 2001, 2). With pride, one of the men displayed his "Native card" to me as he sat on the stoop at Whale Cove on a sunny Saturday morning watching one of the weirs being seined. With glee he told me that he had applied for it a few years back when he wanted to ensure that he would always be able to go moose hunting on the mainland. For others, Aboriginal status is a form of insurance that they feel will guarantee them access to any new special privileges. Still others enjoy the idea of meeting regularly to talk about their connections to Aboriginal culture. Another indication of the significance of Aboriginal connections was illuminated during a casual dinner conversation, in which it was mentioned that their great-grandchildren would be the last who could claim Aboriginal status because they were currently "one-eighth Native" (the legal minimum to claim status).

Most Grand Mananers over seventy years old have stories of regular seasonal visits by the Passamaquoddy, who would arrive for several weeks each summer to trap seals and porpoises. While there is controversy about the degree of permanence of any Aboriginal settlements, there is no question as to the regular visits during the summer months. There are pictures in the Grand Manan Museum that confirm an Aboriginal presence, showing Aboriginals casually leaning against the rocks next to drying seal skins hung over lines. As one islander told me, he remembers that there were also women who carried babies on their backs. They didn't come for the fish, he said, just the seals and porpoises, making the cross-channel trip to Grand Manan as recently as the 1950s. In the early 1950s, their main interest was in collecting sweetgrass which grows along the shoreline, for baskets that they then sold door-to-door on the island.[6] Despite the persistent efforts of several Grand Mananers to prove that the Passamaquoddy were the original permanent residents on the island and that, therefore, they have a claim to the land, the general consensus is that they were always seasonal migrants. But the current situation, which involves the presence of the Maliseet from New Brunswick, has introduced an entirely new perspective to the idea of "Native connections."

The formal organization representing the Maritime Aboriginals is the Atlantic Policy Congress (APC) of First Nations Chiefs. In an important document responding to the *Marshall* decision, the APC was unequivocal: "This policy intends to provide a foundation for the preservation and sustainability of lands, waters and resources within the traditional territories of the Mi'kmaq, Maliseet and Passamaquoddy based on their traditional values and

treaty responsibilities ... The Mi'kmaq, Maliseet and Passamaquoddy continue to assert their exclusive authority over conservation of the natural resources
and the consequential right to direct access to those resources."[7] Through the
years between 2000 and 2006, the growing Aboriginal presence on Grand
Manan represented a double threat: to the community and to control over
marine resources.

New Aboriginal Connections

A licence was first sold to an Aboriginal group in 1996, when a "go-between"
Aboriginal who had worked with reserves and the government met with an
island fisher who had been considering retirement. Provided through East
Coast Marine Brokers, an appraiser had valued the islander's equipment and
boat and, following an agreement with the DFO, the reserve negotiated the sale.
With the first two sales of lobster licences in 1996 and 1997, respectively, there
was little comment on the island. But following the *Marshall* decision and the
government's implementation of its new policy, Grand Mananers became
much more concerned, muttering that the new trucks and traps offered to the
Aboriginal off-islanders was unjust at every level. Suddenly, in 2001 there were
eight new boats plus two older boats, all owned by the Maliseet reserves.

Crucially, however, it was the inflated value of lobster licences that caught
everyone's attention; and in 2001 the construction of a large new $1.4 million
facility at Ingalls Head by the Maliseet of the Tobique reserve became the talk
of the island and the focus of suspicion and resentment. Even the process of
construction itself engendered hostility as islanders commented that trucks on
board the ferry were hauling construction materials from the mainland. In
other words, the Tobique Maliseet were not buying their supplies from island
merchants. The building occupied a prominent seafront position, overlooking
the Ingalls Head wharf. The two-story building, with a floor area of approximately 6,000 square feet, which included a large boat-building garage, was not
inconspicuous. Fourteen large bedrooms, plus a director's suite and modern
bathrooms on both floors provided accommodation for the "trainees" – the
Aboriginals who would be spending weeks on the island learning how to run
the fishing boats. A large conference room and smaller offices as well as laundry
facilities made the entire structure an impressive "school" for the people from
the Tobique reserve. In 2003-04 the Woodstock reserve built a smaller facility,
with five bedrooms, so that its people could move out of their rooms in the

Surfside Motel. By 2006, there were security floodlights and the property had been surrounded by an eight-foot chain-link fence topped by angled barbed wire. On an island where fences were non-existent until the mid-1990s, this enclosure represented more than the protection of Aboriginal fishing gear. Implicit in these security measures were Aboriginal fears of island hostility and an acknowledgment of their outsider status.

One young man with whom I talked in 2001 said that he was being trained to be a trainer. That summer there was talk of setting "dummy" traps during the summer, a training strategy that was quickly abandoned when murmurings from the community suggested that there would be no tolerance for out-of-season "training" for these outsiders. As news spread that eight lobster licences had been sold (between 1997 and 2000), and that the prices had consequently risen from about $100,000 for the 1997 sale to $200,000 to $250,000 for each of the licences sold in 2000, Grand Mananers began to become increasingly apprehensive. The mayor at the time said: "They are selling their heritage. They'll have nothing to offer their children. What will their children do?" A few years later, with the island still uncertain and suspicious of the federal government's ultimate objectives, one member of a fishing couple commented: "People would rather sell to Indians than protect family heritage. Seeing my neighbour benefit is okay; but there's gotta be a reason for it." They were worried that the government's long-term strategy was to ensure the withdrawal of licences from the community. In other words, there was a widespread perception that the transfer of licences to Aboriginals was only a first step in a major restructuring of the lobster fishery, the primary concern of which was the withdrawal of licenses. This would effectively leave the companies that financed the now extremely high-cost licences in control. The market prices for lobster licences had risen as a direct result of the government's Aboriginal policy, from $25,000 in 1995, to $100,000 in 1997, to $200,000 in 2000, to $450,000 in 2005. Obviously, these prices put most islanders out of the market. The only ones who could afford to buy or finance a lobster licence in 2005 were the large companies, such as MG Fisheries and Helshiron. These companies owned or controlled multiple licences that could be "leased" or "rented" to other Grand Manan fishers, who would run the boats.[8] This arrangement was generally accepted because it allowed local residents to continue fishing, and they recognized that, because of the escalating costs, there was no alternative. But there was a growing sense of apprehension. As one observer commented, "The Natives have almost 10 percent of the lobster licences and over 50 percent of the sea urchin licenses. It's going to be a problem." For the community, the new rule that no official fishery meeting could occur unless there was an Aboriginal representative present simply added insult to their growing sense of injury.

For the Maliseet, the challenge was multifaceted and certainly not "the easy road" envisaged by local Grand Mananers. Arriving at Blacks Harbour after a three-hour drive from their reserves, they were then confronted by the ferry line and the ever present tension of wondering whether there would be space for them. On the island, some from the Woodstock and Kingsclear reserves would stay in homes purchased by the reserves; others stayed at the newly built Tobique facility (2001), which had fourteen double bedrooms and a large kitchen and eating area. Before the new facility had been built, they had had to stay in nearby tourist cottages. In 2002, for KaChiMooWin the weekly return drive and ferry trip represented the possibility of increased income for his family and the challenge of learning new skills, but it also meant missing his wife and children for a week at a time. Occasionally, wives would visit, often in pairs. For the weekend they would "do the tourist thing." As KaChiMooWin talked he was fixing lobster traps and trawl ropes that had been damaged during the previous winter. Asked whether they had considered moving to the island, he acknowledged that it had been talked about but that it seemed like a big change. His wife liked the island because it "is so peaceful here." As well as KaChiMooWin and his wife Sarah, Mike Solomon and Virgil had both arrived with their wives (Shelley and Teresa) and four children, ranging in ages from one to twelve. At that time there were eight men from the reserve staying there, but he expected the facility to be full during the summer. While KaChiMooWin enjoyed working on Grand Manan more than being in the woods at home, he was disappointed that the money was not as good.

For the women, the feeling of having a getaway along with the possibility of selling their beadwork at the local market held out the promise of improved lives. The beadwork and baskets that Shelley and Sarah would sell at Foxwood on weekends seemed to offer potential on Grand Manan as well. They were all quite optimistic, despite the long journey and the struggle to learn a new occupation. Solomon had always worked in the woods in northern New Brunswick and was excited about the prospects of the fishery. His wife Shelley, however, was more circumspect. Because she had a good job as the director of Tobique Employment Initiative on the reserve, she did not want to move to Grand Manan "unless [she] could run a business [there]." One of the women had tried to take courses on the reserve about the fishery so that she could bring school children to the island for a week at a time and educate them. For one of the women who considered moving to Grand Manan school was an important issue as it directly affected her children's ability to be integrated into the community. As she described it, in their own community it was difficult for newcomers to be accepted. Even people who move away and return are "still not really accepted." She fully understood the challenges that a move to Grand Manan would entail for the family.

For Mike Solomon, his first weeks on the island had been a positive experience, and he was looking forward to training other Aboriginals. In the meantime, he was working on a boat that had multiple licences, including sea urchin and lobster, and there was also the promise of two mid-bay and two Bay of Fundy scallop licences for the reserve. He complained that the spring fishery did not seem very lucrative and that, for his thirty-three hours work, he had earned only $120. On the other hand, in describing the procedure for dividing the value, he admitted that the winter fishery, which began in November, would be very different.

The division of the revenue was fixed in the following way: 51 percent went to the band council; 25 percent to the captain (inevitably a Grand Mananer

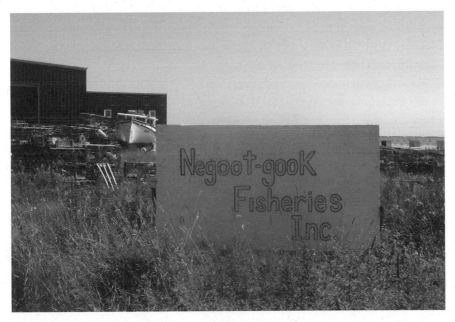

Tobique facility at Ingalls Head

who was supposed to be training the Maliseet to eventually take over the boat); and then 12 percent for each of the two crew. But, said Solomon, after a sixteen-hour day in April, his share had been only twenty-five dollars – not enough to pay for gas for a return trip to the reserve. On the other hand, he admitted that the fall fishery had been extremely profitable. When their boat took six thousand pounds in the first week, which sold for about $30,000, his share for one week was about $3000: "The first three weeks is your season!" Some of the bands devised different arrangements, even having some of their men work on salary. However, it seemed that the incentive of working for a share of the catch was much more effective both with regard to encouraging commitment to work and with regard to being more lucrative for the Aboriginal fishers. According to Solomon, the other bands use different strategies, such as fixed amounts per week (Kingsclear's amount is $450 per week, while Woodstock's is $500 per week). It is the band council that determines the contracts and arrangements with its crews. In the end, Solomon felt that his new job was an improvement: "Now we're not landlocked! It's a nice environment, laid back and relaxed, not like at home, where all kinds of things are going on. And I like the work; we're not feeding the flies. I haven't broken a sweat in two years."

For most of the Maliseet, the biggest challenge is the weekly drive of over three hours plus the ninety-minute ferry trip. They liked the idea of being independent from the politics of the reserve but were concerned about the social adjustments. They felt frustrated that the DFO agreement was with the band rather than with individuals and that, therefore, their futures were tied to that of the reserve. Living on the island permanently entails a few changes in their administrative status, but as long as they retain their ties to the reserve they do not lose their status-Indian claims. Those Aboriginals who live completely off reserve have to pay the same taxes as do other Canadians. Therefore, if these men were to move permanently to Grand Manan, they would either have to pay taxes or have the land on which their facility is built declared Aboriginal reserve land. Indeed, this latter possibility (eventually rejected) was proposed in March 2005 to the Grand Manan Municipal Council in a letter from the Tobique reserve. The office on Grand Manan would act as the conduit to the reserve. The first $7,000 earned off-reserve is not taxable, and when they return to the reserve they can apply for a rebate of taxes paid. The Grand Manan captains who train the Aboriginals and run the boats (supposedly for a period not exceeding four years) are paid by the reserve. Compared to the Tobique reserve's organization of a "lobster business," other reserves, such as Big Cove, which is associated with the northern coast of New Brunswick (and which signed a $19 million contract with the government), has a more individualistic structure and has less control over the boat crews. In its case, the Aboriginals are leased the equipment for five years, after which time they are allowed to have it in exchange for ongoing royalties paid to the band. According to Solomon, in its first agreement with the government in 2000, the Tobique reserve received $7.5 million and this was increased to $20 million three years later in 2003. Throughout the Maritimes, according to a three-year agreement signed on 1 July 2003, there was to be $650 million for thirty-four Aboriginal communities. Asked what changes he would like to see, Solomon was unequivocal: "Not as much babysitting by DFO!" But his wife was more explicit, suggesting that they would like to have processing done on reserve so that there would be jobs at home. There is so little for the women, she explained, and a fish plant would provide needed employment.

There seemed to be a consensus that lack of consultation with Aboriginal communities had produced a program that did not adequately address many issues. For example, according to Mike Solomon, rather than being taught about boat building Aboriginals need to know how to repair traps and how to deal with the day-to-day maintenance of boats. But the Maliseet acknowledged that the problem also lay with their own band council, which knew nothing about the marine fishery and had to rely upon a consultant. Meantime, the money arrived and no one was sure how to spend it. New trucks and boats appeared within months, yet the Maliseet had no training and no experience with the high tides of the lower Bay of Fundy. And they knew that it would take years for them to learn. When a course was first offered, forty-five people enrolled but only eighteen passed. Then, of these eighteen, only four actually continued with the on-boat training out of Ingalls Head. They felt their people's time and money had been largely wasted. The problem of retention was echoed by the (non-Aboriginal) consultant, Dave Bollivar, who was mainly responsible for managing the office and for liaison with the Tobique reserve. In 2003, of the original forty-five trainees, only one was still in the program; and of the seven currently on site on Grand Manan, only one had been in the training program. His analysis was that, due to the long history of government projects, the Aboriginals had no culture of hard work. As well, none of them had ever been in the marine fishery and, without that history and experience, combined with long periods away from their homes, they felt unsupported and uncomfortable. The result was persistent use of drugs and alcohol, absenteeism, and a lack of motivation. Further, the leadership was rotational, beginning with Henry Sapier in 2001 and moving through several other men to Henry Bear in 2003. But always Dave Bollivar was there to fill in during the intervals.

The problem for the Grand Manan fishers involved in the program was that they had been accustomed to working long days, and, as the captains of reserve-owned boats, they did not have the same control over working time. At least one Grand Mananer who had sold his licence and now served as captain with a 25 percent share, receiving it from the band council weeks later, complained that the Aboriginals did not want to work as hard as Grand Mananers were used to, and therefore the islanders were not able to earn as much as they wanted. One Grand Manan fisher suggested that he would like to renegotiate the agreement on a salaried basis so that he could be guaranteed an amount; he did not like having to depend upon the work patterns of the Aboriginals. On the other hand, he pointed out that the lower effort would actually mean a more sustainable fishery. Arguing this, he was not taking into account the proliferation of licences that were being split upon sale to the Aboriginals. Many islanders were reticent to take a public position against the situation, although murmurings continued to reverberate throughout the community during the first few summers. Many felt that the Aboriginal fishery would not last since "they're too lazy." Others pointed to the number of licences held by off-island reserves and worried that if the reserves did indeed withdraw, DFO would retire those licences. At least one fisher who had sold to the Aboriginals argued that, because they were fishing fewer hours, their impact would be less than if a Grand Mananer were running the boat. Moreover, he said, they have freed up money that islanders now have to buy land and build new houses. He felt they would actually contribute to the island economy. But probably the most irksome issue for many fishers was that they had to see their boats rotting at the wharf. Although the later agreements allowed the Maliseet to buy new boats, earlier they had purchased boats and licences from the fishers without knowing how to maintain them. One of the boats was abandoned completely, left tied up at the wharf, while a new boat was purchased, much to the chagrin of Grand Mananers: "A complete waste of money, and there's Melvin [an island fisher] with his small *Carioca* wondering if it'll get back to the wharf at night!"

In the spring of 2002 hopes were high that a summer training program that replicated real conditions might be developed. The plans were for a "catch and release" program, overseen and filmed by the DFO, that would enable Aboriginals who had never been on the sea to gain the skills and knowledge that would eventually allow them to run their own boats. There were to be eight boats, eight teachers, and thirty-eight students. The boats would include two from the Woodstock reserve, two from the Kingsclear reserve, and four from the Tobique reserve. The changing of band chiefs and of those who were managing the bands on Grand Manan created problems for continuity within the system as well as problems for basic communication. With a new chief in 2002, Solomon felt that the system was working better but that there were many issues that had not been addressed. Given the different incomes from the fall and spring fisheries, Solomon felt that they should have two different methods for paying the crew: shares in the fall and salaries in the spring. The other changeable factor that the Aboriginals found difficult was the rotation of different boat captains as the Grand Manan men found their new jobs as "teachers" or "mentors" more challenging than they had anticipated.

The perspective of the Grand Manan captains was quite different from that of the Maliseet. As one lobster fisher described his trainees, "They have no navigational skills. You need greed and drive to be a good fisherman. For the government to think these guys will work is ridiculous!" Moreover, some of the Grand Manan men do not like to go out in bad weather because the Maliseet are lacking in experience and, thus, the risks are too high: "The guys don't pay attention. There's just too many nets and anchors and lines. You've gotta watch them the whole time!" Asked about any regrets with regard to having sold his licence, he said no, except that he might have insisted on some different conditions. He believes that it will take the Aboriginals a long time to adjust, that "they [won't] become commercial fishermen overnight!" He criticized the government for not making enough provision for adequate training. The time line of three to five years, during which Grand Mananers were to serve as captains, was certainly not sufficient, he felt. While it seemed that, as long as the government provided funding, the Aboriginals would stay, he did not believe that they would ever want to live on Grand Manan permanently since too much was provided for them through the reserve system. His experience had been mixed, with some of the Maliseet being willing to work and learn and others being completely unreliable. Some fishers complained about perceived preferential treatment. One of the Aboriginal boats was caught with "a case of bobs" (short lobster) and nothing was done, said one of the men: "If it was us they'd hang us for one!" He complained that they are able to afford larger boats and more powerful engines and, of course, "they get to the grounds before us" on setting day. One of the Grand Manan fishers who usually set traps close to the island rather than out in the deep water was upset because the Maliseet seemed to want to fish closer in than the former licence holders had done. This meant that there was a significant shift in the fishery closer to Grand Manan, creating more competition and the realignment of traditional grounds. For another islander the opposite was true. Having sold his licence and been hired on as captain, he was upset when the Maliseet wanted to take his relatively small boat into the deep waters offshore. He refused, was fired, and was replaced by another Grand Mananer, who was willing to take the boat into rougher seas.

Another of the men was more circumspect in his comments. While he understood that there were centuries of wrongs that needed solutions, at the same time "you can't just throw money at the problem! More than $600 million in Atlantic Canada last year and even more this year! It's not fair when we have to work so hard." Despite such resentment, some islanders acknowledged the dilemma for fishers who wanted to take advantage of the lucrative government offers. In 2003, the news was that the DFO was looking for another seven licences. A typical comment was: "At $300,000 who wouldn't want to sell? If you're sixty or sixty-five and you're looking to retire, why not? No one here can afford that money!"

In August 2003 I met with Hank Bear, a lawyer from the Tobique reserve who expressed his determination to ensure the success of the Aboriginal fishery on the island. Acknowledging that Grand Mananers would naturally be apprehensive about their presence, he was unequivocal about their right to be on the island. "Definitely we feel we have a legitimate right to be here on this side of the island. This side is the traditional outflow of the Saint John waters. The other side belongs to the Passamaquoddy Indians, because of the St Croix watershed," he argued. He wanted to get out a single message: "That we want to fish here and have a meaningful life for ourselves and for our community, and our new community, Grand Manan." According to his account, he was the only Aboriginal lawyer who had argued the case of Donald Marshall before the Supreme Court, and he felt he well understood the many issues of Aboriginal resource use and community relations. During our conversation he was intentionally vague about future plans for the reserve and would not provide any figures related to licences or boats, despite the agreement that had been signed only weeks earlier. When I finally decided to terminate the interview, he relented somewhat, explaining: "We have a policy of keeping a very low profile. We're seen socializing and around, but we don't want anyone to know details or to know our plans. Fishing is a very competitive business … We are a danger here, we know that." Talking about the various problems he was trying to solve, he admitted that the work ethic is a major concern. One solution was to have two crews for each boat, with each crew working three and a half days. Or, "if they work for two weeks, then they have two weeks off." He boasted that they were contributing to the economy and that, in fact, they were "the second largest company on the island." By Christmas he too had left the island, and Dave Bollivar was back as manager.

In the late fall of 2003, the rapid growth of salmon aquaculture and good lobster harvests had continued to give Grand Manan a sense of security. Despite worries about the impacts of the Aboriginal presence on the island, there seemed to be an uneasy truce between the two communities. Nevertheless, stories continued to circulate about drunken carloads of young men shouting at women by the road, and there were complaints of unpaid bills at some of the stores. Tensions were just below the surface, and fights began to break out at the only bar on the island. In November one of the women speculated that just one major incident involving islanders and "there could be trouble." As island patronage of the pizza bar began to drop off because of its growing reputation for rowdiness, the owner finally conceded, closing it in February 2004. It was a victim of a combination of circumstances associated with the Maliseet arriving on the island, not the least of which was a government program that took no account of its potential impacts upon the community. In June, a few months later, the November prediction of violence was realized. An alledged homosexual rape of a Grand Manan man by several Aboriginals at the Ingalls Head facility in May culminated in two of the Maliseet boats being burned out on 29 June 2004. Grand Mananers had demanded that the RCMP investigate the assault within the month. When it appeared there would be no charges laid, islanders themselves alledgedly took action. Despite quiet comments, such as "you can't mess with fishermen," the community held its collective breath. A new Aboriginal manager was brought in.

In August that year a young woman worked at the Ingalls Head facility and tried, as she said, to ensure that the accounting was properly done. The treaty posters were gone from the walls; the kitchen was bare; and the sofa was gone from the common area. The facility was quiet and cheerless. She had arrived with a three-week-old baby and was finding it hard to find adequate babysitting so that she could work. She was unable to get home to the reserve as often as she had expected and was finding it "really hard without the family." But, with thirteen boats on Grand Manan, it seemed that the Maliseet would be around for awhile.[9] That summer another licence had been sold for a reported $450,000. Speculating about the future, Grand Manan fishers repeatedly mentioned several scenarios. The most common suggestion was that the Aboriginals would gradually decide they no longer wanted to be involved, and the DFO would take back the licences and retire them as a way of limiting the fishing effort. A second possibility mentioned by several men was that islanders would be hired to run the boats and that the reserve would be the company to which revenues would be paid. Alternatively, someone wondered if eventually the reserves would give up on trying to fish themselves and would use their companies (Ne-Goot-Gook and TEDCO) to hold onto the licences and lease them back to Grand Manan companies, such as MG Fisheries, being content merely to collect royalties.

Different cultures, different histories of resource use and government relations, different values associated with family lineage and community welfare: all of these and more are at the heart of the boundary tensions between Natives and Grand Mananers. The futures for both communities and for the traditional fishery are still unknown. But, regardless of the eventual outcomes, the transformation will be profound. The government introduced a radical program. Its intentions were good but too narrowly defined, and it failed to consider the implications for the receiving community. Grand Manan will be marked by this failure forever.

Rules, Regulations, Regimes

For the community, the impacts of the Aboriginal agreements with the DFO were not isolated from other aspects of the fishery regulatory regimes. In December 2006, the presentation of a proposed new Fisheries Act to Parliament raised new problems for the island. By February 2007, fishers were expressing concerns about the new act's implications for them and their community. It appeared that there would be a withdrawal of the right to sell or to transfer their licences, thereby negating any inheritance or investment value. The wording in Section 30(1) was the main cause of consternation. According to this section, "A license confers privileges and not any right of property, and may not be transferred." The basic issue was not related to the philosophy of negating property rights per se. The problem was in the historical context of previous practice and government compliance – indeed government participation – in exchanging licences as though they were private property.

In particular, a crucial issue was that this proposal followed immediately upon several years of high bidding for licences associated with the Aboriginal agreements. Moreover, historically families had perceived licences as significant investments in future generations, as capital security that could be passed on to one's children. Just as mainlanders might feel secure in their investments in their homes, so islanders felt secure in their investments in their licences. This was especially so after the huge value increase in licences as a result of the government's Aboriginal policy following the *Marshall* decision in 1999. Through its policy of buying licences and transferring them to First Nations reserves, the government was essentially responsible for driving prices up from $50,000 prior to the first sale in 1996 to about $450,000 in 2006 – an increase of nine times in less than a decade. Now it appeared that, with the signature of the minister of fisheries and oceans, the proposed legislation would withdraw all of that value. In an apparently misguided effort to rebalance the regulatory regime by removing the claim of licences as property, the deleterious impacts of the Aboriginal policy would simply be compounded.

However, the changes to the act in fact did not affect the notion of property rights accorded to licences. Licences had never been accorded rights of transfer and sale, and, as has been mentioned, were actually privileges that, in theory, were supposed to be relinquished to the DFO when no longer being used. That the practices accepted and even abetted by the DFO had not reflected this legal requirement would not be a legitimate reason for expecting continuation of the practice. One DFO official even acknowledged that they had been "advised" by their lawyers not to use the words "sale" or "purchase" in any of their transactions with the First Nations reserves. Lawyers had warned that they must ensure that the written documentation reflect transfers that could not be argued to be sales. For the community the difference was a moot point. They had never considered the transfers anything other than sales; and the prices they were offered and received were set in terms of "buying" either licences or a "complete kit," and for fishers this meant that the DFO accepted their ownership of licences and their right to sell them as "property." This crucial difference in definitions, which involves erroneous perceptions of the notion of ownership, reveals the ultimate control by the government and suggests why misunderstandings and mistrust define so many of the DFO's relationships with the community. In this case, there is also the question of the government's ultimate responsibility to ensure compensation for the exceedingly harmful escalation in the value of licences in the wake of the *Marshall* decision. Asked about this explicitly during a telephone interview (23 February 2007), a DFO official tried to dismiss the evidence of value escalation as anecdotal. When confronted with specific information as to rising values over a period of six years (2000 to 2006), he retreated and took a non-committal position with respect to licences as "property" and the government's role in acting within this perceptual framework.

However, a significant change in taxation rules announced during these discussions somewhat allayed the worries of fishers. As announced in the news media, there would be no capital gains tax payable upon transfer to children or grandchildren of boats and licences. Furthermore, in the case of sale to non-family members, the capital gains tax would apply only to values over $500,000. This taxation change meant that fishers might anticipate being able to retire, leaving their enterprises to children without penalty. Pointing out that the benefits would accrue mainly to owners of forty-five-foot boats rather than the large purse seiners (because the seiners were mainly owned by large companies), Greg Thompson of the Fundy North Fishermen's Association pointed out that, for the individual fishing family, "It's quite a lot of money to come up with and keep your operation going" (*Telegraph Journal*, 27 February 2007, A1, 2). The creation of a "fisheries roundtable" for New Brunswick, initially as a way to discuss the proposed legislation but also as a way to provide long-term co-management possibilities, was another effort by the provincial and federal governments to develop sustainable regulatory regimes and practices for the Atlantic region. As the provincial minister pointed out, in New Brunswick the $1 billion industry accounts for twelve thousand jobs in harvesting, processing, and marketing (*Telegraph Journal*, 26 February 2007, A1, 8).

Not only is the setting of rules a complex problem, but the maintenance and enforcement can also be equally difficult. On the one hand, everyone wants equal access and equal opportunity, and there is a common interest in ensuring that everyone obeys the rules; on the other hand, in every group and community there are those who will bend and stretch rules. While most fishers acknowledge knowing about the various ways in which these rules have been counteracted, only a few actively push the limits on a regular basis. For the DFO, the challenge is to enforce rules in a way that is acceptable throughout the community, ignoring the occasional insignificant lobster lunches and instead catching the bold lawbreakers. While generally most fishers are happy to have DFO enforcement (as this means that they do not have to take action on their own), sometimes there are ambiguous situations. In the winter of 2006 a boat was caught carrying urchins below the permissible size. In the meeting to discuss the case that would proceed at court, the penalty mentioned by DFO was $10,000. However, the GMFA felt that this was not enough. Then it was told that the DFO wanted the GMFA to send it a letter requesting it to take action. There was a sense that the DFO was attempting to have the island enforce its rules in a way that would remove the onus from the government. For the fishers, the dilemma was the potential for community divisions over the long term should such enforcement policies make their way into other sectors of the fishery. The idea of co-management seemed fine, as long as they did not have to enforce the regulations themselves, thus threatening their relationships with neighbours.

The possibility for co-management of fisheries resources, with local communities having significant input in the development of regulations and responsibilities for enforcement, has been widely discussed both in the literature and among policy makers. Communities such as Grand Manan will always be ambivalent about having to police their neighbours themselves. Nonetheless, given the mandate to ensure sustainable stocks within reasonably equitable rules of access, the history of the island prior to the 1970s illustrates the potential for shared responsibility in the context of embedded fishing cultures. Whether or not it is realistic to expect the evolution of effective co-management within the present morass of company-owned licences and growing disparities of access is an open question. Certainly, as Evelyn Pinkerton (1999, 351) has pointed out, "growing problems in fisheries management have outstripped the ability of our current institutions to manage matters effectively." The challenges are further illustrated by the following discussion regarding the relationships between individual fishers, the GMFA, and the DFO.

Boundaries, Politics, and the Grey Zone

The tensions between the DFO and the lobster fishers were reflected in conflicts within the community itself, especially in relation to the structure and mandate of the GMFA. Many lobster fishers felt that their interests were not being protected by the association because of perceived collaboration between it and the large companies and family interests. The growing cleavages between lobster fishers themselves, because of varying degrees of capitalization and number of licences held, became a significant issue for about half the licence holders, who felt increasingly powerless as their votes were worth less because they controlled fewer licences. They argued that the GMFA was not open or transparent in its operations and that they could never obtain information related to financial reports. Allegedly, there were no annual, audited reports that allowed the membership to assess how their dues were being used or even where the finances were obtained. Here the story becomes very murky and is fraught with political innuendo. Not only did two key personnel at the GMFA refuse on several occasions to be interviewed, but they were also not willing to provide any documentation that described the structure and financing of the organization.[10] It appears that the rising values of licences attributed to the Aboriginal fishery policy had exacerbated increasing disparities of capitalization that, in turn, were contributing to growing tensions on Grand Manan.

The conflicts within the community became explicit in 2002, when a group of lobster fishers from Seal Cove proposed the opening up of a new "summer fishery" in the disputed Grey Zone south of the island. In the arguments and debates that ensued, the GMFA took a leading role that involved collaboration with the DFO and, ultimately, led to a significant group of licensees breaking away from the association. The crucial meeting on 9 July 2002 illuminated both the divisions within the community and the ambiguous and problematic role of the GMFA as an institutional watchdog operating in the interests of the community.

In the case of the Grey Zone, created as a new area (38B) within Area 38 but only for a designated season (see map 4.3), international politics and two different regulatory regimes in two different countries added an entirely new level of complexity to the problem of resource access. For the approximately three thousand Maine fishers who fish along the northeast seaboard, the regulatory regime focuses on size limits, and the maximum size allowed is not imposed upon Canadians. However, the American trap limit of eight hundred is more than double that for the Canadians, and the Americans do not have seasonal limits, being able to fish twelve months of the year. Second, while all Grand Manan licensees are allowed to fish in Area 38 (see map 4.3) throughout the period from November to June, encompassing a territory around the island that extends halfway across the Bay of Fundy and south to the American boundary line, in fact fewer than one-third of the licensees keep their traps in the water throughout the entire season. Typically, the fishers who have small boats and who set traps inshore and close to Grand Manan Island pull their traps up in early January because of the poor catches as the lobster move offshore as well as because of the very harsh weather conditions. Further, the scallop fishers, whose season opens in early January, fish in some of the same areas, causing potential gear conflicts. Third, since about 1990 there have been increasing numbers of fishers who have large boats and who set extra large traps near the shipping lanes in the middle of the Bay of Fundy – areas to which the lobsters retreat in the winter months. These men do not lift their traps in January, continuing to fish throughout the season, from November through to

June. The distinction between these two groups (small boats inshore and larger boats offshore) became a significant factor in who would support the summer fishery in the Grey Zone. Finally, as mentioned earlier, island fishers have historically fished in specific areas, related to some extent to the harbours from which they sail. Thus, those boats that leave from Ingalls Head tend to set traps to the east and southeast of the island, whereas those that sail out of Seal Cove tend to fish further south and along a series of ledges that extend to Machias Seal Island.

For maritime areas, setting boundaries and delimiting territories involves conflicting uses and meanings related to resource use, both present and potential. These resources include hydrocarbon (seabed), marine space for aquaculture inshore, fishery (water column), and tourism (both seabed and water column), to say nothing of total ecosystem protection. While boundary setting may seem to be a straightforward process, it is extremely complex, enmeshed in social and political relationships that link global and state concerns to the everyday lives of local communities.

In 1968, the three-mile territorial sea was widened to twelve miles; this was extended again in 1977 to a two-hundred-mile exclusive economic zone (EEZ). Not only was the jurisdictional space over which nations had sovereignty extended, but also the number of overlapping jurisdictions suddenly increased, with a concurrent need for resolution. With the agreement in 1982 of UNCLOS III, there was explicit recognition of five national maritime zones between the land itself, where aliens have no rights, and the high seas, where all countries have equal rights and responsibilities pertaining to fishing, mining, navigation, overflight, and scientific research. Within a nation's "territorial sea," usually extending twelve nautical miles from the coastal baseline, aliens have few rights beyond controlled use for air and sea navigation. Extending two hundred nautical miles out from the coast "and overlapping the territorial sea and contiguous zone, the EEZ grants the coastal state full control over mining and substantial control over fishing. Within the *continental margin* – a geologically determined zone extending sixty nautical miles beyond the foot of the continental slope – aliens may fish and conduct scientific research in the water column but cannot mine or catch sedentary (nonmigratory) species" (Monmonier 1995, 138).

Until the 1984 Georges Bank decision, establishing what is known on Grand Manan as "the Hague Line," the ocean boundary between the United States and Canada had been commonly referred to as the "northeastern boundary," originally settled according to a series of treaties and reports in 1783 (Treaty of

Paris), 1798 (commissioners acting on the Jay Treaty of 1794), and 1817. The creation of the Grey Zone was related to a series of national and international moves to redefine territorial limits. Contestation of the marine boundary began when the United States enacted the Fisheries Conservation and Management Act (FCMA) on 13 April 1976, which extended its exclusive fishery zone from twelve to two hundred miles, effective 1 March 1977. Several months later, on 1 November 1976, the Canadian government responded when it proclaimed its limits as two hundred miles, effective 1 January 1977. These acts upset reciprocal fishing arrangements between the two states and effectively alienated a significant part of the high seas that had previously been regulated by the International Commission for the North Atlantic Fisheries (ICNAF) (Rhee 1980). A new disputed area of twelve thousand square miles over the eastern sector of Georges Bank was suddenly created. Settlement of the dispute in this area was crucial because of the enormous potential of Georges Bank, historically a productive fishing ground and also a potential location of hydrocarbon deposits. There are two options for settlement: the first is by mutual agreement between the two countries, the second is through adjudication, which would necessitate a referral to the ICJ in The Hague.

Despite a bilateral agreement reached in 1979 that would have established a new boundary, final ratification was denied, apparently because of intense opposition from American fishery interests directed through the Senate. Eventually the matter was referred to the ICJ because it was becoming clear that the increase in competitive overfishing by both sides threatened the depletion of major stocks of scallop, cod, and haddock (Rhee 1980, 675). The formalization worldwide of a two hundred nautical mile EEZ occurred in 1982 under UNCLOS III. Then, in 1984, came the ICJ decision establishing the Hague Line across Georges Bank, but only to a designated point to the southeast of Machias Seal Island, thereby leaving a "Grey Zone" still in dispute between the two countries.

In the referral to the ICJ in 1979, the two countries had asked the court to decide the course of the single boundary dividing the continental shelf and the fisheries zones. They stipulated that the boundary should extend from Point A (latitude 44 11 12 N, longitude 67 16 46 W) to a point in a triangle having its apexes at latitude 40 N, longitude 67 W; latitude 40 N, longitude 65 W; and latitude 42 N, longitude 65 W. They chose to exclude Machias Seal Island, the sovereignty of which was being contested by the United States. Placing the final point in the triangular area avoided determining the outer edge of the continental margin (Section 21, Decision 1984). The Americans argued that an equidistant line would unfairly benefit Canada because of the protrusion of Nova Scotia into the Gulf of Maine, despite the fact that the US coastline is four times as long as is the opposite coast of Canada. Furthermore, the United States contended that the Georges Bank is geologically a natural prolongation of New England and, therefore, that the Northeast Channel should be the natural boundary between the two countries (Sections 51, 52, Decision, 1984). In earlier cases, the court concluded that, under customary international law, the delimitation of continental shelf boundaries must be effected by agreement in accordance with equitable principles.

As it turned out, the court accepted neither the American nor the Canadian arguments and determined a line approximately between the two claims. In its decision, the ICJ divided the boundary into three sectors, AB, BC, and CD. These three sectors represented "a combination of a median line and perpendiculars drawn from geometric construction lines" (McEwen 2001, 6). With doubts about the legitimacy of the "natural prolongation" argument, and a related argument that the Northeast Channel also represented a natural divide between two ecosystems, the ICJ rejected a geographically based solution. However, it did give weight to the length of the two coastlines, incorporating a modification into the equidistant line with the result that the Americans were given more than they would have received under a simple equidistant formula.

Boundary definition worldwide remains a contentious issue, largely because of the constraints of UNCLOS III, within which remain significant ambiguities and contradictions associated with the solutions to delimitation. For example, there are no clear definitions for terms such as "island" and "natural prolongation," and there are problems defining normal, or straight, baselines, depending upon the configuration and scale of the coastline. In an attempt to clarify these dilemmas of definition and principle, a major study overseen by Jonathan Charney and Lewis Alexander in the early 1990s examined the factors that have historically affected decisions in maritime boundary delimitations. Whereas Charney points to evidence for "trends and practices" (such as the role of equidistance) that offer some guidance, overall "no normative principle of international law has developed that would mandate the specific location of any maritime boundary line" (Charney and Alexander 1993, xlii). Furthermore, as part of that study, Weil (1993) shows that similar geographic characteristics between places do not result in similar boundary solutions. Most analysts acknowledge the centrality of politics, which "influence the question of whether, and if so when, a maritime boundary will be negotiated or submitted to a tribunal for determination" (Oxman 1993, 39).

Two issues define the parameters of the current debate around territorial claims and delimitation of the US-Canada boundary. One is the claim on Machias Seal Island, the second is the precise location of the US-Canada boundary extending from the end of the 1984 agreement to the middle of the Grand Manan Channel near Passamaquoddy Bay. The issues are linked but separate. There are a number of overlapping considerations that are being applied to the discussions that are seeking resolutions to the island and boundary questions. They include: equidistance, equity, determination of the Northeast Channel as a border between ecological zones, fishery resources, regulatory regimes, and tourism revenues. Ultimately, resolution will need to consider the distinction between linearity (boundary) and zone (frontier), which distinguishes between division of resources and conservation and management of resources. The problem is one not only of negotiation but also of incorporating into the discussions a recognition of different objective outcomes related to sharing the Commons, the long-term resource use in the area.

The Debate in 2002

The disputed Grey Zone and boundary questions that began to emerge in the spring of 2002 necessarily implicated Machias Seal Island. While Canada had maintained a continuous lighthouse service on the island from the early nineteenth century onward, in recent years it had been the summer tour operators, one from each of the two countries, who had exerted their "claims" to the island through their tour groups. The two tour operators had been licensed for boat tours to view colonies of terns and puffins during the hatching season in June and July. If it were not potentially so serious, the story I was told about the strategic manoeuvres on the part of the two countries during the early summer of 2002 would read like an episode of Monty Python. In 2000, the Canadian government had decided that its liability in permitting tourists to land on the treacherous rocky outcroppings of Machias Seal Island involved too much risk, and it notified the Canadian tour operator, Seawatch Tours, that it would no longer be granted a licence for its puffin tours. The company owner, Peter Wilcox, had maintained friendly relations with his American counterpart (a puffin marine tours operator) based in Cutler, Maine, and solicited its support. The Americans gave him a permit to conduct tours to Machias Seal Island, following which he duly informed the Canadian Wildlife Service, which subsequently agreed to issue him a permit. In each of the following years he had to go through a similar "blackmail" situation in order to receive his permission from the Canadian authorities.

In 2002, he heard that, with sovereignty issues becoming prominent again, the Americans had decided to erect handrails along the entrance path, partly as a sign of their continuing maintenance (i.e., "sovereignty") of the island infrastructure. When the Canadian Wildlife Service heard the plans, which were to be enacted on 5 July, the Canadians rushed to build the handrails first and sent out Grand Manan fisher Laurence Cook, with supplies, to do the work on 1 July. When the Americans arrived a couple of days later, Wilcox directed them to another part of the island, where "they might be useful." Then two weeks later another Grand Manan fisher was hired to install even stronger steel rails. According to Wilcox, it was all about "$60,000 sovereignty games." Two years later (in 2004) it was reported that the government wanted to invest about $7 million in new facilities as part of its territorial claim on the small outcropping. In the autumn of 2005, a $620,000 Canadian government construction project upgraded the landing quay and walkways across the small island.

For the community of Grand Manan, an awareness of the renewal of conflict-laden relations with their American neighbours began in March 2002 at the twenty-first annual meeting of the GMFA, attended by about eighty members and twenty guests. In his report for Lobster Sector 38 (Grand Manan), President Laurence Cook summarized "the ongoing work for equitable access to the Grey Zone" (GMFA minutes, March 2002). A working committee (called the Grey Zone Committee) had been established a few months earlier, based on concerns of the Seal Cove lobster fishers that their catches had been declining for several years in the area around Machias Seal Island known as the Grey Zone. They were concerned that the decrease in catches reflected increased American fishing and that, because the Americans could fish all year, by the time the Grand Manan fishers returned to the area in November (having pulled their traps, according to Canadian law, in June) there were few lobsters left. Furthermore, although the Grey Zone had accommodated both groups for years, the decline was being attributed to progressively more aggressive American fishing efforts closer to the Canadian line, causing the baited traps to effectively attract "Canadian" lobsters over to the American side during the summer months when Canadians were not fishing. Initiated by the Seal Cove fishers, who set traps along the ledges in the Grey Zone, the committee lobbied the DFO to support its request that the Canadian government make representation to the Americans to rescind the latter's rights of access to the Grey Zone during the season when it was prohibited to Canadians. During the early discussions, most of the Grand Manan community remained unaware of these negotiations. At the GMFA annual meeting in 2002, the Lobster Sector report brought it to the community's attention for the first time.

In a subsequent meeting of the Grey Zone Committee, it was agreed that Canada's position should focus on two issues: (1) equal access to the Grey Zone fishery, which entailed consideration of incompatible fishing regimes; and (2) the possibility of an alternative management strategy for the area, which might involve the delimitation of a conservation or buffer zone (GMFA minutes, 10 April 2002). The subsequent meeting with the Americans achieved nothing that advanced the case towards resolution. Furthermore, at the 23 May meeting of the Lobster Sector, and indicative of the increasing role of the DFO in the discussions, a DFO representative emphasized that its position was focused on sharing the resource rather than on conservation. This effectively constrained the debate to a single question that, ultimately, could have resulted in forcing the Americans to negotiate the boundary line through the Grey Zone. The government did not seem concerned that the fishers might prefer to consider a "no fishing" zone that could be a conservation zone, and it did not encourage this as an option. A news report quoted a GMFA official: "this isn't about creating a new fishery; our aim is to bring the Americans to the table" (Wilbur 2002). The extent to which the Grand Manan fishers were guiding the discussions, as opposed to the DFO, was already ambiguous. Before the end of June the DFO had made concrete proposals for the active participation of the Grand Manan fishers in the Grey Zone, with trap limits, quotas, and numbers of licences already being negotiated (Lobster Fisheries Association, 38 Advisory Committee, minutes, 4 July 2002). The original DFO proposal was for a target catch of 175 tons, using twenty vessels with 375 traps each. The season would begin on 15 August.

By the time the open meeting of all licence holders was called for 9 July, the entire community was engaged in the debate. At the meeting it became apparent that, with regard to the question of extending the lobster season for Grand Manan into the Grey Zone between 15 August and 31 October, the leadership of both the GMFA and the DFO was united. What was not clear was which institution had the stronger interest in pursuing the challenge to the American summer fishery. In the interviews that I conducted that week, with fishers on both sides of the question as well as with a DFO official and a GMFA representative, it was clear that there was no consensus in the community. There were distinct divisions related to (1) traditional fishing grounds of particular fishers; (2) whether or not the lobster fishers had other licences (such as sea urchin) that would preclude their participation in a summer lobster fishery; (3) a perception that the DFO was forcing an issue not wanted by the community; and (4) whether or not the issue should be raised at all. Many fishers said they would simply prefer the "status quo." One of the fishers said that he would vote for it because he wanted the boundary resolved but that he was convinced "it [would] be a fight" and that Canadians would have their traps cut. The rumours leading up to the meeting generated a great deal of apprehension about the motives of those who were leading the move to inaugurate a new lobster season as well as concern that the imposition of a quota (the DFO specified a maximum of 175 tons) was the beginning of a new regulatory regime that would eventually be applied to the rest of the lobster fishery. Four key concerns dominated the opposition's position:

1 that the new fishery would introduce more dispute and conflict into an area that had managed to maintain good relations over several decades;
2 that the new season in the Grey Zone would effectively create new boundaries and territories, with different regulatory regimes;
3 that the introduction of quotas would be just the beginning, eventually spreading to all seasons; and
4 that divisions were being created within the community as not all lobster fishers wanted or were able to participate.

In explaining the position of the Grey Zone Committee, one of its members said that the American fishing effort had increased significantly over the past decade. When he had started fishing in the Grey Zone in about 1992, there were only four American boats in the area. By 1999, the number had increased to twenty-six; and in 2002, reports suggested there were about thirty-seven American boats fishing in the area around Machias Seal Island. As he pointed out, not only are their numbers higher, but they are allowed to fish twelve months of the year and they can have as many as eight hundred traps, compared to Canadians, whose licence limits them to 375 traps. Furthermore, there is evidence that there are more frequent violations of the line. When this is combined with a minimal DFO presence in the form of regulatory Coast Guard patrols, the Grand Manan fishers have felt increasingly vulnerable.[11]

In explaining why he thought the members of the committee had originally started the initiative, one of the opposing fishers explained that their case was built on an argument of declining catches. Not only were they getting fewer lobsters, but the sizes were also smaller and the Canadians were no longer allowed to keep the three-and-three-sixteenth-inch lobsters (called "Canadians"). With irony, the two men (father and son) pointed out that the men on the committee had recently bought new boats and new trucks: "They can't be doing too badly!" Another fisher speculated that the reason his catch had been "way down" in 1999 was because the DFO had committed all its resources to controlling the conflict over the Aboriginal fishery in northern New Brunswick, leaving the Grey Zone unprotected and open to violations. Quite simply, he said, "There was nothing left."

But the antipathy was not only towards members of their own community. At least three fishers told me they thought that the DFO had fabricated the situation in order to introduce quotas into the lobster fishery. An important part of the problem stemmed from basic distrust of the DFO and a sense that no one understood why there was so much pressure to open up a new season in an area that would inevitably involve conflict with the Americans.

That evening, many stayed home, hoping to avoid angry divisions in the small community where neighbours are important and may well be needed in the future. Significantly for the level of apprehension and the importance of religious culture on the island, one of the fishers who was to keep the speaker's list spoke at the opening of the meeting, asking everyone to remember that the meeting was being convened in "the House of God" (the Anglican Church hall) and that he hoped they would respect that and avoid any profanity. Representing the 136 licences, approximately sixty-four men attended (representing roughly two-thirds of all licences). The meeting lasted four hours and was chaired by the president of the Lobster Sector, fisher Laurence Cook. Of those who attended, only forty-five voted, less than one-half of the individual membership. It was a virtual tie, with twenty-two against and twenty-three for opening the season on 15 August.

The criteria for entry were clearly established, refuting the rumours regarding equity of access that had angered many people in the community. The guidelines were as follows:

1 The 2002 season would be from 15 August to 31 October; thereafter, it would continue from the traditional end (29 June) until October 31.
2 Entry would be by lottery, with up to thirty-seven drawn in each year out of a total 136 licences for Area 38.

3 Each year, those who had been drawn in the previous two years would not be permitted to resubmit their names for the lottery.

4 No other Class A licence (fixed gear) or developing species permit (such as for crab) would be allowed to fish simultaneously.

5 All those who submitted their names for the draw had to advance payment of $4000 to be eligible. This amount would be used to purchase electronic technology for each boat, specifically a "black box" GPS system that would allow their location to be monitored. This money would be refundable if their name was not drawn.

6 Total number of traps would be 7,500, divided equally among participants, with a target catch of 175 tons.

7 There would be dock-side monitors (DMP coverage) to record the catch.

According to several people, many of the mid-capitalized fishers who had arrived prepared to vote against opening up the new season changed their minds after hearing the case presented by the DFO. At the conclusion of the meeting it was decided that there would be a write-in ballot, due within two weeks that would be invigilated by the GMFA and open to all licence-holders. As agreed at the meeting, this vote would have to pass by a margin of two-thirds of those who voted, not two-thirds of licence holders. In an analysis of the prospects for the write-in vote, one official said that the weight would shift to the larger, capitalized fishers who held several licences. This was because the show-of-hands vote at the meeting only allowed for one person/one vote, despite the possibility that the person might actually control several licences. As well, it was felt that there would be a shift towards voting "yes" if neighbours could not be seen to be voting against each other.

What became increasingly obvious was that the DFO wanted the Grey Zone opened in the summer to Canadian fishers. Several fishers who attended the meeting reported that a DFO official had made a clear, unequivocal threat at the meeting: "If you vote against participation in this new fishery, we'll ask Campobello [Area 36]; and if they won't go, we'll ask Cape Breton!" Of course, it should have been obvious that such an action would be rejected by those other communities: it was an empty threat. But it had a significant impact, according to several of the men I interviewed in the week following the meeting. They felt powerless to prevent the initiative. They also realized that the situation that had begun with a few Seal Cove fishers had become embroiled in federal and international politics over which they had no control. When the final tally was in, eighty-six had voted: fifty-three had voted yes and thirty-three had voted no. In other words, the fishery that went ahead did so on the basis of a 62 percent vote. Not quite what the meeting had agreed upon.

As the debate continued in the days and weeks immediately following the 9 July meeting, a new dimension was introduced. There was an assertion, by at least two officials, that the Department of Foreign Affairs (DFA) had become involved. Suddenly, it seemed that the fishers, divided in their own community, were being used by the government to resolve the long-standing international boundary dispute. The Americans were not interested in negotiating, and it seemed that the only way they might be encouraged to begin a new round of discussions was to threaten the viability of their fishery.

With the decision to participate in the new fishery, the media began to report the dispute, with headlines such as "Open Season on Lobster an Issue of Sovereignty" (*Telegraph Journal*, 6 August 2002), and an editorial entitled "Keep Fishermen out of Dispute" (*Telegraph Journal*, 12 August 2002). With the opening of the season on 15 August, it was clear, as reported in the paper, that the "Lobster Zone Divides Grand Manan" (*Telegraph Journal*, 17 August 2002). The reporting, based on information from the GMFA, gave a particular "spin" to the evolving situation that did not quite reflect my interviews. Commenting on the outcome of the 9 July meeting, one story went as follows: "Reluctantly, the Grand Manan fishermen asked for, and received, permission from the Canadian government to fish the Grey Zone starting August 15. But by going out to sea the New Brunswickers hope to force Americans to the bargaining table" (*Telegraph Journal*, 5 August 2002). In the same article, the manager of the GMFA was quoted as saying: "We're having a summer fishery, not to have a summer fishery but to prevent a summer fishery in this area." He said that they wanted to pressure the United States to work with Canadians on an agreeable management plan and a closed season. The next day, the local Conservative MP, Greg Thompson, was quoted as saying that the Grand Manan fishers were "setting off on a mission of national importance" and that "a government decision to send 18 lobster boats out to fish alongside American vessels in disputed waters, could help Canada on the international stage" (*Telegraph Journal*, 6 August 2002). He went on to say that it was "a risky, but necessary, strategy" and that "if it provokes Americans to the bargaining table, it's worth it," In an editorial a week later, the *Telegraph Journal* (12 August 2002) argued that "Sending out fishermen to take up space or fish beside working peers certainly sounds like a recipe for attracting their attention, if not their ire. The new fishery has become a matter for Foreign Affairs to address ... If asserting sovereignty and forcing a dialogue on this territory is Ottawa's goal, then leave the fishermen out of this exercise."

Laurence Cook, the president of the Lobster Sector on the island, added his voice to the media reportage on the eve of the fishery opening: "We have been asking for this since the first time I was on the lobster board ten or eleven years ago. We are going to participate in the fishery and the idea is to recoup our economic losses right now." He went on to say that they would like to see "a blended conservation plan for the area which recognizes that both parties can operate there." Commenting on the different regulatory regimes, Cook said that there are three thousand American licences, all eligible to fish in the area, and each having an eight-hundred-trap limit. "That's 2,400,000 traps they put in, so the 7,500 the Canadians will be adding to that won't make much of a difference" (*Saint Croix Courier*, 13 August 2002). The rhetoric,

Media headlines over the Grey Zone, 2002

with the exaggerated figures (though theoretically possible) was clearly aimed at supporting their case. A DFO spokesperson reiterated the importance of the different regulatory regimes, pointing out that the Grey Zone "is not under any enforcement convention because there is an area between what each of us is claiming as boundary that has not yet been resolved." Acknowledging the possibility of conflict in the area, he said that "no one wants to get into gear conflict … the bulk of the fishermen want to get on with their lives and work with each other to come up with a comprehensive management strategy" (*Saint Croix Courier*, 13 August 2002). The next day, one of the fishers described their tactics as they strove to set their traps: "We're going to have to look for some empty space because I don't expect the welcome wagon to be there" (*Telegraph Journal*, 15 August 2002). Despite some explicit threats from the American fishers, setting day proved to be uneventful. Cook attributed the relative peace to "heavy fog, and a strong DFO and Canadian Coast Guard presence" (*Telegraph Journal*, 16 August 2002).

A New Fishery, a New Sub-Area: 38B

The situation on Grand Manan continued to be divisive. Of the eighteen boats (twenty licences) that had opted to participate, three were Aboriginal licenses operated by non-residents through special agreements with the DFO. Of the Grand Manan fishers who participated, none had multi-licence boats. This was because of the provision that they could not engage in more than one major fishery (Class A licence). There were few of the North Head fishers, who have not traditionally fished in the southern part of Area 38 because of its distance from their home harbour, about sixteen miles (or forty minutes), more than for those sailing out of Seal Cove. According to one fisher, opposition was related to a variety of factors: some simply did not want to give up their summer holiday but, at the same time, did not want others to go; others had multiple licences and were not permitted, and they felt that preservation of the status quo would have been a wiser course; a few felt that it would not be economically viable, given the long distance and the cost of the monitoring equipment; still others did not trust either the DFO or the GMFA: "if they're for it, then I'm against it." One fisher quoted in the media said: "It's not fair that a handful of guys have gotten licenses when about 75% of the fishermen on this island are against it" (*Telegraph Journal*, 17 August 2002). He was one of the signatories to a petition, signed by fifty fishers, that circulated on the island shortly after the opening. Complaining about the lack of consensus, another fisher pointed out that 15 percent of the licences were controlled by seven people (*Telegraph Journal*, 3 September 2002).

At the close of the season, on 1 November 2002, Canada's minister of fisheries and oceans, Robert G. Thibault, called for a bilateral agreement that would define a common management regime in the area. He accused the Americans of fishing "with few restrictions," creating "a potential for over-harvesting" (*Telegraph Journal*, November 1). The local Conservative MP representing Grand Manan in Ottawa, Greg Thompson, reiterated the need for compatible fishery practices, saying that "American fishing rules ignore conservation measures [and] ignore market realities. It's a system that has impoverished a lot of their own fishermen." He went on to acknowledge that the Grand Manan fishers had been pushed into a position that they would have preferred not to be in (*Telegraph Journal*, 1 November 2002). Interestingly, according to several Grand Mananers, following the season's close positions had changed slightly. As word spread that the lobster catches in October had significantly increased compared to earlier weeks, more of the fishers began to express interest in participating in 2003. At the same time, the federal minister of the DFO, Hon. Robert G. Thibault, was quoted as saying that he did not favour continuing the "summer fishery" in 2003.

Nevertheless, in 2003 there were thirty-six of a possible thirty-seven licences allocated, including three Aboriginal licences: double what had been allocated the previous year. Most of the members of the original 2002 group were again participating. One major difference was that, because of the designated total number of traps to be allocated, each licence had only 227 traps rather than the 375 of the previous year. The Americans were unhappy, particularly because the Canadians, under their regime, did not have to throw back the ones with over five-inch carapaces, as did the Americans. By mid-summer, it had become apparent that low catches did not make the thirty-mile return trip worthwhile. Some of the men withdrew until later in the season, when the lobsters would be moving into the area. Said one fisher: "The catch was terrible. We were all mildew in July! There were three weeks we didn't go out at all." But the catch improved late in the season, with daily takes of five hundred to one thousand pounds. However, "it was nothing compared to last year," and he wondered whether many would return in 2004.

Return they did, but in lower numbers. About twenty-seven boats decided to continue in the Grey Zone fishery in 2004, although by the end of the season in late October the consensus was that it was lucrative only late in the season, just prior to the new season's opening in November. One non-participant described the low catches with a sense of vindication: "The Canadians are getting 150 to 200 pounds compared to 1,900 for the Americans. Well, what does that tell you?" I did not know, and asked what it told me. "Well, you know the Big Power don't like it when you push too hard where you shouldn't be!" One of the dockside monitors corroborated the stories of low catches. The year was so bad, she said, that "the fishermen call it the 'Dead Zone.'" The prospects for the fourth year, 2005, did not seem very attractive, with the result that the original fee charged by the GMFA ($4,000) was being lowered to $1,000 in order to encourage some non-participants. What is curious is why the GMFA would want the fishery to continue. Or was the Grey Zone fishery actually being encouraged by the Canadian government through the DFA, as many had suggested? By 2003, it was clear that the politics of the Grey Zone issue were a tangled web of local and global relationships, involving not only the individual fishers and their conflicting views but also their organization, the GMFA, the DFO, and the DFA.

Local and Global Politics in the Grey Zone

At the beginning of the Grey Zone conflict in 2002, it was apparent that two groups of lobster fishers were in disagreement about the wisdom and desirability of requesting a "summer fishery" in the Grey Zone. Inextricably bound up in these divisions was a lack of trust in the leadership and in the role of the GMFA. While the Lobster Sector of the GMFA had operated separately for several years, the negotiations over the Grey Zone made that organizational independence untenable due to the involvement of DFO. The government insisted on dealing through the GMFA, which it saw as the official representatives of the island fishers. Unfortunately, fewer than half of the lobster fishers were members of the GMFA. Therefore, when it became apparent that the DFO would only talk to Grand Manan fishers through the GMFA, a way had to be found to formally bring the Lobster Sector back into the latter. The resignation of Laurence Cook as president of the Lobster Sector in the spring of 2003 began that process. His resignation and subsequent joining of the GMFA to form a new Lobster Sector subgroup essentially cut the former group off from any administrative or negotiating role in the Grey Zone deliberations.

Meanwhile, the position of the federal government was also being modified. Whereas the media had reported at the end of the previous season that the minister did not wish to continue the fishery in 2003, by the next spring he was focused on "developing complementary management plans." He also pointed out that the 2003 fishery "reflects a written formal request from GMFA and its membership" (interview, May 2003). In other words, the DFO was placing responsibility for the continuing Grey Zone fishery on the Grand Manan fishers themselves. The fishers, on the other hand, felt overwhelmed by the pressures from the DFO and the GMFA. Moreover, in the context of so much community opposition, they could not understand the GMFA's objectives. Then, in the summer of 2003, the GMFA sought to consolidate its power and to ensure greater leadership over the Lobster Sector by restructuring, which effectively cut loose the Lobster Sector Advisory Committee (GMFA Board Meeting, minutes, 29 July 2003). The GMFA continued to insist that the DFO was driving the Grey Zone initiative. The DFO, on the other hand, offered two scenarios: in an 11 August letter to George Brown it said that it was only supporting a local request; and yet, in response to one of my requests for information, I was told to contact the DFA, which had "the lead in this file." No one, it seemed, was willing to accept responsibility for or to explain their interest in pursuing the Grey Zone fishery. Thus, when it seemed that only seven boats might go out for the 2005 season, the GMFA announced that it would lower its participation fee from $1,500 to $1,000. Catches that year were as important for the "bycatch" of crab as they were for the lobster itself. One fisher pointed out that being allowed to keep the by-catch enticed fishers into the summer lobster fishery since the poor lobster takes could be more than offset by lucrative crab sales. One young man who had applied for the summer licence said that he had gone out three times and that the catches were hardly worth the trips to the Machias ledges. The Grey Zone summer fishery is all but gone as an issue for Grand Manan, despite the apparent efforts of the DFO and the DFA to engage the Americans in a battle that would eventually resolve the international boundary dispute.

More International Boundary Disputes

The discovery of natural gas on Canada's east coast and the growing prospects for importing liquefied natural gas from Russia and the Middle East created new potential for industrial sites for the reconversion of the super-cooled gas. As the companies involved began to design pipelines that could carry the liquefied natural gas (LNG) there was also a need for terminals for importing and conversion. Communities along the coast of New Brunswick and northern Maine began to hear of plans for refinery facilities in the period between 2004 and 2006, with varying degrees of alarm and hope for economic benefits. The decision as to where to locate the facilities depended to an important extent upon the existence of deep-water harbours and community support. However, there were other problems that required negotiation. Surrounding areas were concerned about the potential environmental impacts and risks associated with large tanker ships that would be moving through local waters. One proposal in particular concerned Canadian communities along Passamaquoddy Bay because the American locations would require shipping through a passage claimed by Canada as sovereign waters. Two proposed sites were Robinson Point and Pleasant Point in Northern Maine, a few miles from St Andrews and Deer Island, New Brunswick. To reach these sites, ships had to pass through Head Harbour Passage, which Canada claimed as sovereign waters.

In a letter from Michael Wilson, the Canadian ambassador to Washington, the Canadian position was stated firmly: "The impact of the proposed siting of the terminals, and the potential passage of LNG tankers through the environmentally sensitive and navigationally challenging marine and coastal areas of the sovereign Canadian waters of Head Harbour Passage present risks to the region of southwest New Brunswick and its inhabitants that the government of Canada cannot accept" (quoted in "Canada to Deny LNG Passage," *Globe and Mail*, 16 February 2007, B3). The two American companies, Downeast LNG Inc. and Quoddy Bay LNG LLC, both filed applications with the US Federal Energy Regulatory Commission (FERC) for site approvals. The letter from Canada's ambassador was intended as a warning in advance of any decision by FERC so that the companies would not invest to no avail. Seemingly at cross-purposes with the federal government, the New Brunswick government was quoted as supporting the American proposals, although Shawn Graham, the premier of New Brunswick, later refuted this in a commentary piece in the *Telegraph Journal* ("NB Supports Canada," 24 February 2007, A9). The premier said that the intervener status of the province in the FERC hearings would be in the "interests of the Province and, in particular, the interests of the citizens who depend on the Passamaquoddy Bay for their livelihoods and who cherish it as an unparalleled natural gem." He went on to point out that it would be unwise to prejudge the FERC process, but that, "if the final result is in any way contrary to the best interests of New Brunswickers, we will act swiftly and decisively." Subsequently, the province made a formal request that the FERC process be terminated on the basis that the Canadian government's refusal to allow tanker ships through the passage essentially precluded development of the entire project. In June 2007, FERC rejected this request (*Courier Weekend*, 8 June 2007, 1).

Throughout the spring of 2007, the argument of whether the Head Harbour Passage was international waters (as the Americans argued) or sovereign Canadian waters (as the Canadians argued) was being debated in the media by lawyers. One factor that was pointed to was the existence of precedents that might uphold the Canadian position, such as when, in the 1970s, Canada refused to allow US oil tankers though the passage. But a problem for the Canadians, one lawyer suggested, was that the United States is not party to the UNCLOS and, therefore, cannot be forced to arbitrate. Another argument held that the waters are certainly internal Canadian waters because it is impossible to reach the passage without going far into Canadian waters, unless one were to take a circuitous route around Grand Manan. In a CBC interview with UBC law professor Ted McDorman, the complexity of the issues for both sides was apparent. While the Americans argued for "right-of-innocent passage," the right is not absolute insofar as "innocent" must be proven. In other words, fishing activities or environmental pollution would not be regarded as "innocent," and risks associated with these might preclude acceptance of the American claim. On the other hand, the passage represents the only route into Eastport, a route that has been used by American boats for generations, thereby establishing a historic right of passage. McDorman did not believe that LNG tanker shipping would be seen as non-innocent, and he opined that the Canadian case would probably falter if it relied on that basis (CBC interview, *Radio Morning*, 26 February 2007). Sharing the waters that provided access to shipping ports was an issue of bilateral negotiations and of potential harm to the fishery and surrounding communities, including Grand Manan. While some brushed off the concerns of environmentalists and fishers as "fear mongering," others were outspoken in their call for strong government action. Janice Harvey, from the New Brunswick Conservation Council, was unequivocal: "What's at stake is the complete destruction of the economy in Southwestern New Brunswick based on fishing" (CBC interview, 5:30 PM, 23 February 2007). It is unlikely to be resolved without representation to the ICJ.

Access to the marine Commons has become increasingly complex in recent decades, partly due to technology that has put pressure on the fisheries resource but also due to government efforts to manage a notoriously difficult resource that is not well understood. As well, the different levels of governance, from international laws to federal, provincial, and community regulations, has contributed to weaving a web of structures that precludes the kind of personal understanding that had governed community relations for generations. Similarly, as the next chapter explains, there have been growing disputes over the boundaries of traditional land-based resources as the community seeks to develop new rules for sharing shoreline access, the woods, and the trails that facilitate movement about the island.

5

Sharing the Commons:
On Shore and Land

Intertidal Zones and Shorelines

It is not only the resources of the open ocean that pose problems for sharing the Commons. Two crucial areas of negotiated access are the intertidal zone and shoreline, both of which have been the arenas of debate and conflict over generations. For two hundred years, as the incubator for key habitat species, the rich intertidal zone has provided many niche fisheries for local Grand Mananers. While traditionally dulse, periwinkles, clams, and nori are the most important of these resources, the most recent addition to this list (1996) is rockweed.

Rockweed
Rockweed is the most prolific of the local seaweeds, covering the rocks all along the shore across the entire intertidal zone area. Despite its easy accessibility, without known markets rockweed had never attracted islanders until a collaborative arrangement in 1996 between the New Brunswick government and Nova Scotia-based Acadian Seaplants initiated a pilot project to harvest it. It would then be shipped off the island for drying on the mainland. The company had explored the idea of harvesting around Grand Manan as early as 1969, when it had approached a local carrier captain about the possibility of being a transporter of rockweed that would be harvested using mechanical rakers. A dense brown algae described as a bladderwrack (*Ascophyllum nodosum*), rockweed is protected from the sun by its own mucilage when the tide recedes but

floats upwards with the rising tides because of its leather-like bubbles, which act as buoys. Among its freely floating fronds are the eggs, hatchlings, and spawn of hundreds of species of marine life. Over one hundred invertebrate species and thirty fish species are directly associated with rockweed habitat (Berrill and Berrill 1981). Rockweed is a descendant of the most ancient plants, and only along northern coasts where fog protects against the sun can it survive in shallow waters. Stem, leaf, and root are all of the same material, the holdfast (i.e., the root) acting only as an anchor, while the nourishment is absorbed through all the seaweed's exposed areas.

In the late 1980s, the New Brunswick government was aggressively exploring all possibilities for diversifying the provincial economy. Knowing that rockweed was being harvested elsewhere, notably in Nova Scotia where the stocks were reportedly being depleted, the New Brunswick government felt there might be an opportunity for a new venture along the Fundy coast, where jobs were chronically scarce. Thus, in 1991 New Brunswick signed a Memorandum of Understanding (MOU) with the federal government, clearing the way for the provincial government to license private companies to harvest rockweed off the shores of Grand Manan. Considering the eventual outcome of this story, the five objectives of this initiative seem rather ironic. They were:

1 to maximize the number of continuing full-time employment opportunities for New Brunswick residents;
2 to ensure a sustainable harvest;
3 to promote the development of a commercially viable industry founded on sound business principles;
4 to integrate the rockweed industry with other users of marine resources; and
5 and to ensure that rockweed harvesting and processing are undertaken in an environmentally acceptable manner.

By November 1991 a draft management and development plan had been prepared by the Department of Fisheries and Agriculture (renamed after 1994 to include Aquaculture), New Brunswick (DFA NB), and in January 1992 applicants were invited to submit proposals for a single harvesting licence.[1] The licence was to be for a pilot period of three years.

Following the signing of the MOU, the government consulted with a marine biologist, Dr Robert Rangeley, to assess the viability of a rockweed harvest. His report, which included strong arguments against the rockweed harvest as it was being proposed, clearly indicated that, in going ahead with its plan to license rockweed harvesting, the government was misinterpreting or consciously ignoring the available scientific information (Rangeley 1991). He pointed to two key weaknesses in the government's position. The first concerned the strength of the negative evidence. While the government argued that there were no data demonstrating negative impacts, equally, Rangeley said, there were no scientific data to suggest positive impacts. However, he concluded, there was overwhelming evidence that the extant biomass of rockweed is an essential component of the ecosystem. There was a fundamental need to adopt the precautionary principle. The second point concerned the burden of proof. Implicit in the decision to allow the rockweed harvest was an assumption that the cost of being wrong regarding no negative impacts was low. Rangeley, on the contrary, argued that such costs would be unacceptably high, in which case the burden of proof should lie with the company or government to demonstrate the benign effects of the harvest.

Not only did the government ignore Rangeley's report, but its so-called public consultation on Grand Manan in March 1991 was so poorly advertised that only two people appeared. A booth at a trade show in St Andrew's at the famed Algoquin Hotel in 1991 was the only other opportunity to publicly consult with companies involved in sea harvesting, and of course it, too, failed to reach members of the community on Grand Manan. By 1993, the Conservation Council of New Brunswick had begun to take an interest and demanded assurances that an independent environmental impact assessment would be conducted. The government did not respond. In 1995, it granted a harvesting licence for a three-year pilot project to the Nova Scotia company Acadian Seaplants Ltd. On 16 June 1996, the company announced that it would begin harvesting within two days. By the next day, under the leadership of three Grand Manan women, through phone chains and word of mouth, the community was galvanized in protest. A meeting at the Legion Hall was attended by about one hundred people, including Eric Allaby (local MLA), Joey Green (island mayor), and Klaus Sonnenburg (head of the GMFA).

People were concerned about the impact on the spawn and nurseries that the rockweed protected. Referring to the weir industry, one fisher said: "If the rockweed harvest employs five people, every time I seine a weir I take five with me and I don't know how many women work at the plant." Someone else pointed out that the periwinkles bring in over $160,000 annually: "That is more than this rockweed industry will benefit Grand Manan for many years to come." Jobs and the environment were a constant refrain. Another speaker,

Blockade of Ingalls Head wharf, 1996

Janice Harvey from the Conservation Council of New Brunswick, argued that communities must be able to retain control over their resources and that "Islanders should be able to say yes or no to economic projects like rockweed harvesting." Allaby defended the harvest, saying that it represented an opportunity to diversify the economy. Despite the protest meeting and letters to newspapers and the government, the harvest went ahead. A blockade of the wharf in July that forced the company to bring in its own boat rather than use the ferry was a final, short-lived protest. The harvest continued for four years as a pilot program.

Then in March 2000 it was announced that the licence was about to be renewed, and DAFA NB called a meeting on Grand Manan that was to bring information to the community to reassure islanders that the environmental impacts were being monitored and that a continuing harvest would not threaten the marine resources around the island. There were five key speakers: one from the DFO, one from the University of New Brunswick, one from Acadian Seaplants, one from the Department of Marine Research (in Maine), and one from DAFA NB. Among those who were there to listen were three harvesters, a municipal councillor and manager at Atlantic Mariculture, seven fishers, and three others, including the man who would later be hired as the island manager.

In reporting on the four-year harvest period, the main speaker was Dr Glynn Sharpe of the DFO, who presented a wealth of data in a twenty-five-minute talk that described in detail the scale and intensity of the harvest, the status of the resource, the exploitation rate, and what new knowledge had been gleaned through the pilot project. With regard to this project, called a "regional assessment process" (RAP), Sharpe explained that a wide representation of stakeholders had been involved in the review and production of the summary report. He said that the average exploitation rate had been less than 10 percent, "well below the possible," and that, because the annual production was only 13 percent to 16 percent of the total, he argued that it "was not eating into standing stock." According to DFO studies the rockweed around Grand Manan has five to seven inches annual growth, with exposed sites having lower biomass production. Overall, Sharpe was convinced that the harvested biomass was replaced within the year and that the canopy height had remained unchanged in areas where there was a 44 percent incidence of harvest. Later in the evening, Dr Ugarte, the scientific advisor for Acadian Seaplants, reiterated the assessment of low rates of harvesting, pointing out that they had taken only about 11,000 tons from around Grand Manan, an area that was producing about 66,000 tons annually. As Sharpe continued presenting his massive amount of data, one began to feel overwhelmed and to wonder what the impacts would be over the long term. He acknowledged that eider ducks could be affected but that the harvest season would begin only after mid-July to avoid their reproductive period. (This particular promise was never kept.) They had found many fish species within the rockweed canopy, but these were mainly pollock (85 percent), which relies upon the rockweed as a refuge from predators. According to Sharpe, the wide range of invertebrates found in this area was not threatened by harvesting practices.

Questions remained. While the pilot project had involved production well below possible limits, there was no discussion of impacts at higher rates of harvesting. Moreover, as Sharpe admitted, there were many knowledge gaps with respect to measuring how ecosystem complexity would be affected by the harvesting. As well, an understanding of how water fowl might be affected over the long term due to the destruction of rockweed habitat was unknown, while assessing the impacts upon fish populations would necessarily involve high-intensity sampling and lengthy studies. In assessing the evidence that would preclude a rockweed harvest, Sharpe concluded that none of their data provided evidence of negative impacts in the short term. The changes in habitat structure that they had noted were "patchy and limited in extent."

The recommendations contained in his report were specific. Crucially however, they also alluded to potential environmental degradation. These recommendations included:

1 varying harvest rates according to habitat conditions and wave exposure;
2 ensuring that holdfast content is low by protecting the area with friable substratum; and
3 continuing to gather information on potential impacts on the ecosystem.

Sharpe asked: "Is the present rate of harvesting appropriate?" To which his reply was that it may be appropriate for the resource as a whole but that it was not appropriate for specific areas. Therefore, a fourth recommendation concerned the possible need to manage some areas differently than others.

The biologist from the University of New Brunswick who attended the March meeting had been chair of the RAP Working Group and a member of the Canadian Atlantic Fisheries Scientific Advisory Committee (CAFSAC). She described the many complexities of ecosystem management in terms of the lack of scientific knowledge related to such topics as the spatial relations of organisms within rockweed habitats, the degree of dependence of organisms on rockweed, the trophic links, temporal and seasonal variations, and even the importance of detached rockweed. She summarized the discussions and papers that had been presented at a symposium the previous December, noting that there was a broad consensus that, under conditions of uncertainty, any harvesting should proceed on the basis of the precautionary principle. Once destroyed the rockweed would take decades to be re-established. There was explicit recognition of the distinction between ignorance and uncertainty and arguments for "policed compliance" as a crucial component of any management plan. While the Acadian Seaplants scientific advisor, Dr Raoul Ugarte focused on biomass rather than on habitat structure, most other participants in the workshop were concerned about the more complex and infinitely more important problem of ensuring a sustainable habitat for multiple species protection. One speaker, Dr Chris Finlayson from the Department of Marine Research in Maine, made a series of specific recommendations related to harvesting rockweed under conditions of uncertainty:

• establishment of site specific studies;
• rotation of harvest areas;
• community representation on management committees;

- monitoring of spatial effects;
- establishment of no-take zones;
- specification of minimum by-catch amounts;
- studies of nutrient cycling;
- establishment of integrated coastal zone management committee; and
- assessment of biomass baseline prior to approval of harvesting plan.

For the attending vice-president of Acadian Seaplants, Steven Spinney, who did not make a presentation but who responded to questions, the rockweed harvest had been successful – "full-scale development at the pilot scale level" – meeting the objectives of business as well as the objectives of the New Brunswick government. Enumerating these objectives, he pointed to a sustainable harvest, commercial viability, assuring the maximum number of jobs, and the integration of the harvest with other marine activities. For Grand Manan the reality certainly did not reflect this optimistic view. Total jobs, by a conservative estimate, would be fewer than three in any year, with additional labour having to be brought in from Newfoundland's Fogo Island and Nova Scotia. Ironically, with the exception of the manager and two or three other young men, there were no islanders involved in the harvest. Finding reliable, steady labour for the back-breaking work of raking the tough seaweed up into small dories that had to edge in along treacherous rocks at low tide had proven to be difficult. As for the goal of sustainability, this, as was clearly acknowledged by all of the scientific advisors (with the notable exception of Dr Ugarte), was a long-term question mark in terms of volumes, ecosystem complexity, and species protection. Further, there had been no integration with other marine activities. Finally, the company explicitly ignored the promise of avoiding the eider duck reproductive season, beginning its rockweed harvest in early June. The only government/company objective that had been met was that the harvest be commercially viable. In part, this was explained by the number of jobs created at the mainland drying facility in Pennfield and by easy access to American markets (compared to access from their previous location in Nova Scotia).

Not only could many of the questions from the Grand Mananers who attended the 16 March meeting not be answered (due to insufficient scientific study), but there were also problems with regard to assuring a continuing long-term commitment to monitoring subsequent harvesting practices. A question about the baseline knowledge of periwinkle populations, for example, met with the response that it was unknown. A question related to the hiring of local people to monitor ongoing harvest practices was not answered. For several islanders, the presentations had provided substantial information on the pilot project without actually addressing key questions about the future. While one islander (subsequently hired as the local manager) offered the opinion that the harvesting of rockweed should be compared to forestry practices in that the seaweed could benefit from planned culls, most of those attending the meeting were concerned about ecosystem structure and impacts upon species habitat. While Tom McEcheron from the DFA NB argued that the meeting constituted a valuable basis for the development of a research plan, Grand Mananers were sceptical. Despite McEcheron's feeling that the meeting had been a "consultation," islanders felt that it had been an informational meeting that did not actually intend to be receptive to critiques or to ideas such as local monitoring.

One group of islanders that would be immediately affected would be those involved in the niche sectors – dulse and periwinkle. The representative from Atlantic Mariculture, Michael O'Neill, pointed out that dulse harvests had declined and that there was speculation that this might be related to changing tides as the periods for extreme drain tides were longer than they had been, thus providing longer harvest times and allowing for greater depletion of the resource. As well as being concerned about the impact of rockweed loss upon dulse, he wondered whether these tidal changes might affect the regrowth of the former. While the rockweed harvest only occurs on the eastern side of the island and would therefore not affect the Dark Harbour dulse, a major part of all Grand Manan dulse and most of its periwinkles are harvested on the eastern side, adjacent to areas being subjected to rockweed harvesting. With an increase in the amounts to be harvested, there were legitimate concerns that the March meeting did nothing to allay.

For those involved in the harvest the work was gruelling. While they were able to make up to $150 to $200 per day (more than twelve dollars per hour), in fact they could not sustain that level of work because it was too arduous. One man said that he had tried wearing a back brace but that his neck, shoulders, and legs were still just as sore following a day's harvest. He described it as "the toughest work you will ever get," and he had a basis for comparison as he had also worked in other sectors of the wild fishery. The harvesters were paid thirty-eight dollars per ton (a boat would take 3.5 tons when overflowing) and, on average, would make between $600 and $800 per week. The men had to provide their own boat and fuel. In 2000, there were only two full-time harvesters, although four other men worked part time when the "dulse was off." They had expected Nova Scotians to arrive but they had not been able to find accommodations on the island. The harvesters felt that as long as the practice was restricted to hand harvesting this would not jeopardize the sustainability of the resource. However, the harvester I talked with admitted that the calm areas were fully harvested – areas such as Grand Harbour, where the limits were supposed to be at 17 percent of the total and data showed that this had been exceeded by 2 percent. By 2004, the harvest was growing. On 5 July that year (during eider breeding season) three trucks left loaded with about thirty tons each. On one day alone, ninety tons were shipped from the island: the equivalent over an eight-week season of six days would be 4,320 tons.

Many questions remain unresolved. Even after ten years of the harvest, there is no concrete information as to amounts leaving the island, the areas being harvested, or the impacts on other marine resources. There are still no data available on long-term effects. Nevertheless, local observers noted disturbances that have not been tabulated. One islander living in the Ingalls Head area described seeing thousands of small crabs washed up dead along the shore immediately after they had begun the rockweed harvest. A year later, he said, there was none, and he assumed that the young crab had been obliterated. Moreover, regardless of the five objectives in the original MOU, the government neither diversified the local economy nor increased the number of jobs. While a few jobs were available, in fact the concurrent development of aquaculture resulted in there being few workers, thus necessitating recruitment from Newfoundland. And, as has been mentioned, despite specific claims that they would avoid the eider duck breeding season, Acadian Seaplants was actively harvesting in early June 2006, with full truckloads of rockweed seen leaving the island in mid-June. Finally, for the community there is only a vague awareness that the harvest has continued, and while some islanders retain concerns for its impact on a multitude of species, the government has made no effort to reinstitute any consultations. As Harrison and Burgess (1994) have shown in their study of the Rainham Marshes in the United Kingdom, the concepts of nature and conservation become embroiled in conflicts that, ultimately, are more about political expediency than about environmental protection. One could add that, to the government as much as to private companies, commercial values have precedence over the community's control of its own resources.

The Dark Harbour Seawall

Perhaps the most actively used (and occasionally contested) intertidal zones are those associated with the Dark Harbour seawall. What captures the imagination of any first-time visitor to Dark Harbour is its many changing moods. The grey, dismal light on rotting timbers, soaked nets, and dirty ropes may not at first seem beautiful (as my daughter told me quite unequivocally); however,

Rockweed at ferry terminal, 2004

transformed by the setting sun those same artefacts can become objects of beauty. And if one is in tune with the rhythms of the tides and the dulsing community, it is possible to capture something of the urgency and determination of the dulse harvest. As the tide retreats, about two hours after flood, trucks appear down the steep hill, following each other onto the beach around the pond. With only a few cryptic words the men and women unload their baskets and bags, pull on their boots, grab jackets and lunches, and then head for their dories. Within minutes they are all motoring across the pond to the seawall, where they attach the boat to an electric hauler that draws it up and across the wall. Unhooked from the hauler, the dory becomes like a giant unwieldy toboggan that is to be guided down the other side, with several men and women controlling its descent on the slippery rocks.

The first time I walked out onto the seawall an old truck caught up to me, bouncing around on the rocks of the mud flats that had been left bare by the low tide. It slowed down, and two women with warm but curious smiles peered out and invited me to have a ride. I declined, but a half hour later I found them on the far side of the seawall already engaged in gathering the purple seaweed. As I watched I marvelled. These women were, I later discovered, mother (Carolyn) and daughter (Dorothy), seventy-five and fifty years old, respectively, and in their hip-high rubber boots on the slippery rocks they were collecting pounds of the wet dulse by hard pulls and yanks. This was not light work! Eventually,

Top: Dark Harbour: Seawall and camps
Bottom: Dark Harbour: Sliding dories down the seawall

as I saw that they were about to return to firmer high ground, I offered to carry Carolyn's burlap bag, which turned out to contain almost fifty pounds of wet dulse. Scrambling up the steep rock wall with the bag I wondered at the strength, perseverance, and non-complaining attitude of these women. Then they invited me into their small camp on the seawall to have a sandwich with them. Carolyn had made egg sandwiches, one for each of them. She would not accept my "no thank you": I had to share her single (delicious) sandwich. The camp, Carolyn explained, had been built when she was just a year old, and they had spent entire summers here for many years during her growing up. Such kind, generous, and hardworking people became for me a touchstone of Grand Manan life. Despite the many problems and contradictions, islanders continue to watch out for others and always reach out to help.

Along the inside of the seawall is a terrace that has stretches of evenly arranged rocks, on top of which are nets. These are the drying grounds for the dulse, the nets allowing for easy turning. Today most dulse is taken directly to the large, often sloping drying grounds next to people's homes or to Roland Flagg's drying grounds. Occasionally, small amounts can be seen drying out at the harbour, but the dampness of the environment, held in by the soaring cliffs and lack of sun until after midday, does not make for ideal drying conditions. One of the many changes to the island working relationships and sense of the Commons as shared work areas occurred in 2006 along this seawall. As described by Carolyn Flagg, her son-in-law had burst into the house the previous day (26 June) exclaiming about the dastardly destruction of their drying beds along the seawall next to their eighty-year-old camp. Apparently some of the "Indians and Newfoundlanders" (hearsay) did not want to sell their dulse through "the company" (Atlantic Mariculture), which maintained the trough (or ramp) in the wall and provided the machine hauler for the boats. Therefore they had contracted one of the island machine operators to dig another ramp. The problem was that it went right through the Flagg drying grounds, which had been established there decades ago. While known to be Commons, until now accepted by everyone as "belonging" to them, even their small seawall drying grounds had been claimed by newcomers and had been destroyed during the building of the new ramp. Carolyn explained that, out on the seawall, none of the camps paid taxes since it was common property, although it was acknowledged that the pond itself "belonged" to Buddy McLaughlin on the strength of a ninety-nine-year lease. While the camps along the shore were on Anglican-owned land, and paid taxes, no such arrangement had ever been required of the seawall camps. With the incursion of new people on the island, who felt they had equal rights to the seawall, there was a fundamental change in the sense of community shared spaces.

Beaches

The beaches too have become zones of contestation: between islanders and new residents, between traditional and new activities, and between islanders and tourists. Islanders, used to wide access to the beaches as Commons that could be used for preparing top poles or weir stakes, assumed that the building of plastic cages for the aquaculture industry would be similarly accepted. Beaches where women were accustomed to taking their young children for an outing seemed to have become sites for the industrial assembly of large fish cages or storage areas for anchors awaiting transfer to their ocean locations. It was not only islanders who were upset with these intrusions onto their beaches. In fact, there were widespread protests, sometimes including letters written to the municipal council, about the proliferation of black plastic pipes and rings on the beaches of Seal Cove, Woodwards Cove, and Stanley Beach in North Head. While islanders tended to be less opposed to this because some of their family members were involved in the industry, tourists and summer residents had no such inhibitions.

Salmon cages on Seal Cove beach

Compared to the complaints of islanders who tended to mutter about the intrusions, those of tourists and new summer residents were much more explicit, public, and focused. They wrote letters to the municipal council (July 2003) and threatened to explore legal action if the companies responsible did not remove the plastic pipes and cages. However, this did not mean that islanders welcomed the articulate opposition from outsiders. Indeed, for the most part, islanders sided with the aquaculture companies. They did this for two reasons: (1) they believed that "outsiders" had no right to "tell them what to do" and (2) nobody wanted to challenge the industry that had brought so many jobs to the island. Despite acknowledging that the cages were an unwelcome nuisance for families and children wanting to enjoy the beaches during the too brief summer weather, islanders did not claim any "rights" to their usual beach playgrounds. Generally, the industry attempted to limit construction work on the beaches to the spring and fall seasons so that tourists and families wanting to use them would not be affected. Boundaries are not unambiguous, and spaces of meaning can be created differently according to who is making a claim.

Also harvested along the shore, in areas where there are meadows of tough grasses, is the small floral plant known locally as sea heather. Professional florists dry this plant and use it in their arrangements. While several islanders gather sea heather in late August (when they are blooming) for use in bouquets and wreaths to be sold at the Christmas bazaars, there is understandably some tension when they realize that visitors come to the island to gather them for their floral businesses on the mainland. One person said to me angrily: "Imagine, bags of it stuffed in the back of her truck!"

Through the 1990s, with a strong economy and weak Canadian dollar, there was an increasing number of summer residents, especially Americans, who were buying large tracts of land, thereby causing rapid escalation in housing and land prices on Grand Manan. Shorefront acreage was particularly vulnerable to higher valuation because of demand, with the result that many islanders began to feel that they were being alienated from their own waterfront properties. Without any provincial legislation prohibiting outside purchases, there was no legal protection, and, of course, those islanders who had originally owned the land stood to benefit from the sales. Compared to a price of $55,000 in 2000 for a 1.5-acre waterfront lot in Castalia, by 2006 the asking price for an equivalent sized lot (1.25 acres) in Deep Cove was $95,000. Shorefront land had become as valuable on remote Grand Manan as it was in many urban areas across the country. In 2007, a tiny quarter-acre shorefront lot with a superbly sited cottage sold for $180,000.

Access to beaches quickly became a contentious issue, inevitably involving rights-of-passage. In one case, access to a beach that had been a favourite island family retreat for generations was suddenly cut off by the contiguous shoreland owner. The ensuing dispute, which was eventually tried in court, went on for years (from 1998 to 2003) when the path was blocked because the owner had established new tourist accommodations. In refusing to allow the general public across his land, despite evidence of historical rights going back two hundred years, he was engaging in confrontational relations that were new to the island. In another case, the public road that gave access to a large gravel beach that islanders had enjoyed for generations as the site of barbeques and picnics was closed off by a summer resident who would no longer tolerate people crossing her land. Most islanders simply ignored her signs and continued to use the narrow lane.

In-migration of working residents from Newfoundland has caused other types of shoreline tensions. Island dulse harvesters complained that New-foundlanders, newly arrived as migrant seasonal labour, were not abiding by informal rules of territorial access and were threatening the sustainability of the harvest. But the harvest of rockweed proved to be even more threatening when a mainland company in 1996 embarked on what was termed a "pilot project" supposedly for only three years. Despite community opposition, the harvest was still continuing in 2005, providing jobs for only two islanders and about twelve Newfoundlanders. Nine years after this project had started, only a few islanders continued to voice objections.

The increased presence of provincial regulations in the lives of islanders affected the lobster fishers' ability to use the rocky shorelines for sources of weights for their traps. Strict enforcement of regulations prohibiting removal of rocks from the beaches led some to surreptitious searches, while others bought the cement weights commercially. For all fishers, use of the beaches and shorelines for the preparation of weir stakes, mending of nets, and access to the sea across slippage ramps have been crucial for both work and social relations. The intrusion of aquaculture companies, tourists, and summer residents have all played a role in creating new spaces of meaning. At Whale Cove, for example, where fishers launch their dories and skiffs and check herring catches in the weirs, competition for space from a new kayak company (after 1999) provoked antagonism from island fishers. Over generations these slipways have been gathering places where the men assess the weather and the tides, trade stories, and set out for their work on the sea. Unlike the fishers, who pay an annual fee to the harbour board for the maintenance of these launch slips, the kayakers have no such obligation. When the men arrive in their trucks, expecting to back up to the slip so they can easily unload the heavy air tanks for the diver or repair items for the weirs, if the brightly coloured kayakers and their boats happen to be taking up space, they are seen as usurpers. The appearance in 2001 of a NO PARKING sign, on an island where there has been no history of formalized landscape rules, was a matter of comment around the wharves. As with aquaculture, the intrusion of modern activities such as tourism has brought new challenges to social relations on Grand Manan.

Another issue is the new uses to which the wharf facilities are being put, including storage of nets and the requisite tools for pen manufacture as well as net manufacture. This inevitably involved old sheds being converted and new large trailer trucks making deliveries in places and along roads that, historically, were only able to accommodate pickup trucks. As the traditional fishers vied to keep their spaces intact, there were frequent debates about who was encroaching and who should have access to common roads that were designated as Crown reserves. While these debates were gradually resolved over the early years of aquaculture development, they presaged many similar disputes involving new uses of areas that had always been considered as Commons. For example, the trend to buy portable prefabricated homes that were brought to the island on special barges rather than constructing one's own new homes involved the need for holding space near the North Head wharf. As a result, the large area used for preparing weir stakes and top poles, or repairing lobster cars, or painting boats became, for days at a time, the transit point for the new homes. Again, the men tended to express annoyance and to mutter about the inconvenience, but for the most part they shrugged, pointing out that it was a "sign of the times."

More than contests for space and social relations are being played out along the shorelines of Grand Manan. The humour and creativity that emerges unexpectedly across the island can also be found here. One spring I discovered an iridescent pair of waterproof working gloves carefully mounted on two pieces of driftwood, seemingly reaching out of the grasses atop a sand dune, stretching towards a stark blue sky. Several months later, walking along the same stretch of shoreline, I found nine more gloves all grouped around the original pair, again carefully arranged as though in conversation, like so many of the men on the wharves. On the opposite side of the island, along three different and equally isolated natural pebble bars that enclose brackish ponds, there are camps of varying degrees of luxury and home comforts that may include propane lighting, television (in one), and elevator pulleys that transport food and weekend staples up and down the 325-foot cliff. While some proudly fly a Canadian flag, others display an enviable stockpile of split wood; one is surrounded by newly made rabbit traps, while another has a veritable yard of creative driftwood furniture decorated with plastic flowers. Clearly, the survival mode of the hard-working islanders incorporates moments of respite and fun.

The shore has been a Commons that has been shared over hundreds of years with regard to the harvest of dulse and berries; the preparation of stakes and the mending of nets for the weirs; and, more recently and controversially, the building of salmon cages. The shore is more than a boundary between sea and land: it is a space of evolving meanings and shared activities that contributes to a sense of community and belonging. Personal identities are illuminated in the furniture and landscaping around the camps, and social relations are shaped by the unspoken agreements and acknowledged renegotiation of who has claim over these spaces.

Camps on Indian Beach seawall

Indian Beach camp

The Woods: Camps, Trails, and Development

Ordinary space resists traditional patterns of narrative. Space is a fragmentary
field of action, a jurisdiction scattered and deranged, which appears to be negotiable
or continuous but is actually peppered with chasms of economic and cultural
disjunctions. ~ (Yaeger 1996, 4)

In telling the stories of the woods, camps, and trails one is tempted to focus on
the sweet-smelling grasses and brightly coloured meadows of Indian paint-
brush and sparkling yellow buttercups; or the soft spongy bogs that harbour
delicate orchids, cloudberries, and pitcher plants; or the large swaths of white
bunchberries, enveloped by high fronds of light green ferns; or the sounds of
the ever-present black throated green warblers or the screeches of a hawk.

These and so many more wonders of the Grand Manan woods and bogs bring visitors back year after year. It is an island that offers rich varieties of scents, sounds, and views along the beaches, through the woods, and along the steep cliffside trails. To a frightening extent many of these marvellous encounters are threatened, both by the outside forces of speculative development and by the islanders' lack of awareness of the inherent fragility and vulnerability of their natural landscape. Indeed, in the October 2007 newsletter of the Grand Manan Trails Association it was announced that the new trails guide book had deleted any mention of the "Blue Trail" from North Head (Tatton's Corner) to Money Cove because of clearcutting, which had obscured large parts of the trail: "It was impossible to follow the markers. Clear cutting may become more and more of a problem" (*Grand Manan Trails Newsletter*, October 2007, 1). In a conversation that same week an islander described his trip out to Money Cove on the back of an all-terrain vehicle (ATV). "The trails are a mess," he said; "hikers couldn't use them because of the destruction from ATVs. Tourist operators should be concerned."

Until the 1990s, land had always been taken for granted, with families owning many acres that would gradually be given to children and grandchildren as they began their own families. But in recent years the incursion of buyers from the mainland, especially from the United States (partly due to an attractive exchange rate and partly due to an inexorable movement up the coast of people seeking "isolated" second homes), has forced up land values, changing the perceptions of what land means to individuals. Conflicts between tourists and private residents, between fishers and tourist operators, and between Grand Mananers themselves began to surface. PRIVATE and NO PARKING signs began to be seen and boundaries were challenged. Fences that had never existed on the island began to appear. When a new summer resident decided to erect boulders across a traditional path, he found them gone when he returned several weeks later, with a more ominous warning left on his doorstep.[2] Islanders did not take kindly to barriers that prevented them from gaining access to traditional hunting land, especially when the perpetrators were on the island for only a few weeks every year. It was generally acknowledged that, if everyone actually fenced the land they claimed to own, the area of Grand Manan would have to be extended by at least 50 percent.

The result was increasing numbers of challenges and feuds as neighbours sought to benefit by subdivding waterfront land or prohibiting passage along traditional rights-of-way. In the municipal council meeting of April 1998, four boundary issues were discussed: (1) a major dispute at Deep Cove, (2) an easement between the house of an islander and two people from-away who complained about sewage run-off, (3) rights-of-way between houses down to Whale Cove, and (4) a stream embankment that had suffered erosion near the site of an old cemetery. In the latter case someone argued that something had to be done because "caskets had been floating down the stream!" One sale on Bancroft Road in 2000 unexpectedly involved not only the two negotiating parties but also a grandmother who was in a nursing home, an American, and a son who had power-of-attorney. The process was extremely complex but certainly not unusual. Many lovely century-old homes can be seen along the main road, all of them abandoned because of problems with "heirship" – contested claims that cannot be resolved. A common expression on the island is, "Oh, that's an heirship house." One report about the rising prices of land suggested that real estate on Grand Manan was the most expensive in the province. When I asked the local MLA about provincial legislation that might constrain the large purchases of land by off-islanders, his response was that it was not an issue anywhere else in the province and therefore it would not be easily passed in the House. In about 1991, he said, the government had briefly looked at the PEI model but ultimately rejected it because it "would raise more problems than it solved." Nonetheless, a report in the media in 2006 suggested that the problem was becoming more widespread. "Campobello Island Dream a Tax Nightmare" read the headline in an article that described the perceived invasion of American buyers over the decade of the 1990s. "Empty oceanfront lots, which five years ago fetched about $40,000, are selling for $250,000 ... Foreign investors buy up choice land, driving up the property values and taxes, pricing some locals off their land" (*Globe and Mail*, 5 July 2006, A3).

As well as growing conflicts between islanders and new arrivals who were looking for summer retreats or just a few weeks vacation, there were contentious issues around the long-term consequences of some traditional practices that were no longer deemed to be environmentally safe. Notable among these was the debate about a new method of garbage disposal on Grand Manan. Until 1998, garbage had been disposed of at three open dumps: in North Head, Grand Harbour, and Seal Cove. One simply took anything and everything there and left it. There was a continuously burning, foul-smelling pile of garbage that attracted rats and gulls. While tourists had been complaining about the unsightly and hazardous situation for years, it was ultimately the provincial government that enforced a new law prohibiting open dumps, forcing the island to consider other options. Just before the dumps were to close, apparently, "people were panicking, throwing everything away, tons of stuff," according to one observer. Islanders, even those who recognized the rightness of the government's position, regretted the impending loss of an easily accessible and anonymous recycling system as people who wanted to dispose of useless but essentially functional items (such as bicycles or sewing machines) would simply leave them at the dump, setting them off to the side where those who wanted them could pick them up. Another, less supportable, activity involved going through individual garbage and finding letters or bills that would reveal interesting family secrets. I had been warned never to throw away confidential information in case it was picked up.

One man had dump scavenging down to a science. He and his wife regularly visited the dumps, each carrying a sack. In a one-hour walk, he said, they had each collected twelve dollars worth of aluminium cans. He said that one of his friends had collected $105.40 worth of bottles and cans in eight and a half hours, which is "pretty good money." While the price for aluminium is "low right now, at only thirty cents a pound," he felt that the work was worthwhile. When he found old water heaters he would take out the internal heating unit for the copper and brass parts; and he collected batteries for the lead. He figured he had made about $1,000 worth of supplemental income from his scavenging activities. There were many valuable things that could be retrieved from the old dumps, and many islanders were nostalgic about losing them.

After several public meetings to consider the various options for a new method of garbage disposal, eventually they settled on a transfer station, which was to be located in the middle of the island just outside of Grand Harbour. Garbage, papers, recyclable glass, and plastic would all be accepted there. Heavy machine parts were to be deposited in a separate area. At first, there were many complaints about the distances people had to travel to get there and about the fact that its operating hours did not enable fishers to get there easily. But within a couple of years there were no more muttered comments about its foolishness, and garbage was no longer seen floating off the end of wharves in protest. In the meantime, all island garbage was being trucked off the island to a mainland dump.

One of the most contentious claims, which pitted an individual against several tourist operators and islanders who sought to retain traditional right-of-access to the beach and access to a private gravesite, involved land in Deep Cove. Because the beach area had always been a favourite spot for islanders during the brief warm months of summer, the paths that provided access to it were of special concern. With experts consulting old maps and historic documents going back two hundred years, and lawyers asked to consider legal rights with respect to acquired or riparian rights, the debate continued for several years. During the course of the debate letters were exchanged with council and between lawyers, boulders were placed in strategic locations, a ditch was dug that would discourage all vehicles, signs were removed and repainted, and families stopped speaking to each other. While the individual wanted to protect his new tourist rental cottage from walkers, in fact there would always be access to the beach from the lower end. This conflict was never fully resolved, with the pathway being blocked but with people continuing to visit the small private gravesite beside it en route to the beach. This was indicative of the land-related battles that were threatening to become more common during the 1990s.

Water, too, was a source of contention as development threatened to expand into new areas, both residential and industrial. The addition of a second shift at the Connors plant in Seal Cove caused several debates related to assured water supplies. Because many villagers have dug wells as opposed to drilling them, there is great sensitivity to the water table, especially during dry years. In 1999, several families reported water shortages. In one case the fire department arrived with its truck to replenish a dug well, which almost immediately emptied down the hill. When Connors added its extra shift, it needed additional water for the processing and tried to tap into existing wells above the plant. Strong opposition from neighbours forced it to search in other areas, notably along Red Point Road. Eventually, before it had to make expensive infrastructure commitments, the second shift was discontinued due to restructuring, and Connors was finally sold in 2004, in a complex agreement with the American-owned Bumble Bee, which created an income trust. At the same time as Connors was searching for water, its manager developed a trailer park to accommodate the migrant labour arriving from Newfoundland. Despite its being a four-hundred-foot drilled well, people in the village claimed that it affected their residential supplies. Similar problems were envisaged when an aquaculture company proposed a hatchery that would draw water from a hillside near Cedar Street; and someone else proposed a disposal site for salmon gurry, which was seen as a possible contaminant of the water table.

The ownership of many camps had been an ongoing source of tension for some families, whose claim went back many years on the basis of informal "squatters rights." In the late 1980s, one of the high school teachers had embarked upon a series of projects with her students, the point being to sensitize them to the ownership patterns on the island. While she was unable to recall many of the findings, she said that in 1990 non-residents owned about 48 percent of the thirty-two thousand acres of Grand Manan.

Of this off-island ownership, about one-quarter was American owned, while almost one-third was owned by a German-Swiss national (known by everyone simply as "Kaufmann") who owned about ten thousand acres around Miller's Pond. In 1998, a mainland buyer expressed interest in this large piece of undeveloped land. The Crabbe Company had been exploring the possibility of buying the land as a source of high-quality (very dense) wood for its mainland operations. That the province might be interested was also rumoured, although some people expressed dismay that the government might build a hiking trail and did not want their tax dollars used for something like that. In an undated letter to the Anglican Church, which also owned significant acreage, another company, Bestwood Lumber, indicated its interest in buying either land or the rights to cut on the land. An important concern for islanders was that their access to winter wood supplies was being threatened. They had been used to being allowed to cut on the Anglican land for a stumpage fee of twenty dollars per cord. The new company arranged a lease for five years on the Anglican land, which precluded islanders having access to it for their wood cutting. A letter suggested that it might be mutually beneficial to undertake major cutting: "On March 31, 1998 we spent the day cruising some of your holdings to obtain a visual perspective of the current condition. The woodlot consist [sic] primarily of mature Spruce, Balsam, Fir and Birch. From our initial observations there are significant problems in your woodlot."

The interest in cutting rights on the island, even given the problem of access and the ferry, was considerable. The wood is valued on the mainland because of its density (due to salt water spray) and small growth rings (due to short growing seasons). For many years following these two expressions of interest, despite the provincially mandated attempts of the municipal council to enact a zoning and planning process, active wood cutting escalated on the island, resulting in large areas of clearcut (which could be seen only from the air or by those skirting through the cutover areas on their ATVs. For some islanders the wooded land was a source of income. Commenting on the extensive cutting that was being done in Deep Cove in 1999, one of the landowners shrugged and said, "Well, she's a widow and needs the money I guess." Several large flatbed trucks left the island that evening, fully loaded with thirty-foot-long logs that were eighteen inches in diameter. While much of the wood leaving the island was clearly for lumber or pulp, there was a concern that the vast acreages owned by off-islanders might be developed for housing.

The contradictory responses to this possible threat were interesting because most islanders inherently opposed any form of legislative interference. Even the hint of a formal plan was enough to bring several hundred people out to a meeting, despite the fact that a plan might very well have been in their interests with regard to protecting the long-term viability of woodlands and hunting territories. The idea that the municipality might buy the property was supported by some islanders, but in the end this never happened. When Crabbe offered to buy the land, the Kaufmann family accepted, and the transfer was made in 2001. For some people it was an opportunity lost. John Cunningham felt that if the issue had been properly presented to the island populace, they would have supported the municipality's purchase of the land because of the high percentage of hunters on Grand Manan. Apparently the island has the highest ratio of hunters to population of any community in North America.

For most islanders the woods do not really represent a retreat into the wonders of bird and plant life; rather, the woods are important first as a source of firewood, second as a place to hunt deer, and third as a camp getaway for men. For generations firewood has been an essential fuel for all homes across the island, cut during the winter months and delivered to homes during the spring for drying and, later, stacking. Many families have their own land for cutting, while others buy from local cutters. As well as heating fuel, the woods are a source of material for home building and for the top poles for weirs. The men seem to love working in the woods, often trekking in between boat trips during the winter lobster or seining seasons, usually alone.

As already mentioned, the sheds, wharves, and garages are important daily gatherings places for the men. Equally, the camps in the woods and around the ponds and on the gravel bars have an evolving significance for them. Names such as Pig 'n Whistle, Shangri La, High Chapperal, Kentuck, and Xanadu are clearly not about a connection to the sea. Many of these camps have propane lighting, while some have generators and even television sets. But the personal comforts stop at running water, and all must rely upon traditional outhouses. Some go into their camps in late February when the maple sap starts to run, and they spend days collecting the sap and boiling it down for the syrup, which is to be given as gifts to friends. For others, the first spring trips are in the trouting season, which opens on April 15.

The woods have traditionally been a gendered environment, one in which men have explored all of the hills and streams, often naming them without reference to official map designations. Fox Hill, Snake Corner, and Quigley Hill do not appear on maps but certainly are known by the island men. Their knowledge of the woods has probably been developed more by their hunting of deer than by anything else. While some have been known to set rabbit traps, a few trap muskrat on the outer island, and one hunter apparently devastated the beaver population by trapping and then selling pelts on the mainland, most Grand Mananers restrict their hunting to deer. Because they are allowed only one deer in a season, the hunters may see to it that their wives also have licences so that, if they are careful about it, they can get a second one. No one walks in the woods between the end of October and late November during the four weeks of the hunting season, which is a source of frustration for the women who are then beginning to pick spruce "tips" for their wreath making. Deer stands can be found throughout the backwoods, and truckloads of fallen apples are often brought to Grand Manan to provide tempting bait, which makes the hunt much more reliable. One of the Newfoundlanders was quite dismissive of this practice: "It's cheating," he said, "that's not real hunting!" But for many Grand Mananers it is not just the sport that is important but the food. The men are expert butchers. It is fascinating to watch as they butcher the quarry, which they hang in their garages. They wield sharp knives with the expertise of surgeons, slicing, cleaning, and cutting the meat into meal portions for the freezer. The liver is taken out to be soaked for twenty-four hours in baking soda and water to get rid of some of the gamey taste. Rabbits, on the other hand, are usually jugged, packed in Mason jars, and sealed in boiling water.

According to the older men, even the deer hunt is beginning to see signs of change. While "you might see a few women carrying shotguns," the younger men do not seem to want to go out into the woods. "When I was a kid I'd be out there with my father. But you don't see that now." For the older men, however, there is immense pride in being able to bring home a 150-pound deer that will provide meat for the winter. Ninety-one-year-old Smiles Green had just returned from the woods on the third day of the opening of the hunting season in 2004. Not only had he shot the deer near his camp (the Pig 'n Whistle) unassisted but he had also managed to get it out of the woods on his ATV – all 150 pounds of it. By the time I saw him in his garage he had already cleaned it and half butchered it, organizing the steaks, slabs, and cubes in the freezer. Two years later, at ninety-three years of age, he proudly described his successful hunt at 8:30 AM on opening day of the season. Following the close of the 2004 season, Smiles headed out just before Christmas to another camp on Miller's Pond with fifteen men who had organized an all-male birthday party for him and for

Curtis Brown, who was turning eighty-one. The gendered nature of these cel-ebrations is a common feature of island social events, although among the twenty- and thirty-year-olds this seems to be changing.

Historically male spaces, since the mid-1990s the woods have been opened up for women and the camps have become more oriented to weekend recre-ation for couples and families. The social spaces of the woods have been profoundly changed with the changing social relations, technologies, and eco-nomic rhythms of the island. Women and girls are no longer satisfied staying at home, and, increasingly, the woods represent the empowerment of women who, riding in on their own ATVs, have invaded the previously male territories of the woods and gravel bars, from Bradford's Cove to Indian Beach (even though men continue to be the key decision makers around the social spaces of the camps). Sometimes the women go trail riding in groups, sometimes they visit the camps.

ATVs have taken over the woods, enabling easy access to the camps as well as making the winter wood supply much more accessible. One real drawback has been their devastating impact upon the trails, which have become mud trench-es in many areas. Whereas the ATVs have not used the sea-front hiking trails, which have remained narrow and relatively free of ATV damage, the interior trails have been made almost impassable for walkers during any wet season. Indeed, in the case of the steep trail down to Money Cove, the ATV damage required considerable repair, to the extent that several hundred yards of flat stone paving was constructed to protect against further destruction.

One of the most interesting individuals who relied upon the camps for his own way of being in the world was Philip Parker. Upon retiring from the ferry crew he spent increasing amounts of time in the woods, until he had built five camps (two in the woods and three on different gravel bars at Indian Beach, Money Cove, and Beal's Eddy) to which he could go every day. He would leave his house in North Head just before sunrise and return every night at sunset – a return trip on his ATV of about an hour and a half. Even at seventy years old he would use a self-designed rope system to clamber down a steep cliff, bring-ing in food, water, and assorted tools. His travels had created a fairly decent road between Money Cove and Beal's Eddy along a very steep rocky and grav-elly beach. His days would be spent dulsing, building rabbit traps, splitting wood, and devising interesting and innovative solutions to the problems pre-sented by the environmental challenges at hand. In one notable case he con-structed two differently designed rafts that were tethered to the shore but that had roof coverings. They floated on the high tide but would remain easily accessible at the mid- or low-tide points. His rationale was that these vessels would protect dulse on the days when it could not be dried. The rafts provided storage areas that would ride up and down with the tides and offer cover from the sun. Parker's undertakings were essentially private, and even his family did not know about them. When I first met him he was invariably accompanied by his lovely golden retriever, Pal, who died in 2002. Parker was a quiet, peaceful soul whose camps were both retreats and places of incredible energy. They were places of being and becoming.

Trails and Development

One of the most significant tourist attractions on Grand Manan has been the coastal trail system. Originally occurring because of the need for fishing families to scout for boats and to check herring in the weirs in the early morning dawn, the trail virtually encircles the island, overlooking weirs, lobster buoys, tidal currents, and soaring gulls. Varying from only a few feet above the beach to several hundred feet, the trail is a spectacular way of experiencing the natural landscapes and physical beauty of the island in its complicated moods. No two walks or arduous hikes will be the same, depending, as they do, on winds, temperatures, tides, and seasons. One must carry the tide tables or at least be informed about them, especially if any beach access is involved. One of the problems that became an issue in the 1990s was the fact that all of the land traversed by the trails had become private. In the past, even people from-away who owned shorefront land did not spend enough time on the island over extensive enough areas to preclude hikers using the trails.

But in the 1990s this changed. Beginning in 1997, when a private landowner developed Hole-in-the-Wall Park, the trail system began to be interrupted. By establishing a fee for entry to the park, Basil Small effectively restricted any access to the trail that was not at either end. Because of the large acreage involved (over two hundred acres), this meant that, through the park area, where tent sites were created, the trail became both difficult to follow and somewhat intimidating because it appeared to be intruding on private property. Thus, for the area from Swallowtail to Whale Cove, the trail was barely accessible and difficult to find.

More significantly for local fishers, however, the Hole-in-the-Wall Park presented an important problem affecting their work. Whereas this area had been the local airfield until the late 1980s, with vast expanses of open flat land, and the owners had welcomed the men who wanted to dry or repair seine and weir nets, the current owner did not. Indeed, he charged a rate per foot of twine (fifteen cents in 1997) that, universally, the men felt was exorbitant and unfair. At that rate, with an average weir being about 1,100 feet around and requiring two sets of twine for both top and bottom, the cost would be about $330 for twenty-four hours, compared to the fifty dollars they were used to paying. A traditional working area had been removed from their space, and the island fishers were not pleased. Everyone talked grumpily about "the changes" over which they had no control.

A slightly different situation existed for the trail between Seal Cove and Bradford Cove along the western side of the island. An extremely remote area but with spectacular views across the strait to the American shore, this trail was constantly being challenged by local residents, who removed the signs. Because of the many ATV trails and wood-cutting trails, the trail signs are essential for tourists, especially if they move away from the shoreline. In protecting what they deemed to be their rights to private land (although some suspected the problem was illegal meadows of marijuana), those who removed the signs were endangering tourists and discouraging tourism. The conflicts between residents who operated tourism establishments and other islanders were often muted, but they were, nevertheless, significant sources of tension. When a large piece of land overlooking Whale Cove was subdivided in 2002, there were again concerns about trail access. It was reported that the owner had assured council that he would not bar hikers and that land sales would include the provision that the trail be kept open. But the land sales and house building along the Whistle Road seemed destined to threaten public access to the historic trail. The most hopeful sign of new forms of relationships between tourism attractions and land developers was the creation of a nature preserve along the trail between Long Eddy Light and the Bishop (a small pillar of rock north of Ashburton Head). The generosity of an American landowner named James Monroe explicitly established a preserve through the New Brunswick Nature Trust that protected the trail from incursions by residential development. This was an important step insofar as the future of the island's tourism industry depended to a large extent on the continuing existence of the trails, which were, in any case, maintained by a small group of dedicated volunteers.

Interestingly, while the trails had originated as access to viewpoints for the fishers, their existence since the second half of the twentieth century had depended upon the leadership of people from-away, usually people who had moved to the island as retirees for the very reasons that tourists enjoyed visiting: the scenery, outdoors, bird life, and hiking and cross-country skiing. In finding volunteers who would agree to take responsibility for clearing particular segments of the trails and contribute to the guidebook, Bob and Judy Stone (since 2001) assumed an important but quiet and generally unacknowledged role in the life of the island. While a few islanders enjoyed the hiking trails and volunteered to adopt specific portions for maintenance, for the most part both maintenance and hiking were the purview of people from-away. Again, however, there is evidence of change. In 2006, an organizational meeting in early June was attended by almost an equal number of islanders and people from-away. At a closing dinner a year later, about one-third of the forty-five people were native islanders, although that percentage is much higher if it is based on being a permanent resident. For tourist operators it is crucial that the trails be maintained as, without the trails, summer visitors will not come.

The woods are at the centre of the island, both physically and emotionally, even if they do not define the culture to the same extent as does the sea. The woods have provided heating fuel for generations as well as basic foods such as meat and berries. But almost as significantly, the woods have provided spaces of retreat, escape, camaraderie, and exploration – a Commons that islanders could share in ways that were not applicable on the sea. Today, the ways in which the woods are implicated in island lives are beginning to change due to the incursion of outside lumber companies and residential development as well as changing gender relations. When the museum curator was mounting an exhibit about the entire Grand Manan Archipelago, she sought out people who had visited all of the islands, something that proved to be more difficult than she had expected. It was suggested that she ask older men who had been both fishers and hunters and who, therefore, would have explored the outer islands for gulls' eggs and muskrat. At one time the high prices for muskrat pelts had encouraged active trapping throughout the archipelago. A few people still seek out gulls' eggs, but the worldwide boycott of wild furs destroyed the prices for pelts and the incentives for trapping.

For generations, the cultural meanings attached to the woods have been rooted in both sustenance and recreation, and they can be discerned in even the small clues left by gambolling wanderers. One of my favourite personal photographs is of a precarious piece of ephemeral art that was captured on film one early afternoon in 2000 and that was disassembled less than two hours later. I call it the Grand Manan inukshuk. It had been constructed on the base of three beer bottles. Three layers of very large rocks had been carefully balanced on the bottles, and the sculpture was both elegant and imaginative as well as fun.

Grand Manan "inukshuk"

There are other examples of humour that show islanders' relationships with their land. For example, also recorded on film is a cut-out rectangular "frame" in the middle of a tree that is nine inches in diameter. Approximately ten inches by four inches, the frame is completely through the tree so that the scene on the other side looks like a picture hanging on the tree. Impromptu sculptures along the beaches – of driftwood, bait pockets, and crab shells (June 2006); an evolving diorama of orange and blue gloves and ropes and twine (2004-06); a log chair created in the middle of the woods for the park warden (1998); a tribute to a teenage love (Ashburton 1996) – these and more provide some lively clues to relationships with the landscape.

A Future for Sharing?

The Commons of the sea, the shoreline, wharves, and woods are all spaces of changing relationships and meanings. Over the past decade the growing number of conflicts generated around issues related to sharing these spaces is the direct result of increased connections to the mainland, involving movements of people, capital investments, and new perceptions of social and economic networks. For Grand Mananers, these conflicts occur as isolated events. Tourists seem to presume rights-of-access to kayak launching areas; prefabricated homes take up space in common work areas where weir stakes must be sharp-

Shoreline hands

ened; a new campground precludes use of large fields for mending nets; trails cross private land that is claimed for cottagers; and aquaculture companies build their cages on the beaches. These problems and more cause tensions and conflicts that tend to be resolved one problem at a time. But they are symptomatic of a broad trend towards the privatization of common spaces, and they raise the spectre of the possible alienation of work areas that are crucial to maintaining a working fishery.

An article in the *New York Sunday Times* in early 2007 might presage what is to come for the island. It was reporting on a similar range of pressures spreading up the eastern seaboard of the United States. Headlined "In Maine, Trying to Protect an Old Way of Life," the story described the "rapidly shrinking working waterfront" that is threatened by the "hot demand for waterfront property" and a real estate market that is driving community leaders to seek solutions that can protect the working landscape of the traditional fishery. It was estimated that the mileage of working waterfront had decreased by 20 percent in only a decade (Katie Zezima, *New York Sunday Times*, 25 March 2007, A12). "It's a crisis in slow motion" said one observer who was working to find solutions. A real estate agent acknowledged the problem but suggested that "it's really an unstoppable force." The recognition of the enormity of the long-term implications, with regard to the loss of traditional access to the water, galvanized the village of Harpswell, which was able to raise US$1.5 million "through pancake breakfasts, corporate donations and door-to-door solicitations to buy a dock and businesses at Holbrook's Wharf." It was hoped that this would help to maintain a sense of community and a working waterfront. Meantime, the State of Maine provided matching grants to six communities to support local initiatives that would preserve open spaces and working waterfront areas. One example of local initiatives involves the Town of Machiasport, which used a small grant to buy a fifty-foot right-of-way leading to public clamming beds. There are lessons to be learned and warnings that must be heeded if Grand Manan is to avoid similar incursions that will threaten both its landscapes and its way of life.

As discussed at the beginning of this chapter, the evolving nature of property rights is significantly linked to the functioning of local economies and the social relationships within them. The role of government, at all levels, is a crucial factor in establishing regulatory resource regimes and the community administrative structures that determine issues of land distribution and use. How communities are able to share their Commons, whether of the sea, shore, or woods, is part of a complex amalgam of formal government policies and regulations and a historically based network of informal negotiated understandings. Common property is not a clearly defined institution; rather, it exists according to a continuum of evolving rules and normative guidelines that reflect social relationships as much as changing global markets and technology.

6

Globalization and Restructuring

PETER GRIMES: I am native, rooted here
By familiar fields,
Marsh and sand,
Ordinary streets,
Prevailing wind ...
And by the kindness
Of a casual glance.

CAPTAIN BALSTRODE: You'd slip these moorings
If you had the mind.
~ (Britten 1963 [1945], 108–9)

Contradictory feelings around whether to stay or whether to go are growing among Grand Mananers. Evidence that women and youth, in particularly, are beginning to want to "slip these moorings" is discussed in chapter 10. By 2006, the insidious inevitability of globalization was touching all lives on the island to a greater or lesser degree. In chapter 2, I describe the historical importance of the herring fishery, both weirs and seiners, and the significance of the fish plant and smoked herring sheds as sites of social and productive activities. Of all fishery activities on Grand Manan these have most defined island identity. It is, therefore, not surprising that fundamental changes in this industry will have profound effects on future social relations and community expectations. When we consider the many other areas of change that have been described so far, including the rapid growth of salmon aquaculture, the degradation of groundfish stocks, and changing patterns of migration, we begin to get a sense of the magnitude of change, most if not all of which can be linked to the forces of globalization.

Globalization has been variously defined, but at its heart is the notion of connectivity (Tomlinson 1999). While the notion of connectivity is complex as it involves concepts of distance and networks as well as an understanding of time-space compression that implicates dimensions of local cultural experience within the daily rhythms of life, its relevance to the changes on Grand Manan must be acknowledged. As Tomlinson (4-9) points out, increased connectivity means that we experience distance in different ways. One measure of globalization is "how far the overcoming of physical distance is matched by that of cultural distance" (6). From an instrumental point of view "connectivity works towards increasing a functional proximity," creating globalized spaces and connecting corridors that ease the flow of capital (7). Globalization can involve the internationalization of legal norms (e.g., liberalization of trade through the World Trade Organization) or of human rights (e.g., through the United Nations); it may mean the dominance of English as the language of business or the spread of liberal democracy or the internet. Planes, boats, satellites, and fibre-optics, as well as ideas and people, are all drawn into a single web that "is physical, biological, electronic, artistic, literary, musical, linguistic, juridical, religious, economic, familial" (Appiah 2005, 216). But these complex interconnections do not imply a true "global village" (McLuhan 1964) since face-to-face relationships are not possible on such a scale. But the knowledge imparted, the sense of new experiences to explore, affirms different realities that gradually transform expectations and views of the future, creating new cultures. For Grand Manan collectively, the impacts of technology (including the internet), the degradation of the fishery, the growth of aquaculture, and in-migration can be summarized in terms of new connections and dynamic new possibilities. Incorporating the metaphor of boundaries and, later, framing the discussion in terms of the concept of habitus (see chapter 9) enables me to focus on some of the dynamics that are occurring within these complex areas of change. The connectivity of globalization is usually understood as beginning with economic activities and culminating in cultural transformation. As Tomlinson has pointed out, whereas actual proximity has not changed, there is a new experience of being connected that affects perceptions of places as being accessible (4). However, the compression of distance as a measure of the accomplishment of globalization also incorporates cultural distance.

Another feature of globalization is that individuals and communities are seen to lose control over their lives. For the agro-food industry the implications of this are economic, cultural, and ideological (Watts and Goodman 1997). Within the fishery, a key feature of globalization has been a profound loss of control over every aspect of the production process. In other words, globalization has meant that local fishers no longer "determine what, how much, by what method and for whom" fish are caught (Whatmore 1995, 37).[1] The change in Canadian fisheries policy through the 1970s and 1980s fundamentally altered the relationships of fishers to their resource by putting in place mechanisms that increasingly disembedded them from their communities and inserted them firmly into state and market structures (Apostle et al. 1998). The patterns were established for the active promotion of new productive capacity that would alleviate problems associated with unemployment and a degraded marine environment that was producing fewer fish with fewer fishers. Following the intense period of privatization through the imposition of individual transferable quotas and company-controlled licensing, the next rational solution was the encouragement of aquaculture development.

Four factors have been especially important for the globalization of fish markets (Arbo and Hersoug 1997). The first is the establishment of the new ocean law regime in the 1970s and the delimitation of the two-hundred-mile exclusive economic zone, which radically altered the distribution of rights to fish. In effect, there was a nationalization of resources that meant that trade in fish became crucial, especially for those countries that had lost access to previously accessible fishing areas. The second factor is the development and diffusion of new technology, which enabled large-scale harvests and processing to complement advances in communication and information technologies. The third factor is a rise of international corporations, whose power through markets and pricing mechanisms enabled them to use strategies such as global sourcing to effectively control seafood markets. And the fourth factor is the liberalization of trade and movement of capital, which has been concurrent with the closing of the resource system. As Arbo and Hersoug show, these factors have generated increasingly competitive world markets that are focused on the retail level represented by corporate interests. In other words, there has been a de facto switch from reliance upon the production end of the process to the consumption-driven determination of demand. All food production has become increasingly integrated into global networks, but perhaps no sector has been more profoundly changed than that of the wild fishery.

On Grand Manan, the globalization of economic relationships and productive activities has not been matched by cultural change. The long history of isolation and independence has resulted in a level of resistance in Grand Mananers that is only gradually melting away, allowing for a new community identity. This chapter examines the ways in which economic and productive activities are being restructured and transformed into global networks and how these are affecting boundaries of cultural difference.

Herring and the Weston-Connors Connection

The two key areas of new globalized relationships are found within the herring fishery and salmon aquaculture. Representing traditional and modern economic sectors, respectively, these two staple activities dominate island sensibilities and understandings. They define identities and community relationships. As already described (chapters 3 and 4), the herring industry has been radically transformed because of both changing market demand (declining demand for smoked herring) and new technologies (purse seiners). And the major restructuring of the agro-food industry has profoundly affected the processing of herring as well as the ownership patterns within the industry. This underlies many of the changes on Grand Manan and explains some of the tensions apparent in daily conversations and behaviours. The restructuring that has occurred has essentially reoriented the sardine business from the production end of the process (the weirs) to the retail-consumer end of the process. It is no longer a question of whether or not there are fish in the weir; it is now a question of whether or not there is a market, and whether or not Connors will continue to require fish for its local packing plants. Corporate owners want to control every step of the process, beginning with the harvest of fish.

For Grand Manan weir fishers, the business strategies of the Weston Corporation became the driving force for change on the island beginning in the early 1990s. The closure in 1992 of the Weymouth plant in Nova Scotia was the first indication of structural changes in the industry. This closure was soon followed in 1994 by the closure of the Welkspool plant on Deer Island, with that production being moved to the Back Bay plant on mainland New Brunswick. Then in 1999 the Back Bay plant was closed, putting pressure on the remaining plants to increase production. The immediate result for Grand Manan seemed to be positive insofar as it meant there would have to an extra shift at the Seal Cove plant. This extra shift required labour recruitment that sent the plant manager to Newfoundland in the springs of 1999 and 2000 looking for workers willing to migrate to Grand Manan for the approximately five months that the plant would be open.

At its peak in 2000, the Grand Manan Seal Cove plant had two shifts and employed over two hundred workers from May to November. In 2000, the Weston Corporation bought its main competitor in the United States, the Stinson Seafood Company, which owned five sardine plants. What was especially crucial in this acquisition was Weston's agreement to invest $12 million in Stinson upgrades in order to guarantee the continued operation of its plants.

The governor of the State of Maine had reportedly intervened (interview, May 2005) in order to make the sale contingent upon a legally binding agreement to upgrade the Stinson facilities. Presumably, the executives at Weston knew that their guarantees would not be an issue as the company apparently had already decided that the purchase of American-based Stinson was merely a stepping stone in the strategy to establish an east coast monopoly, thereby enhancing the value of its assets, which were to be sold the following year. Unfortunately, that agreement became legally binding even for the purchaser of the company a year later. However, while the $12 million could be allocated to upgrading the Prospect plant, that did not preclude the immediate closure of three other plants at Lubec, Belfast, and Rockland, leaving only the Prospect and Bath plants operating.

Then, only one year later, Weston put the entire company up for sale, separating Connors sardine production from its salmon aquaculture Heritage division. In step with an agro-food industry that was restructuring globally, Weston sought to rationalize its product lines by increasingly focusing on the bakery business. In its efforts to buy Best Foods (at a reported cost of $260 million), it planned to divest itself of the entire east coast fish business. In 2001, the publicly traded Connors Brothers Income Fund was created, with its subsidiary being Connors Brothers Limited, "one of Canada's oldest food producers and the largest producer of sardines in the world" (Annual Report, December 2001). In its first annual report, the Connors Brothers Income Fund Company announced: "In recent years, the company has invested over $20 million in a plant modernization program. Additional capital projects have been identified which will further improve efficiency and reduce operating costs. The integration of Stinson Seafood (purchased in March 2000) also continues to produce operational savings" (Annual Report, 2001).

The next sad chapter in this story shifts decision-making responsibility to local executive officers and directors in New Brunswick. Through 2002, the Board of Directors at the Income Fund (which included maritime executives Derek Oland and President Ed McLean; two former politicians, Doug Young and Bernard Valcourt; Paul Marion, a Canadian Imperial Bank of Commerce representative; and Don McLean of Suedon Investments Ltd.) apparently considered various options for future business directions. Within months they entered negotiations with their largest competitor, the American company Bumble Bee. The proposal for a "merger" with a company that had global sales six times the amount of Connors Brothers should have alerted Canadians to the probable outcome.[2] However, in late 2003 negotiations with Bumble Bee led to the sale (i.e., merger) of what was left of Connors to the American company in February 2004. Within months, it was apparent that there were problems for the local New Brunswick communities. Soon, one of the four remaining plants was closed. The temporary shut-down of the Grand Manan plant for "inventory" in August 2004, at the height of the herring season, raised questions about the intentions of the new management team.

Then, in December 2004, it was announced that the Grand Manan Seal Cove plant would be closed immediately and permanently. Islanders began to consider the possibility that within three to five years the Blacks Harbour plant would also close. This would mean the eventual demise of the two-hundred-year-old herring fishery and the permanent end of the weirs around the island. Most people calculated that about three hundred people would be directly affected by the closure, without considering those who depended upon what fish plant workers purchased in local stores. The closure would affect not only the plant workers (about 180 on a shift) but also truckers, seiners, weir fishers, and even ferry revenues.

The possibility of major changes in the nature of the business was supported by the news that, in October 2004, the company had ordered a large new seiner at a reported cost of almost $8 million. Then, in April 2005, Bumble Bee sold the can manufacturing plant that had been part of the Blacks Harbour operations. This seemed to confirm fears that the cannery itself was under threat as well – a suspicion apparently corroborated by the concurrent investment in additional freezer facilities. Speculation was reaching a consensus that the company was planning to move from being a canning operation to being a transhipment point, its purpose being to ship frozen fish to an offshore, low-wage country where they would be packed. The American company, with the purchase of the seiner, effectively controlled and owned the fish product; and the plant at Blacks Harbour seemed set to become a transhipment point. Should this scenario be played out, job losses in the order of 60 percent to 70 percent will occur in both the weir sector and in the packing plant. That this is a strong possibility is supported by the experiences of towns in Newfoundland and Nova Scotia, where fish processing plants were threatened with closure in favour of packing and processing at overseas locations, where wages are much lower. The CBC News on 3 June 2006 reported on the lockout of plant workers in Glace Bay, Nova Scotia, where Clearwater had demanded concessions of the fish plant workers. According to company sources, unless they were willing to accept pay cuts from twelve dollars per hour to ten dollars per hour, the plant was to be closed and processing moved to China (CBC News from the Maritimes, 1300 h, ADT, 3 June 2006). In the struggle to sustain the traditional fishery, it is not only the men at sea who are being affected but also the women and men whose lives, for generations, have been tied to the packing and processing of fish. Globalization and restructuring within the agro-food business are profound, and they are changing the future trajectories of Atlantic coastal communities.

For Grand Manan, the importance of the closure of the Seal Cove plant can hardly be overstated. For some families it represented the major source of income; for others it provided crucial money that allowed the purchase of children's clothing. With the announcement of the closure in December, one energetic employee embarked upon a series of enquiries regarding the possibility of obtaining summer work in the tourism industry cleaning cottages. But, as she said to me, "there's not nearly so many weeks available, and the pay isn't as good." In 2005, there were three markets for herring. The first and largest was as sardines, which were now packed by Connors only at a plant in Blacks Harbour; the second was as lobster bait, both on the island and down the eastern seaboard of the United States; and the third was as salmon feed, the "wet" feed given during the first few months that the smolts are in sea cages. The capture of herring, either by large $1 million seining boats (increasingly owned by large companies) or by coastal weirs, engages about twenty families as owners (of seiners and weir privileges) and another fifty men who work as crew. While the weirs capture fish from about mid-June until mid-October, the seiners operate over longer seasons outside the areas defined for weirs. Gradually, through the period from 1995 to 2005, the number of island seiners declined from seven to two. One of the last was the *Polly B*, which ran afoul of government taxes and repayment of capital loans with the result that it became a carrier for transporting smolts to the salmon cages during the winter and spring of 2007-08. By 2006, the only one still operating was the *Fundy Mistress*, which is owned by MG Fisheries and supplies herring to its own feedplant for salmon. However, it too became vulnerable when, in the spring of 2007 it was announced that the Woodwards Cove feed plant was closing, a victim of the cutbacks in the aquaculture industry related to ISA-infested cages. At its peak, the feed plant had been providing herring feed to twenty-two sites around the island.

Interestingly, the *Fundy Mistress* was retired in October that year but only after an order had been placed in Newfoundland for the building of a new seiner. It is to be delivered to the island in July 2008, with the expectation that the government will have granted the company, MG Fisheries, an increased quota for herring. Running counter to the company/Connors-dominated trend, MG

Fisheries (seen as an "independent") appeared to be determined to continue seining herring. The other herring seiners had suffered because changing market prices for herring could not compensate for the growing maintenance costs, which exceeded revenues from the fish. Two of the last three individually owned island boats were confiscated in November 2004 for non-payment of capital loans. These million-dollar boats required continuously large harvests in order to operate profitably, an impossibility when fish volumes were declining and prices paid did not rise as quickly as operating costs.

The threatened loss of the herring fishery was exacerbated by the continued intense seining at the top of the Bay of Fundy, where spawning beds were known to exist. One fisher was adamant that if Connors continued to send its large new seiner, *Brunswick Provider,* into those waters the herring would not survive (interview, 17 June 2007). Meantime, the men on Grand Manan who were thrilled by their unexpected "Christmas presents" when herring appeared in the Whale Cove and Pettes Cove weirs were not about to negotiate with Connors when an American market for lobster bait offered lucrative sales prospects. They sold directly into the American market knowing that the Blacks Harbour plant was not open and that Connors would merely act as a broker into the same American market. Whereas the weir fish typically bring $120 per hogshead, bait fish can be much higher. One American carrier captain with whom I talked in early June 2007 said that he would pay $17.50 per bushel for bait herring, which would be the equivalent of about $75 per pound or almost $300 per hogshead. Obviously, the Grand Manan fishers had incentive to sell directly into the American market when they were not under the stringent understandings of the summer months, when they sold to Connors for the Blacks Harbour packing plant. But their strategies did not win them friends with the management team at Connors. Given that the management team dominated by Bumble Bee executives was now located in Seattle, and that the "head office" in Blacks Harbour remains a shell of its former self, it appears that there is less and less loyalty towards the idea of protecting a relationship with Connors.

The changing nature of the herring fishery also affected women. The loss of the smoke sheds due to decreased market demand meant a significant loss of social connections for island women. Even those who were able to continue working in the fish plant described the conditions there as much less amenable than they had been before. The plant was noisy, so they could not talk; the hours were strict, unlike those associated with stringing or boning herring, when the women could leave for short intervals to look after children. Such comings and goings were not an option in the fish plant. Nevertheless, the work was important, and when the plant closed islanders were unanimously upset, knowing that the repercussions would affect everyone. Almost immediately a volunteer at the food bank reported an increase in the number of families arriving at the weekly depot for supplemental food. The socio-economic implications of the loss of herring processing were exacerbated by the lack of options within other fisheries. Unfortunately, it seemed that even salmon aquaculture was threatened at precisely the same moment. As an industry that had been resisted, it had gradually come to be seen as "insurance" against the complete loss of the traditional fishery, and most islanders had reluctantly come to accept its benefits as a viable industry for Grand Manan. With the removal of the fish and gurry pipes from the Seal Cove wharf on 9 June 2006, it seemed that the final chapter had been written for fish packing on the island.

Salmon Aquaculture

The growing dependence upon salmon aquaculture and its global ties pose special risks for Grand Manan. It is apparent that salmon aquaculture is not a fishery; rather, it is an industrial process. In its organization, technology, and links to global food markets, it is more similar to the poultry agro-food industry than it is to traditional fisheries. The similarities to poultry extend to the nature of the grow-out process as well as to the organizational and ownership links between feed companies, hatcheries, and salmon grow-out sites. But in two important ways salmon aquaculture has a distinct history on the island. Its development has been unlike either the gradual evolution of agriculture or the modernization of the wild fishery, both of which allowed time for their integration into social networks and community relationships. Moreover, the high level of integration within world market structures and the crucial role of technology implicate the aquaculture industry in a complex network of global institutions, resulting in loss of control and access for local communities.

As wild fish stocks have been decimated around the world, the growth of aquaculture has been encouraged as an alternative source of protein and, as argued by state and provincial governments, as a source of employment. However, scholars have questioned the motivations of the government for embracing aquaculture as the panacea for food and employment problems (Milewski et al., 1997). In their study of shrimp farming, Stonich and Vandergeest (2001) argue that government goals aimed at enhancing the economic bases of coastal areas were in fact subordinated to profits for producers and input suppliers. Moreover, as they show, there has been limited attention given to ecological dangers and a distinct lack of concern for changing property regimes associated with marine access. On the other hand, according to scientists at the DFO, who have written extensive reviews of the research, many of the environmental impacts are unknown (Burridge 2007).

From an environmental perspective, the unfortunate paradox is that the use of wild fish in feed mixes negates the real benefits of increased protein food sources, and this is due to the withdrawal of the wild fish resource (e.g., herring). In Norway, fish meal and fish oil used in the mixing of salmon feed are regarded as globally tradable products. The conversion ratio of feed to fish weight gain is estimated at 3:1, meaning that approximately three pounds of herring are withdrawn from wild stocks for every pound of salmon produced (Pillay 1999). In Norway, over half the total tonnage of herring caught (2.2 million tons) is directed into feed for the salmon farming industry, leading to concern that the aquaculture industry may face a fish meal trap as a major resource constraint. Already added to the feed are food colouring (to ensure pink flesh) and antibiotics to limit bacterial and other infections that can destroy thousands of fish within weeks. Another problem is that not all the feed is eaten. While the early evidence that approximately 15 percent to 20 percent of dry and 20 percent of moist feed was not consumed, resulting in this food entering the marine ecosystem, there seems to have been a reduction a decade later, according to local site managers.[3] Ecological interactions that are of concern include suffocation and hypoxia of benthic habitat due to the production of carbon dioxide, hydrogen sulphide, and methane in the sediments. Also, species diversity is reported to decrease in nearby benthic communities, and there are effects on the water columns (such as increased turbidity). Included with this is plankton mortality and possible increase in harmful algae blooms. Furthermore, the use of antibiotics poses the risk of influencing the evolution of bacteria in the sediment so that it becomes more resistant (Hansen et al. 1993). The impacts of these interactions are determined by the size of the farm, the density of the fish per pen, the duration of the farm operation on a particular site, water currents and tides, native species in the region, and the capacity of the environment to assimilate wastes. As Hansen has noted, a strong flow-through of water may mitigate the problem of organic wastes. Ongoing studies include investigations of the impacts of these chemicals on other wild species. For example, laboratory studies indicate that the chemical wastes used to combat infestations of sea lice are lethal to shrimp and lobster (Haya, Burridge, and Chang 2001).

In the last twenty-five years, globally, salmon aquaculture production has nearly matched that of wild salmon catches (FAO 2006). While other forms of aquaculture, such as shrimp, are important globally, in Canada aquaculture is dominated by finfish aquaculture, including trout, char, and salmon, and this accounted for 91 percent of total national sales in 1999. The gross value added to the Canadian economy grew by 22 percent from 1998 to 1999 alone. In New Brunswick, where farmed salmon accounts for almost one-third of Canadian production, salmon aquaculture is concentrated in Charlotte County, where production doubled every five years between 1985 and 2000 and where there are 1,600 direct jobs in the aquaculture industry. By 2007, in New Brunswick salmon aquaculture was the province's "largest agrifood sector, (with) a production value at $225 million and employing some 4500 people" (*Courier Weekend*, 8 June 2007, A2)." For Grand Manan, the growth of the industry in the period between 1989 and 1995 saw sales revenues increase to equal those from the combined wild fishery at $10 million (FGA Consultants Ltd. 1995). They increased a further tenfold over the next decade, to approximately $100 million in 2004.

The increased number of sites (by 33 percent to 24 sites) that the new policy in 2000 precipitated did not mark the final chapter in industry restructuring. By 2003, world prices for salmon were beginning to decline, linked in part to an American scientific study that linked farm-fed salmon to higher than normal PCBS (*Globe and Mail*, 17 May 2002, A7). As a result, pressure on the already narrow profit margins caused salmon farmers to hold on to their fish hoping for increased prices. This, in turn, meant that the sites were not being harvested in a short one- or two-month period and allowed to rest fallow before the introduction of new fish. In other words, the policy of separating year-classes was being abandoned, with the government turning a blind eye to the broken rules. As a result, by 2004 the structure of the industry was being completely transformed, culminating in the spring of 2005 with the sale of Weston's Heritage sites and the Stolt sites (acquired by Marine Harvest in 2004) to the New Brunswick company Cooke Aquaculture as well as the forced sale of Fundy Aquaculture (to avoid bankruptcy) to Admiral Sea Farms, which is owned by local Grand Mananers Glendon, Billy, and George Brown.

This major restructuring had been precipitated by three major factors: the first was the flexibility the large companies wanted for financial leverage; the second was the declining market price for salmon, which severely threatened the resources of local investors; the third was the restructuring of the global food industry, which made vertical integration and control of every step of the production process necessary for maximum profitability. In April 2005 it was announced that Stolt/Marine Harvest had been losing money and that a letter-of-intent to buy its salmon sites had been received from the New Brunswick company owned by Glen Cooke, Cooke Aquaculture. The communications director for Cooke was unequivocal about their search for new opportunities: "We're looking at any opportunity for growth" and a priority would be to turn Stolt around and make it into a profitable company (*Telegraph Journal,* 13 April 2005, C1, 4). The successful purchase of all Stolt sites in New Brunswick would increase the number of Cooke's New Brunswick sites from sixteen to thirty. The announcement caused significant apprehension on Grand Manan related to the further announcement that "the company's culture is to have lean and mean operations, eliminating waste and being as efficient as possible. We have to be a low-cost producer in order to survive in this very competitive industry" (Nell Halse, in *Telegraph Journal,* 13 April 2005, C1, C4). Recall that this announcement came just four months after the closure of the Seal Cove fish plant.

In the weeks following this statement, everyone on Grand Manan had opinions about the state of the industry and what it might mean for the island. One young man working for a machine shop decided to delay an extension on his home; another man who worked as site manager for a company not implicated in these negotiations felt that the jobs would continue, although the working conditions might change; and another person expressed the opinion that most people were worried that they would lose their "perks" – the health benefits, floater suits, boots, and, for a few, the treasured vehicles. It seemed, in May 2005, that the island was holding its collective breath, waiting for the "other shoe to drop." Another related change that affected the structure of the industry and ownership patterns on Grand Manan was the government's modified policy for its Aquaculture Bay Management Area (ABMA) system, which had originally been developed on a two-year cycle as described in the 2000 document. With the reappearance of ISA in 2005, it was decided to enact a three-year cycle that necessarily put additional pressure on individual owners to have three sites in different areas in order to ensure continuous labour and supply streams. This ABMA system was devised as a further guarantee of environmental and fish health, according to Dr Jamey Smith, executive director of the New Brunswick Salmon Growers Association (*Courier Weekend,* 8 June 2007, A2). Its impact was to add further pressure to consolidate the industry.

A major characteristic of the new agro-food industrial structure has been the increased demands for timely delivery. The demand structure for salmon is changing such that it is acting as a commodity in the global market rather than as a luxury. This makes for demand-driven rather than supply-driven growth. From a retailer's perspective, the market advantage of salmon aquaculture is its ability to provide fresh salmon throughout the year. But the shift in industrial focus from the supply/production side to the demand/retailing side of the process has meant increased pressures on the aquaculture industry, and these are not easily accommodated within the cultures of small fishing communities and by distinctive environmental regimes. Its growth has driven the price of all salmon downwards, as well as the price of other wild species, because they are considered market substitutes. The rapid growth of the agro-food industry in the context of globalization has been a major factor facilitating the growth of aquaculture. Transportation and access to overseas production zones and markets, and efficient communications have fuelled the rapid expansion of salmon aquaculture across large geographic areas, effectively removing control of production from small communities.

As well, there are severe restrictions on the extent to which the production process itself can be manipulated in order to produce timely products according to demand. While the poultry industry has tight control over every stage of the production process, the salmon industry does not. In salmon aquaculture, there are many areas that cannot be controlled, notably the rate of growth of both fry and salmon, which is related to site conditions such as light and water temperatures as well as the methods and timing of feeding and types of feed used. Water temperatures can be affected by global sea environments as well as by specific site conditions related to (for example) site proximity to the main island, which can experience cold water influxes during the spring freshets. The winter of 2005–06, for example, was particularly warm, and, according to the fishers on Grand Manan, water temperatures were correspondingly warmer. Affecting the moulting of lobster (see chapter 4), the warmer temperatures also meant that the salmon were feeding more often, growing faster, and requiring higher labour inputs than normal. The timing of every phase of salmon grow-out is a critical factor in the profitability of a site.

In 2004, by contrast, the fish on several sites were below optimum weights due (it was speculated) to lower water temperatures. This meant that sale prices per pound would be lower, squeezing an already minimal profit even further. Following the devastating losses attributed to ISA in 1997 and 1998, the new government policy in 2000 had aimed to ensure that single-year sites would be grouped in "bay management areas." This meant that a site could not have both new smolts and one-year-old salmon in adjacent cages. Therefore, the harvesting had to be accomplished over a relatively short period in order to ensure that new smolts would be quickly introduced, decreasing the time in which cages were empty and non-productive. The dilemma for the salmon farmers was twofold: (1) they had to ensure their salmon were at prime weight to maximize the sale price per pound, and (2) they needed to clear out their site completely during a harvest if they were to conform to provincial regulations.

Nonetheless, one site manager acknowledged that the 2000 policy did not always work the way it should and that, in fact, the pressures of market prices increasingly compromised the decisions to harvest complete sites as required. "The Bay-Management system is dead," he said, because prices were so low no one could afford to sell their salmon except as a way to maintain cash flow. It was only three years later that the dangers of ignoring the separation of year classes became apparent with the reappearance of ISA, leading to a modification of the government policy with its new three-year program. The hazards of various ownership structures became apparent to a local owner who had leased his weir site to Stolt but had also decided to have cages of his own fish. The problem was, despite being able to "piggyback" on the larger company's harvesting capabilities during good times, when the markets were poor Stolt would accept only fish from its own cages. In February 2005, Stolt harvested its own salmon from a Seal Cove site, leaving the local owner with seven cages of his own that, at the low prices prevailing in February 2005, would have entailed a loss of almost $800,000 if he were to have sold immediately. The result was that many sites were being harvested over a period of six months and gradually being restocked throughout that same period, clearly precluding separation of the year-classes. Initially, provincial officials ignored this practice in the interests of preserving competitiveness for the area.

The pressures on local entrepreneurs were not related only to the management of the salmon farm sites. The risks and problems for small investors also extended into the businesses that had been built with both forward and backward linkages within the industry, incorporating a hatchery and a processing plant that would allow for greater control of the supply chain. The commitment of John l'Aventure to the industry had gone well beyond his pioneer investment in 1980. Having survived several financial downturns in 1991, and again in 1997–98 due to ISA, his determination to assert control of the total production process meant that he had invested in sectors that would provide the basis for an overall vertically integrated company. He thereby could control the timing of each phase of the operation, from the production of smolts to grown salmon to processing for market. After a brief, and ultimately unsuccessful, two-year attempt to grow smolts from fry during the mid-1990s, in 2002 l'Aventure launched a major new operation in expanded facilities that included a building dedicated to brood stock and two buildings that housed recirculating tanks for the raising of up to 500,000 smolts. Attracting an experienced manager from Ontario for the hatchery was crucial, and it allowed him to successfully develop a facility that not only provided smolts for his own sites but that also allowed him to sell to other sites on the island. With very low profit margins dependent upon low mortality rates (of less than 5 percent), the hatchery program was an important part of his overall business plan.

Arriving on the island from the mainland where she had worked in the industry for fourteen years, the new hatchery manager represented a new direction for island businesses. The very fact that l'Aventure advertised across Canada for a technically competent and experienced individual was an unusual step for Grand Manan, which had historically relied upon on-the-job learning and creative flexible skills when it came to learning new technologies. As several islanders were to discover through the early years of growth, the aquaculture industry was unforgiving with regard to such strategies as on-the-job learning. As the newly hired manager discovered, there were not only technical problems to be overcome but also many personnel difficulties, not to mention organizational challenges associated with the globalized agro-food industry, which demanded timely delivery at every stage of the production process. The first 200,000 fry were brought into the new facility in late June 2003, at the same time as about twenty-five brood-stock fish were being raised in another building for egg production in November of that year. It was a bold venture that required expertise and intuition, patience and diligence. The feed imported from Europe and the need to kill the brood stock after egg production in order to ensure that bacterial kidney disease would not spread both indicated the high levels of technical knowledge required and the financial risks that were being absorbed.

A second local hatchery on Grand Manan was less successful, mainly because it was not formally connected to any salmon company. When several individually owned hatcheries on the mainland collapsed in 2004 because of disease and inadequate production timing, the resulting shortage of fry meant that the large companies (such as Stolt) had supply priority, thereby preventing the small hatcheries (such as the one on Grand Manan owned by Terry Brown and his son) from acquiring fry in time for spring delivery of smolts to salmon sites. In other words, the small hatchery owners had no leverage in obtaining fry during the summer and autumn of 2004, with the result that three years of successful production came to an end in late 2004. This case clearly illuminates the high-risk and corporate controlled nature of the industry.

Funded initially by his brother-in-law in 2001, Terry Brown and his son and son-in-law had invested heavily to build a small hatchery that produced its first smolts in 2002. These were delivered to salmon sites in April and June of that year. The business arrangement between the families disintegrated over the following six months, apparently due to the loss of one cage of fish due to an inappropriate feeding regime. Nevertheless, when the next stock of 250,000 fry arrived as barely one-third-of-an-ounce fingerlings, it seemed that they were still on track for another successful year. Indeed, sold to l'Aventure in 2004, 180,000 smolts were reportedly "the best ever," with extremely low mortality rates (less than 1 percent). The problem that Brown had encountered with early smoltification had been successfully overcome by a process that involved the addition of magnesium and calcium chlorides to the feed mix. At a cost of $50,000, he said that this was "a lot for our operation." One of the problems is that the anti-smoltification process also speeds up the growth rate of the young fish. Over the extra two months he held the fish they grew to between 1.3 and 1.5 pounds. They would be sold for $2.75 each, for a total sale revenue of almost half a million dollars. He was optimistic that he was using all the most recent technology to ensure healthy smolts and maximum profitability. He had vaccinated every tiny fish a few months earlier, at a cost of twelve and a half cents each, costing another $25,000 altogether, and he was looking forward to a break-through year in this, his third year of production. The unfortunate outbreak of disease on the mainland in the summer of 2004 dramatically changed the prospects for this small enterprise. Unable to replenish his stock of fry (because of the loss of mainland stocks due to disease) before December 2004, Terry Brown was forced to close his Inland Venture operation because he would not have had saleable smolts in time for the next season, and he could not afford to wait out a year. By February 2005 he had to return the assets to his brother-in-law; and, sadly, his home was up for sale two months later (it was later withdrawn from the market). The possibility for high profits seen in the early 1980s tended to obscure the reality of the very competitive environment of the late 1990s and the high risks of this industry for companies that did not have secure financial resources and formalized links to other sectors of the industry.

The major single-cost factor in the raising of farmed fish is feed. Indeed, so important is the feed conversion ratio (FCR) that both Stolt and Cooke offered their growers (salmon farmers) bonuses for the best FCR. While there has been ongoing and unresolved debate about exactly which feed is optimum – wet (grains with herring and oil) or dry (a mixture of grains and animal residues) – generally the sites favour a schedule of about two months on a wet feed regime, when the smolts are first introduced to the marine environment, thereafter switching to dry feed. The total amounts fed to the fish depend upon the season (water temperature) and size. While the fish are fed about three times a week during the winter months, they are fed three times a day during the summer. For a five-month-old salmon (about just over two pounds), being fed in late summer on a site having twenty-eight cages (with 15,000 to 20,000 fish in each cage), the daily amounts of dry feed would average about fourteen tons. With the cost of feed at about $1,200 per ton, feed costs would be $550 per cage per day, or over $15,000 per day per site. With very narrow profit margins the cost of feed can be critical. One site investor said that Heritage (Weston) charged seventeen cents per pound more than feed supplier Moore-Clarke was charging Cooke Aquaculture. The highly variable prices charged were between fifty-one cents and seventy-one cents per pound. In other words, feed costs could be $330 per ton higher, which would represent about $4,500 per day just for feed. The feed costs excluded the costs of the transporter scow, blower, and labour, and represented over 90 percent of the operating costs of the salmon site.

The other two main costs are for smolts and for labour. At $3.00 to $3.25 for each quarter-pound smolt, or approximately $1.5 million to stock 500,000 fish, the quality of the smolts and the transfer process into the sea cages are crucial factors in the mortality rate. Site operators aim for initial mortality rates of about 1 percent but have been known to have to deal with much higher rates, especially in the early years of the industry. Overall, during the complete grow-out cycle, salmon farmers feel successful if their mortality rates are less than 5 percent over the two-year period. They have been as high as 15 percent, which represents a significant drain on profit margins.

Labour is the third major cost input. During the peak summer feeding period, a site will hire six or seven full-time people, at rates varying from nine dollars to twelve dollars per hour. These young people (for they must be young in order to sustain the strenuous work that is demanded) obtain health benefits and safety equipment such as floater suits, but they do not have pension contributions. Cold water temperatures during the winter months significantly slows fish metabolism and, therefore, the amounts of feed required. As a result, during the winter months the workforce is cut by approximately 75 percent on all the sites. Rapidly changing technologies have also affected labour requirements. Compared to the 164-foot cages that were common in 1995, the newer

230- and 328-foot cages[4] involve an entirely different set of organizational and technological requirements, including anchoring designs, fish capacity, feed blowers, and harvest schedules. As mechanical feeding has replaced hand feeding, labour requirements have been further reduced. Most sites have about six or seven full-time workers during the summer and two during the winter months. As well, regular dives are necessary to check the nets and moorings and to collect and monitor any mortalities (known as "morts"). In 2004, divers were paid seventy-five dollars per hour. The sites attract mainly young men, although four or five women have worked on the sites in any one season. Indeed, one site manager suggested that "some women are better feeders" because they have "more patience." Most of the female workers do not continue for more than two years because of the strenuous physical labour required.

The local economy of Grand Manan benefited in more ways than simply jobs on the sites. Economic activities in which major expansion was directly related to the aquaculture industry included trucking, boat and scow building, diving, net and cage manufacture, feed mixing, and even new wharfage due to the need for increased docking capacity. The variety of both semi-skilled and skilled jobs was enhanced by these activities, and the amount of money generated was significant. Divers, for example, could earn up to $1,500 to $2,000 per week throughout the year. One diver purportedly earned about $300,000 annually, a fact that would be constrained by the physical demands of daily dives. While most of the jobs directly associated with the sites are low-skilled, in fact there have been multiple benefits to existing local companies, notably trucking (and truck repairs), marine equipment, and boat building.

As well, local initiatives have included the invention of new designs, such as the wet-feed blower developed by Jimmy Dexter, who worked for General Marine. With the increased size of cages, hand feeding, which had been normal practice until the mid-1990s, was no longer possible. The dry-feed blowers constantly clogged when wet feed was used, causing a problem that was resolved by adapting a technology based on the idea of the meat grinder in order to maintain the friability of the feed. According to local site managers who used the unique blower, it was very efficient and well designed, and, by 2005, General Marine had sold forty of Dexter's wet-feed blowers. Another example of island ingenuity may be seen in the development of a timed feeder for the fry being raised at the Fundy Aquaculture hatchery. With constant feeding required hourly during the first few critical months of growth, from the fertilized egg stage to the one-third-ounce fry stage,[5] some automation seemed necessary. Consequently, two young men working at the hatchery devised an eight-inch disk divided into sections that held the special feed. As the disk rotated on a timer, an arm would push in the feed at regular intervals. It was a simple, creative, and inexpensive solution to a basic problem.

Following the pause in 1999–2000, the industry seemed unassailable, poised for continued growth, and the community of Grand Manan was scrambling to take full advantage of its prospects. The appearance of a large scow in 2003 built on Deer Island to transport smolts directly to sites indicated a prosperous future. "All oak and brass," it was described as the forerunner of new modes of transport for smolts and grown salmon. But not everyone was happy about the growth of the industry. On Deer Island in December 2005, the cages at four different salmon sites were destroyed by vandalism. The usually reticent Conservation Council of New Brunswick sided with the industry, saying that the releases were equivalent to terrorism – an assault on the environment. The estimated value lost to Cooke Aquaculture was about $3 million, and the reward of $250,000 failed to bring the perpetrators to justice. Furthermore, compared to the high prices of 2003, the prospects for the industry dimmed as the competition from Chile and the scientific report from the United States combined globally to affect prices.

Only a year later, prices for salmon were down significantly, almost below cost; and by 2005 major restructuring was well under way. The sale of l'Aventure's company in April that year was a clear sign that the industry was in trouble. Large investments and narrow profit margins meant that there was no cushion for unexpected hazards. A bank calling in its loans, a storm that destroyed several cages of grown salmon, the reappearance of ISA, poorly trained workers whose feeding schedules went awry, disease in the hatcheries that forced up the cost of smolts: these factors and more had profound implications for the ability of smaller companies to survive. In the spring of 2005, Cooke Aquaculture announced a series of purchases, including the sites of Stolt and Weston, which owned Heritage. By the end of the year, the New Brunswick government announced the closure of the entire upper zone of the island for a period of a year in order to allow fallowing to clean out and restore the sites after an outbreak of ISA. In the spring of 2006, for the first time in six years, no shipments of smolts would be arriving on the island, although some smolts were delivered from the island hatchery to the Admiral site in Seal Cove as the government sought to protect the newly expanded company, which was struggling to ensure that its investment in buying Fundy Aquaculture was sustainable.

Following so closely on the closure of the Connors plant a year earlier, the news for the island was bleak. By the end of the summer 2005 it was reported that over two hundred houses were up for sale. By the spring of 2006, there was talk of houses being foreclosed, seized by the bank for non-payment of mortgages. For his part, the bank manager said that this talk was a product of panic-induced rumour and was not substantiated by his records. But the spin-off businesses were certainly suffering. The successful net company from the mainland, Cards Aqua, closed in early 2006; Sea Nets, originally part of John l'Aventure's integrated company, closed temporarily. It was reported that the trucking companies that had expanded so quickly, especially in the period after 2001, were in trouble. While those with long histories in business were able to sustain some losses and, in fact, to buy some of the assets that began to be offered for sale, others were operating on fragile margins that allowed for little flexibility during the tough times. Trucking was also affected because of Cooke's decision to operate two boats (one owned and one on contract from a Grand Mananer) to transport salmon and smolts directly to and from the sites. Rather than relying upon trucks and the ferry, two herring seiners, the *Polly B* and the *Lady Cavell* (which Cooke's bought for this use) were brought into service, each able to carry eight to ten truck equivalents in their holds. Everywhere on the island people were concerned about the future. In the spring of 2007, smolts again were being transported to Grand Manan to restock the cages of the upper island zone, this time via the old seiners rather than containers on the ferry.

The new three-year regime mandated by the government in 2006 and implemented in 2007 was having repercussions as potential investors sought to have sites in different zones. The reason for this was to assure fish at all growth stages as a way of maintaining continuous labour and other input costs and revenues throughout the entire two-year grow-out cycle and fallow period. The two areas defined as year-class zones were divided by a boundary running from Ingalls Head to the west and south of White Head. For the zone to the south (lower end of the island) Cooke's appointed a new area manager (Michael Ingersoll) to ensure that the entire area was cleared of fish by late 2006 (to be restocked in 2008). The zone to the north (upper end of Grand Manan), managed by Area Manager Matthew Ingersoll, received smolts in eight of the twelve sites in the spring of 2007, will begin harvesting in 2009, and will be ready to receive new smolts again in 2010. This implied a dramatic change in the number of fish being raised on the island in 2007. Compared to the peak in 2004, when all twenty-three sites were producing over $100 million worth of salmon, in 2007 the value (based on estimated sale price of $3.50 per pound) was estimated at $60 million for the fewer than 3 million fish on eight sites.

But the transitions were not only about fallow sites and changing ownership between the large companies. Most islanders who had substantial financial interests invested had decided to sell to Cooke's, losing both their ownership and leasing interests and essentially becoming hired labour or managers. The final chapters on the evolution of Grand Manan's aquaculture industry have yet to be written. But as the consolidation and restructuring continue through 2007, even those who were only peripherally involved in the industry through leasing their weir sites have begun to withdraw. Because of the need for real capital rather than the leasing arrangements that mainland companies had initially contracted, Cooke's set out to buy sites from the fishers who had been previously leasing them as former weir sites. In effect, the privatization of the Commons was being solidified, and the loss of control by local fishers was becoming ever more certain. With the government's mandate in 2005 that the upper zone of Grand Manan be cleared of all salmon by 2006 for a year of fallow, the low profit margins were further eroded. For Bothwick's company, Limekiln, empty cages meant two years with no income. Threatened by bankruptcy and with an offer from Cooke's to buy the site, Bothwick decided to renege on the lease with the Grand Manan fishers, opting out of the final two years of a six-year agreement. For the fishers the choice was limited: sue and be forced into bankruptcy with no promise of compensation or accept the buyout.

For Grand Manan fishers, the loss of their weir site and the loss of any control over the marine Commons that was being alienated for aquaculture represented a fundamental loss of elements of their identity as Grand Mananers. The men who had worked as labourers on the sites recognized that security within the industry would always be only a dream and that, unlike the unpredictable vagaries of the wild fishery (which had always held the promise of a better year), aquaculture held no promises for the future of the average working person.

Globalization in other areas

Of course, globalization is not only about the agro-food industry: it affects all aspects of daily lives – economically, socially, and politically. For Grand Manan the centralization of village stores and increasing reliance upon chain stores such as the IGA and Save Easy was the earliest indication that global influences were beginning to intrude on island structures. But all areas subject to new government rules, such as municipal amalgamation and the legal requirement for a municipal plan, reflected the increasing involvement of global trends in local affairs. The fact that the government encouraged a Nova Scotia company to harvest rockweed on Grand Manan reflected the sense that there were no boundaries to resource exploitation and that corporate interests were the government's interests.

As basic as it is to Grand Manan culture, dulse, too, is part of global markets. Despite its niche role in the lives of many islanders who pick the seaweed in between seasons, dulse represents an important business, with links to markets in France, the United States, and Australia. For Sandy Flagg (son of Roland Flagg), who employed seventy people at peak season, the international markets were a window on the world of business that demanded computer access and sophisticated accounting skills and that drew on multiple family talents.

Always ready to take advantage of new opportunities, the businesses of Bev Fleet and Everett Dakin, based on trucking, construction and heavy machinery, benefited by the rapid growth of aquaculture and the need for on-demand transport. Between 1995 and 2005, a conservative estimate would put the growth rate of both companies at more than triple – an enviable increase by anyone's assessment. Unfortunately, the very rapid expansion was followed by the sudden death of Everett Dakin in 2006, just as the downturn in aquaculture was gaining momentum, and this caused significant problems for Dakin's business. By spring 2008, according to the new owner in a letter read out at a council meeting (5 May 2008), the business had been significantly downsized and was even in jeopardy of closing down altogether. Another major company, General Marine, began as a small engine shop, servicing the fishing fleet. In the decade between 1995 and 2005, its business more than quadrupled as it diversified to build scows and feed blowers for the aquaculture industry and invented new technologies that solved problems related to blowing wet feed.

Perhaps one of the most interesting ventures involved one entrepreneurial islander's investment in a sea urchin aquaculture experiment aimed at commercialization. Taking advantage of a postdoctoral student who had funding from the Natural Sciences and Engineering Research Council, Kenny Brown agreed to supplement the research initiative as a two-year industrial contract investigating the ways in which the growth of sea urchins might be enhanced, with particular focus on the valuable roe. The advanced research was guided by Dr Chris Pearce and was aided by two assistants, who themselves had degrees in science. Relying upon their own designs for tanks (both indoor and outdoor) and filters, plus a computer imaging spectroscope, they worked with larvae, trying to increase their rate of sexual maturation from four years to about two years. In testing different grow-out environments they were trying to optimize both biological and economic efficiency. According to Pearce, "The knowledge is there. It'll just take someone with guts to commercialize it."

On Grand Manan one finds that risk-taking courage. The problem for the urchin initiative was to optimize the roe-to-body-weight ratio. In the Bay of Fundy, the average percentage is only 4 percent to 5 percent productive, whereas to be commercially viable at least 10 percent is needed. With wild stocks already showing a decline in 2001, the support of aquaculture production seemed to be a good investment. Pearce felt that the preferable option would be to develop an enhancement program that would entail hand harvesting from the wild and then enhancing the roe to 15 percent. Unfortunately, two years later Pearce and his wife left the island to pursue new careers on the west coast, and Kenny Brown's venture came to an abrupt end. Nevertheless, as one of ten sites worldwide to have engaged in this innovative study, clearly individuals on Grand Manan were beginning to bring the island more fully into integrated global networks.

I have already described the harvest of rockweed as an enterprise brought to the island from a mainland company. Two other unusual enterprises were developed by people from-away who wanted to take advantage of the fishery resource and accessible labour. One that lasted for only one year (2002) involved an eel company, the idea of a Saint John man who engaged one of the local fishers to trap eels using specially designed cages. He saw in the small eels an opportunity to take advantage of a leather market in South Korea. After an initial cleansing, which was done by about ten women, the eels were shipped live to Korea for use in making colourful imitation leather goods. The second company, Fish-On-Bait, was developed and 50 percent owned by Tim Monroe, an American from Florida, with a partner in Seattle. They had found a market for speciality packed and flash frozen whole herring, which were sold as bait to the exclusive longline sport fishery in the southern United States. After a tentative start a year earlier in Nova Scotia, Monroe came to Grand Manan in March 2003. With the help of a local fisher, he situated the packing company at Woodwards Cove, with an installed assembly line for handling and packing the fish and a large freezer unit. Although disappointed because the herring were late that year (they did not appear until mid-July), he was satisfied with the availability of labour and the cooperation of the fishers. He hired thirty-five employees who were packing five thousand trays per week by the beginning of

August, and he was optimistic that this could be increased to ten thousand. Flash frozen at minus four degrees Fahrenheit within three hours, the fish were then vacuum packed twelve to a tray and shipped to Texas.

In 2005, after three years of working on Grand Manan, Monroe seemed satisfied with the quality and quantity of weir herring he was able to negotiate from weir fishers. They, in turn, were pleased because his requirement for small amounts of high-quality fish meant that they were able to sell their small catches (up to ten hogsheads), which were of no interest to Connors, and to receive a much higher price. Compared to Connors' price of about $120 per hogshead, Monroe would pay $300. One of his major problems was continuity of supply, which would ensure that his employees would have steady work on the packing assembly lines. Notoriously unpredictable, the herring runs not only varied according to the beginning and ending of the season but also with regard to where they arrived and in what size schools. Monroe sought a solution to this by building a cage similar to those used in salmon aquaculture, which would act as a holding pen for the herring and could be towed to the Woodwards Cove site. His dilemma was ensuring the high quality of bait, which meant the herring could not tolerate excessive handling (e.g., being put through the sluices on the herring carriers). It was a multifaceted problem that he was still struggling to solve at the end of 2005, his third year of operation. He did not return in 2006.

For many business enterprises on the island, the impact of global price fluctuations was a major factor in their ability to maintain profits. While lobster and scallop prices were directly affected by changes in the American dollar,[6] for the most part fishers were able to absorb the fluctuations because of the steady demand for these luxury products. Nonetheless, in the spring of 2006, with the Canadian dollar up to almost ninety cents, an increase of over 40 percent from five years earlier, the American buyers were becoming much more demanding, refusing entire lots if they found any damaged lobsters (e.g., missing a claw) in a shipment. For companies such as Fish-on-Bait, which relied upon competition with sources closer to their American markets, the profit margins were more fragile. New Grand Manan businesses were also vulnerable to global currency markets. When Linton's Lobster opened in late 2003 as a land-based lobster pound that had installed the most advanced circulation system, the Canadian dollar was still as low as US$0.78, about 12 percent lower than it was in 2006. For a new company vulnerable to increasing interest rates on capital structures, this margin represented a significant loss of possible profits.

Economics is about the distribution of limited resources, and for this small fishing community the economy has increasingly become both dependent upon outside ownership and subjected to increased disparities among community residents. As the market prices for various licences increased beyond the ability of the average fisher to buy them, island companies bought them. In the decade after 1995, fishers were increasingly dependent upon the financial support of the large companies, which could "rent" them a licence or lend them a boat if they agreed to market all their catch those companies. In the period between 1995 and 2006, the loss of individual control over the small enterprise represented by a boat and licence was a significant change. This change was the result of a combination of the following: government regulations, the restructuring of the agro-food industry and the concomitant growth of salmon aquaculture, and the incursion of an Aboriginal presence that caused an escalation in the value of licences. The implications for community relations and collective identities can scarcely be overstated.

Tourism

In a sense, tourism, through the annual migratory visits of Americans to small summer cottages, offered Grand Manan its first glimpse of globalization. Ever since the late nineteenth century, wealthy New Yorkers and artists have sought refuge in the beautiful landscapes of Grand Manan. Of these, the most famous is Willa Cather, who arrived with an entourage of other young women writers and artists and stayed at the Whale Cove Cottages for a month or more each summer. Their contribution to Grand Manan history, and even its collective identity, cannot be overestimated (see chapter 9). Although they had very limited contact with local fishers, these women had a distinctive culture, and the fact that they continued to return to the island for over two decades lent a particular sensibility to the land around the Whale Cove Cottages – a sensibility that seems to remain today. Currently ably owned and managed by Laura Buckley, the cottages and the restaurant are open from mid-May until mid-October, usually offering a sprinkling of community-type dinners that create a sense of collective spirit and amiability that is appreciated throughout the island. In October each year, the Grand Manan Trails Association's closing dinner is held at the restaurant and functions as a fundraiser. For the approximately forty-five fortunate people who reserve tickets at twenty-five dollars each, a sumptuous dinner of Cornish game hen or scallop ravioli rewards them for volunteering to keep trails clipped and free of brambles. The entire proceeds of the dinner (approximately $1,200.) is given to the association as an unsolicited gift from the Whale Cove Cottages, thereby contributing an added layer of meaning to the trails for tourist operators and summer visitors.

Despite the perception of many first-time visitors that tourism is a high-value contributor to the island economy, according to one source it is probably not worth more than about $1 million annually (interview, Allaby, September 2001). Reflecting the perception of a different reality, in an article in the *Island Times* a young observer wrote: "The main industry on the island is tourism. The island is so eye-catching and so beautiful that Grand Manan tends to earn lots of money from the tourism business" (Connor Ross, "Grand Manan Island," July 2007, 12–13).[7] Not only has tourism not been a major economic factor but its seasonality and the impact of a declining American dollar have also combined to lessen its attractiveness for investors. The increasing value of the Canadian dollar in relation to the American dollar – by 15 percent in one year alone (September 2006–07) – was a major factor in a sustained decline in tourism between 2003 and 2007. By 2007, a Toronto couple's twelve year-investment in the Compass Rose Inn was being terminated, with the two buildings up for sale in September of that year. Across the road, the owners of a bed-and-breakfast had also decided to sell their business. Partly, of course, there is a desire to retire from this very demanding business, but there are wider concerns related to the changing nature of tourism. Cottage owners complained that visitors no longer wanted to stay for weeks at a time as they had during the 1960s and 1970s, thus causing business owners to question whether or not people were seeking new kinds of holidays. Indeed, there was evidence that there were more visitors but that they were changing their patterns of consumption. The Grand Manan Trail Association, for example, noted that the number of guide books sold in 2007 had more than tripled compared to the previous year, perhaps indicating an increase in the number of campers who enjoy hiking (Bob Stone, introductory remarks, annual Grand Manan Trail Association closing dinner, 9 October 2007). Another concern in the spring of 2008 was the impact of rising fuel prices on those visitors who drive many miles from homes in the United States or Ontario.

Historically, two groups of visitors have defined the tourism industry: (1) those who arrive for one or two weeks, staying in rental units and enjoying bird watching, hiking, and other nature-oriented activities; and (2) those who own their own cottages and come for the summer. These summer residents see themselves practically as islanders and will claim such status in conversations with the short-stay visitor whom they encounter at the market or in the stores. Somehow, the notion of being an "islander" is important to people who have long-term relationships with Grand Manan.

What changed during the 1990s was the ratio between these groups of visitors. Whereas the summer residents continued to return each year (and, in 2006, there was an increasing number of them), there are fewer of the two-week visitors and more who visit for only a few days. This poses a problem for the nature of tourist establishments, which, while seasonal, nevertheless depended on long-term stays that required lower costs as well as time for cleaning and laundry. In 1997, based on a tourism survey, almost two-thirds of the visitors to Grand Manan had stayed for three days or fewer (Tourist Survey, Marshall, 26 June–6 July 1997 [285 returns]). Out of the seven establishments that were created through the Atlantic Canada Opportunities Agency grants in 1986, only two were still operating in 2006. Similarly, of the six bed-and-breakfasts operating in 1995, none was left in 2006 (although there were three new ones). One factor in this decline may have been the provincial government's promotion of its Day Adventures program, which tried to encourage tourism through emphasizing day outings such as whale watching. For Grand Manan, this posed a particular problem since a day-only visit was severely constrained by the ninety-minute ferry trip, which cost about fifty dollars for a couple in a car. Moreover, the escalation in weekly rates for cottages, from $600 to $1,000, while not expensive for Americans when the Canadian dollar was low, was expensive for Canadian families who wanted a two-week stay and became increasingly expensive for Americans as the dollar rose at the same time as did fuel prices. World energy costs and currency markets had a profound impact upon the Grand Manan visitor population. Indicators such as museum visits (down about 20 percent in 2005 from the previous year) and reservations at the provincial campground (down 30 percent the same year) suggest that important adjustments will be required of Grand Manan's tourism industry.

One important feature of Grand Manan tourism has undoubtedly been the visits by Elderhostel groups who stayed at the Marathon Inn, mainly during June and September, providing valuable "shoulder months"[8] support to other tourism establishments around the island (such as craft stores and the museum). Usually staying for four to six days, these groups were led by inn manager and investor Jim Leslie (originally from Toronto), whose intimate knowledge of the island was based upon his thirty-plus years as a hotelier, fisher, dulse harvester, and erudite guide. The importance of these guided tours for Grand Manan was that they provided introductory visits that might entice tourists back in future years.

Probably the most obvious type of tourism on Grand Manan comes in the form of whale-watching tours, which have been offered since the early 1980s. These tours were first offered by Preston Wilcox out of Seal Cove. Using his lobster boat following the close of the lobster season at the end of June, he took out groups of twelve to fifteen in his forty-foot boat and was always able to guarantee spectacular views of the rare right whales, which usually arrive in the Bay of Fundy in late July for their breeding season. By the early-1990s, Preston's son Peter had taken over his business, and a few years later there were several other competing enterprises running out of North Head. With the impact of the new ferry boat in 1990, the capacity for an increased number of day visitors was a boon to whale-watching operators. In 1999, direct-line telephones were set up at the North Head wharf for individual operators, and grand new signs appeared. Two new boats – impressive sailing yachts – were introduced, and visitors now had a choice of tour operators (which was not the case a decade earlier). Between 1995 and 2005, the number of tour groups rose from three to six, before slipping back to two in 2006. Peter Wilcox kept his business going throughout this time, despite his initial concerns that perhaps he should have relocated to North Head in order to take advantage of the day trippers. The provincial government's Day Adventures program, which initially appeared to offer impetus to the Grand Manan whale-watching enterprises, in fact provided negligible benefits since visitors did not stay overnight. People even tended to bring their own picnics, eschewing the island's restaurants. For Grand Manan, the Day Adventures program was a distinct failure from any perspective.

Throughout the period of my study, restaurants were beginning to respond both to tourist complaints about the quality of food on the island and to the changing tastes of islanders themselves. In 1995 there were only two dining rooms that offered a variety of freshly cooked food and the possibility of a glass of wine, and neither of these offered family-oriented prices. All other restaurants depended upon frozen and fried meals such as fish-and-chips, hamburgers, and tacos. Attempts to have summer-only take-outs for lobster rolls tended to last for only one or two summer seasons. This was in part due to opportunities for more long-term employment in other ventures. Despite the fact that islanders were beginning to eat out more often as incomes rose in the 1990s, and their support of new restaurants was strong in the first months of operation, many of them also wanted greater variety. One key factor in the success of a new restaurant venture was opening early in the morning (6:00 AM) for the fishers, who liked to stop for breakfast, a situation that owners found difficult to continue because of labour costs. A second factor was the seasonality and different tastes of tourists, who tended to want lighter dishes and fresh ingredients. A tourist survey in 1997 indicated a consensus that visitors wanted fresh fish meals, and these were not offered by most restaurants as standard fare.[9]

The increasing number of Grand Mananers who were travelling, mainly to Florida but also to Europe and other parts of Canada, was beginning to have an impact upon tastes in a variety of ways. There was a great change in restaurant food and foods available in the local supermarket in the decade between 1995 and 2006. For example, whereas in 1995 one could not find fresh herbs (such as dill and rosemary) or plain yogurt or soy products in the local Save Easy, by 2006 these were available on a regular basis.

Tastes are also changing in areas such as art. Several Grand Mananers, especially young women, have ventured into the competitive world of art and photography following the completion of their formal educations at art colleges in Canada and the United States (Tia Wormell, Sara Griffin, Nicole Green Wolfe, Lori Morse, and Emaly Green). As well, several artists who have summer homes on the island have become increasingly well known to islanders, who have begun attending their art or photography exhibits at two local galleries that open for the summer months (Ossie Schenk, Peter Cunningham, and David Ogilvie). Similarly, music programs, inaugurated during 1996–98 by a young woman from-away, attracted tourists but were rarely attended by islanders. In 2006, a folk singer and composer from the mainland performed at the Grand Manan Museum for about a hundred people, half of whom were islanders. And that same year a new community choir was formed, intentionally ecumenical. These were important initiatives that marked cultural change on the island. The creative and energetic museum curator, who had been hired in 2002, was unequivocal in saying that one of her challenges was to get islanders into their own museum both for exhibits and for the weekly talks during the summer months. The gradual shift in islander tastes towards a broader understanding of culture reflects several factors, including fewer working days on the water, increased contacts through travel, higher incomes, and increasing levels of education.

Tourists or Summer Residents?

There is a new category of from-aways that has become increasingly important during the period of my study. This consists of the people, mainly Americans, who have bought large areas of land for development as second-home properties. With an increasing number of land sales over the period since 2000 has come an escalation in land values and taxes that has begun to worry the community of Grand Manan. Several houses on the market in 2006 were valued at $300,000 to $450,000. Between 2004 and 2006, a $1 million-dollar home that was built at the southern end of the island became a frequent Sunday drive destination for curious islanders. For the contractor who had been hired for the two years it took to build, it was an adventure in new design and high technology that required enormous diligence, a challenge that he had accepted after carefully weighing the risk of quitting his salaried ferry job as a crew member. He took the risk and was still happy about it two years later. Owned by a New York filmmaker who may be in residence only a few months of each year, the house may be a harbinger of more incursions of this type. Islanders have a growing sense of losing their land as well as their fishery, and they see this as a threat to their identity. Frequent references to Cape Cod and Martha's Vineyard reflect growing concerns. Only ten years earlier, housing prices had averaged about $35,000, while in 2006 they averaged closer to $100,000. As taxes mounted, especially with so many questions about the economy, residents began to ask why there were no laws prohibiting out-of-province sales. But it was too late. Whereas in 2003 a 1.5-acre lot in Castalia sold for $55,000, in 2006 a 1.25-acre waterfront lot in Deep Cove was advertised at $95,000. Some waterfront properties were asking $100,000 to $150,000 per acre – prices that mirrored those around many large Canadian cities and that were well out of the reach of most Grand Mananers. One small house in a superb waterfront location that sat on a tiny lot of less than a quarter of an acre was valued (and reportedly purchased) at $180,000 in the spring of 2007 – a price that no islander would ever consider paying, especially in view of the old house's maintenance needs.

Not to be underestimated as one of the important impacts of globalization and new communications technologies is the new library, which was opened in 2001. Despite its relatively small collection, it provides rapid access to books available in mainland libraries as well as an efficient library staff that creates reading programs and meetings to discuss books. It is not an exaggeration to say that the new library and its enthusiastic staff, despite initial grumblings about the cost, have transformed attitudes towards books and what they have to offer. It has sponsored special readings, had book launches, and has a rotating series of interesting posters that provide a lively and welcoming atmosphere for students and adults alike.

Where to with globalization?

As pointed out earlier, highly competitive global markets and increasingly powerful market mechanisms have left coastal communities more open and vulnerable than ever before. These communities are especially vulnerable because they
are dependent upon a combination of unpredictable harvests and even less
predictable international markets. For communities such as Grand Manan,
the challenge is to mitigate this intransigent combination by finding ways to
strengthen the potential for creative change and long-term sustainability.

There is evidence that indeed globalization, through the advent of salmon
aquaculture, is accelerating "the erosion of community solidarity" (Apostle and
Barrett 1992, 328), which ensured support during times of crisis, and creating
"competitive, atomized, and dependent" entities (333). Within the wild fishery,
generations of flexible strategies and seasonal adaptations created cooperative
patterns that cannot be replicated within aquaculture. This is not to deny a historically strong culture of competitiveness, which Apostle and Barrett do not
seem to appreciate. Nevertheless, although the village-based histories and competitive work habits of islanders certainly affected the long-term willingness to
generate island consensus around many issues, the wild fishery was the source
of shared experiences that forged a sense of mutual understanding. There has
been a historic integration of socio-economic relations that has resulted in a
sense of community. With the recent loss of the security provided by the wild
fishery and the growing importance of aquaculture, socio-economic relationships are being stressed to the extent that shared understandings are being
muted and families are feeling increasingly atomised. Richard Adam Carey's
(1999, 7) suggestion that "agri-business-style industrial fishing and/or aquaculture – strictly regulated of course – is simply the next logical step up on the car-
payment culture's evolutionary ladder" must give us all pause. Today, islanders
see their men leave for work on the salmon sites with little knowledge of or real
sympathy for what they do. Relationships between workers and company officials are formal, underlain by a constant feeling of insecurity. Within the traditional fishery there are also tensions. One fisher suggested that the younger
men, feeling the stresses of meeting payments on large new boats in a context
of declining wild stocks, are less cooperative and even engaging in destructive
behaviours such as cutting trawls and stealing lobster on days of thick fog. As
Peter Sinclair discovered in his studies of villages in Newfoundland that were
suffering from the cod moratorium, the "socially disruptive impact of locally

uncontrollable events" can have a devastating effect upon community solidarity and collective identities (Sinclair et al. 1999, 338). The government has taken a powerful leading role in the ideological shift towards prioritizing globalized market forces, and this has been done to the detriment of community relations and identities. The privatization of marine Commons and the loss of effective community control over marine space and economic livelihoods represent significant changes in property regimes and an ideology that ultimately undermines generations of cultural meanings for the small fishing communities on Canada's east coast.

7

Belonging

The constitution of belonging – of any belonging, whether it be institutional, disciplinary, national, regional, cultural, sexual or racial – has never been an exclusive function of its shared terms but also of its shared exclusions.
~ (Rogoff 2000, 5, citing Michel Foucault)

No narrative describing the community and people who live on Grand Manan is possible without reference to the idea of "belonging." The identity of being an islander is central to everyone's sense of belonging and to the informal rules and norms that guide daily lives and activities. Experience in the traditional wild fishery underpins most of the structures, both formal and informal, that define social and economic relations on the island. As explained in chapter 3, boundaries and edges of transition between social groups – "insiders" and "outsiders," residents and visitors, adults and teenagers – are significant mediators of identities and provide a consensus regarding collective meanings. While boundaries may not always be explicit, they have been established over generations, and it is only in 2006 that they began showing signs of becoming permeable. There are, for example, unacknowledged rules about who can hold certain jobs or run for elected office, and these are based upon family lineage and connection to the island. As Foucault points out, it is not merely a question of "belonging," it is also a question of exclusion.

The exclusions on Grand Manan are important because, through resisting new ideas, they have functioned as effective barriers to change and have inhibited adaptation. Exclusionary practices have affected women, spouses born off-island, new residents from-away, migrant workers, summer residents, and tourists. Without the shared stories of generations and intermingled families, the sense of belonging, even for those who move to the island as spouses of native islanders, is diminished. The issue is not simply who is "in" and who is "out"; rather, it concerns "an active form of 'unbelonging' against which the anxiety-laden work of collectivities and mutabilities and shared values and histories and rights can gain some clarity and articulation" (Rogoff 2000, 5). While there are signs that the boundaries are becoming more permeable, the zones of transition continue to exert significant constraints over all aspects of social, economic, and political life on Grand Manan.

Featherstone has explored the importance of the taken-for-granted and repetitive nature of the everyday lives of individuals with regard to defining meanings that underpin local cultures. Insofar as local cultures incorporate "rituals, symbols and ceremonies that link people to a place and a common sense of the past" (Featherstone 1993, 178), they also rely upon these rituals and shared meanings as their defence against perceived threats to their identities. These shared understandings become an exclusionary medium for maintaining both the integrity of their culture and a semblance of control over their lives. In the struggle to maintain identities directly tied to shared histories and experiences of the island, people from-away are marginalized in several ways. That their views of the world tend to be based upon both different and more varied experiences than do those of the islanders means that their language, values, and norms of behaviour all set them apart. This is true both in terms of their own choices and in terms of their acceptance by native islanders. Organizations such as the museum's board of directors, the Boys and Girls Club, and the Rotary Club tend to be dominated by one group or the other, for example, but certain jobs and voluntary positions will always be held exclusively by native islanders. The impact of this division, which effectively marginalizes people from-away, permeates all relationships. One from-away owner of a small inn acknowledged that she had not tried to integrate into the community but that she and her husband certainly did not feel ostracized. However, she said that they felt that they would not be welcomed in any kind of leadership capacity and that, in fact, should they try to take one, "We might be thrown off the island."

The concept of belonging implies a sense of power and access to decision-making mechanisms. People who are "in" are able to exercise powers of decision making for the entire community in ways that are precluded for those who are "out." Women have traditionally felt excluded. In virtually all the interviews I had with women, there was a consistent theme of powerlessness. Their self-identity contradicts my own perception of them, which I base on the strength they seem to have derived from two centuries of surviving while their men were away for weeks or months at a time. Their dominant sense of power-lessness seems to be related to several experiences, beginning, for many families, with the changing markets for smoked herring, which undermined traditional jobs that had provided both income and crucial social support networks. But it is the exclusionary practices of a patriarchal culture that provides the context within which women see themselves as peripheral to decision-making networks on the island. Definitions of boundaries have become increasingly blurred over the past decade as the changing nature of work available to women and the possibility of greater mobility have combined to diminish women's sense of dependence upon men. Boundaries have become zones of transition, with evidence of men's resistance to women's tentative moves forward. The clear-cut role distinctions that defined past relationships are today being renegotiated. Typical comments from women (and note the ages) are: "Oh, but we're nothing!" (age about sixty-eight years); "There's nothing for me to do [no job opportunities]" (age about twenty-seven years); and "My husband told me, 'Nice girls don't go out at night'" (age thirty-seven years).

In past generations, belonging within the community occurred within the context of long male absences, with women maintaining familial and community relations that were focused on village networks. In addition to their home-centred roles, most women were also integrated into the fishing economy through their work in the wharf-side sheds stringing and boning herring. This work provided significant social spaces of female connectedness that underpinned community relations and strengthened bonds of shared experience. Outside the homes and herring sheds, the villages also provided significant spaces of meaning for women, even becoming part of their identities in ways that persist today for the older women, especially in the most remote village of Seal Cove. That men no longer spend extended periods away from home has a double impact, offering new opportunities for enriched family relationships and also creating problems associated with threatened identities and renegotiation of spousal roles. Furthermore, while men have retained and continue to nurture their own social networks and spaces of interaction, for women the loss of the herring markets has meant the closure of the stringing and boning sheds and fewer opportunities for meaningful social interaction. During an informal guided tour of some of the activity centres around the island, a man pointed out an old prison at the top of Mill Hill that used to be a place for the men to gather for cards and cribbage. That was a time "when the women ran everything," he said. "But things are changing."

In her richly evocative description of a small community in the Scottish Hebrides, Sharon Macdonald (1997) refers to the idea of "reimagining" communities as a way of highlighting the continuing process of reworking that defines them. For women who were born on Grand Manan and who "belong" to the community, historic relations of exclusion are being renegotiated, often in subversive ways but increasingly with a growing awareness of equity and independence. As described in chapter 5, "Sharing the Commons," one way that women have created new boundaries of participation, re-embedding their place of belonging, has been by expanding their territories of spatial activity. As others have pointed out, spatial mobility creates an important claim on space and, by extension, on power relations (Katz and Monk 1993; Massey 1994, 1998).

Central to this discussion about the notion of belonging and evolving relationships is an understanding of how personal and collective identities are related to their community contexts. George Herbert Mead's classic discussion of the construction of the self in relation to social context is the foundation for understanding the dialectical process of identity formation. In his view, the individual acquired an identity by means of self-formation within a socially interactive framework of mutual recognition and adaptation represented by the structure of community norms. The notion of process is especially important because it emphasizes the essentially dynamic nature of identity and the fact that identity is inherently unstable, dependent upon relations of difference. Layers of economic, social, and political relationships within particular contexts of history, migration, and mobility create the "webs of significance" (Geertz 1973, 5) and lifeworlds of meaning that are constantly being created and recreated through social interactions.

Other important elements in people's sense of who they are their family roots and, crucially, the "ascriptive nature of family characteristics" (Cohen 1985, 104). Almost no conversation on Grand Manan can occur without reference to someone's family connections, intermarriages, departures, and deaths. Young people soon learn that they are marked by their family associations, whether in school in their relationships with teachers, among their peers, or even upon entering a local store. For anxious teenagers, the expectations of particular behaviours can overwhelm personal goals and personalities in ways that are both intimidating and stultifying. Several people characterized entire families as: "they're trouble," "good, hard-working folk," "no ambition," or "snooty." Referring to a community characteristic, one islander talked about the independence that is "bred into them" and how hard it is "to get them to pull together unless it is life-threatening." Many islanders may demonstrate a defensiveness that can be expressed in unexpected ways. A question about family lineage can suddenly lead to a response that is effectively a denial of inbreeding and an argument that some people seem to think they are backward but that, in fact, islanders "are more resourceful and well-read than many who come here." This response was typical of those I received several times in the first two years I was on Grand Manan.

The problem of modern identity has frequently turned on the idea of alienation, understood as a loss of human powers and the creation of multiple categories of being (Dunn 1998, 55–6). Given the loss of traditional sources of identity and the possibility for new forms of social networks, there is an inevitable tension.

> The points of reference used by individuals and groups in the past to plot their life courses are disappearing. Answering the basic question "Who am I?" becomes progressively more difficult; we continue to need fixed anchor points in our lives but even our personal biographies begin to fail us as we hardly recognize ourselves in our memories. The search for a safe haven for the self becomes an increasingly critical undertaking, and the individual must build and continuously rebuild her/his "home" in the face of the surging flux of events and relations. (Melucci 1996, 2)

The shifting balance in the relative importance of various roles and relationships is associated with changing relations of production, the growth of a consumer society, and changing lifestyles. With economic restructuring, these represent a series of "shifts away from cultural roles rooted in traditional belief systems toward new economic and social roles" (Dunn 1998, 58). New patterns of interaction and different views of the world have changed the ways in which we communicate, socialize, organize our institutions, and work together: "It is above all in situations of crisis that our identity and its weaknesses are revealed – as for instance when we are subjected to contradictory expectations, or when we lose our traditional bonds of belonging, when we join a new system of norms. These conflicts constitute a severe test for our identity, and they may also damage it" (Melucci 1996, 30). Furthermore, the idea of the relational basis of identity means that "the 'social' is constituted in relations amongst persons who interact, belong to groups or communities, and function within institutions" (Dunn 1998, 31). Identity and difference are complexly constructed through the interpenetrations of many forces in a web of social relations.

An understanding of the links between identity and community is intricately bound up with the nature of social and economic relationships, especially in periods of radical change. While some have argued that the forces of modernity and globalization have actually widened the scope of community (Hewitt 1989; Gusfield 1975), others have pointed to a decline in the significance of community life. On the one hand, there is an emphasis on the stretching of connections, and, on the other, there is an emphasis on the decreasing importance of place-based relationships. According to Dunn (1998, 61):

> The expansive possibilities of community contained in modernity represent the source of a potentially fulfilled self, but a self always under stress of conflicted definitions and meanings. Weakened forms of organic community survive in such institutions as the family and religion and are reproduced in such locally based structures as neighborhoods, schools, workplaces, and voluntary associations. Such forms offer a degree of stability and security in the face of rapid social change but can also incur major obligations on the part of the individual.

For a community, Melucci argues, collective identity ensures the continuity and permanence of the group. For rural areas, the impacts of economic restructuring and the intrusion of exogenous forces of change will have particular importance because of the less complex nature of their social institutions. Resilience will be more problematic as there is a lack of flexibility in economic spheres, which creates insecurities in social domains. These insecurities may be manifested as defensiveness and resistance among the more vulnerable segments of the population.

While there has been a broad consensus about the cultural and social content of identity in various periods of early history, there is more ambiguity concerning modern relations. The problem for individuals and communities experiencing profound changes with regard to the grounding of reference points and the changing nature of the "mirrors" of social control is in being able to move from internalized to externalized modes of perception. Faced with the increasingly externalized alternatives for identity referents, rural communities experience fractures that can be both invigorating spurs to growth and disruptive causes of pain and isolation.

With particular significance for the period after 1995 on Grand Manan, migration studies challenge us to understand the processes of changing meanings of community and place as well as their inextricable relationship to identity. The variety of types of migration (stepwise, return, circulation, chain, seasonal) precludes an all-encompassing migration theory (Ogden 2000, 504). But its significance in contributing to social and cultural change assures its centrality in the development of social theory, with particular relevance for understanding sense of place, community, and identity (McHugh and Mings 1996). While migration flows have tended to be seen in relation to labour markets, there is increasing consensus that migration must be seen not only in terms of economic causation but also in terms of its being a social process. The move itself expresses a particular worldview, infusing it with meaning as a cultural event through which Bourdieu's (1984) concept of habitus becomes central to an understanding of the changing dynamics of community life. This concept is explored in chapter 9, which focuses on different population groups (from-aways, summer residents, and in-migrants) as they attempt to establish homes on Grand Manan.

When George Ingersoll shared stories of his growing up in Seal Cove he was opening a door to some of the shared experiences that defined his life and his identity. In his recounting the challenges faced by children whose responsibilities included driving the cattle to slaughter or, in one amusing case, driving their cow south to Deep Cove for "servicing," he was describing some of the most formative years of his life. His hilarious description of trying to get the cow across the bridge at the "crick" vividly recalled a momentous episode in his life. "It refused to go across," he said, "until someone twisted the tail and convinced it to get moving!" Listening to the men talking on the wharf one hears a powerful sense of belonging, even in their expressions, as they mutter about some new regulation: "Give your head a shake!" Or as they describe an individual as being "three bricks short of a load" or, alternatively, as being "as smart as a Wilcox." Describing the storytelling abilities of one island wag, another man exclaimed: "Yep, he's got a tongue in the middle of his mouth!" Expressive local dialogue is one lively indicator of islander status and claim to belonging.

As I describe in the next chapters, new contacts with migrants to Grand Manan, both visitors and workers, is creating an entirely new milieu. With the in-migration of Newfoundlanders, two communities are being brought together, instigating new processes of cultural formation that will eventually achieve some resolution as an evolving whole. The decision to migrate and the processes of developing new networks reflect the identities of those involved and shape the receiving community's conception of them. In the details of church attendance and the patterns of religious participation, for example, we see evidence of conflicting meanings of "belonging" that become part of the consciousness of the newcomers. For Grand Manan, migration was a non-issue until after 1996. Thereafter, the expansion of the aquaculture industry and the restructuring of Connors, as well as increased rockweed harvesting, new lobster pounds, and new business opportunities associated with the expansions of trucking and boat building, created a job market that, for the first time, attracted large numbers of mainlanders and Newfoundlanders. As I describe in chapter 10, the struggles of the Newfoundlanders and the responses of wary Grand Mananers reflect two strong cultures with equally persistent historical foundations attempting to adjust to new community realities.

Islanders can also face questions of belonging, especially if they have spent any time away from Grand Manan. In talking about their seven years away from the island one couple described their return in 1985 as "reverse culture shock." They had been hearing about the increasing prosperity and optimism that the scallop fishery was bringing to Grand Manan and the pull of home drew them back. Their time in Halifax had been a wonderful experience for them, and their two daughters had been born there. Describing their adjustment process upon returning home, both agreed that they had not realized the extent to which they had adopted new norms and expectations, especially in their social relationships. Both had parents and siblings on the island, and they had maintained connections through frequent visits and cards, so the idea of having to adjust had not occurred to them.

They began to question their move following their first efforts at entertaining some friends. Sharon sent out invitations for dinner a few weeks ahead of time and waited to receive replies. None came. The evening approached, and they wondered what they had done to offend people. With dinner prepared and ready at the announced time, they waited. Two hours after they had hoped friends might arrive, a few people came. Then more arrived. By eleven o'clock, everyone was there, an overcooked dinner was served, and the drinking started. At that point no one wanted to leave, and finally in the early morning hours the couple found themselves sending people home. Looking back on their experience they realized how much they had changed in the intervening years and how different social norms and behaviours were on the island compared to what they were in Halifax. It is not a question of "tradition" and "modernity" but, rather, of a cultural distinctiveness born of generations living within parameters defined by a narrow range of choices. The accoutrements of modern lifestyles – computers, internet, and cell phones – have not transformed deeply entrenched relationships and social behaviours. As Pocius (1991, 287) noted in his sensitive exploration of Calvert, Newfoundland, "having up-to-date goods does not mean that people act in ways that characterize modernized cultures."

Another couple experienced their return quite differently. Having left in 1969, when most of his classmates had also left the island, the husband described his coming back twenty-five years later as a return to roots and family. His wife had been brought up in lighthouses, while he had worked on herring carriers from the time he was seventeen years old, so their sense of belonging was strongly linked to traditional lifestyles and cultural values. He said that, in being met at the ferry upon their arrival, he knew his prayers had been answered. So it was something of a surprise to him when, having been elected a councillor, someone questioned his Grand Manan identity.

Belonging is a complex idea, and it depends upon more than family names or fishing experiences. Its meaning is nebulous, and islanders themselves find it a challenge to provide a concrete definition. Even if one applies a "self-described" criterion, there will be questions. Many people from-away will argue that they belong by virtue of time and contribution to the community. For islanders, however, it is one's lineage over generations that is important. For Ina Small, who had arrived on the island in 1926 as a nineteen-year-old teacher and who ultimately married a fisher, her fifty years of marriage and living on Grand Manan were not enough to remove the designation of being from-away. In the final line of her autobiographical book, which describes her years on the island, she acknowledges the difficulty of acceptance. Leaving in 1977, she watched the ferry pulling away from the wharf and wrote: "The ferry pulled slowly away from its slip. I gazed as long as I could at the shrinking steeple of the Anglican Church of the Ascension and the cemetery back of it. Tears filled my eyes. It would be my resting place also, beside Robert. Soon I will come home, at peace, my wanderings over. Only then will I no longer be a stranger from away" (Small and Mutimer 1989, 114). Islanders who knew her agree: she was never really accepted as a Grand Mananer. Her belonging was always compromised by her own sense of difference as well as by the way in which she was perceived by the community.

Even when people return following years away, their situation is often problematic as many islanders view them with scepticism. For those who return, questions of belonging, experience, and family connections all become relevant. The attitudes, perceptions, and values of these in-between islanders are forever changed, and their acceptance on the island ultimately depends upon their ability to reintegrate on island terms.

8

Identity: Place and Community

"Wood Island: A Special Place Being Reimagined"

Fundy waters run deep
Then ebb away

From across the waters
Men and women
Came to settle here

Like so many generations
That come and go

And now, their children
Their childrens' children
Scattered all across the lands
~ (Wood Island, Reformed Baptist Church window)

The villages of Grand Manan have been important arenas of identity formation through generations, both as spaces of historical meaning and memory
and as centres of relationships rooted in family, work, and church. The villages
are places that have historically defined personal and collective identities in
ways that continue to resonate today. There is an important "manner in which
notions of kinship are tied to notions of villageness," and perceptions of class
and status feed into feelings of relatedness (Cohen 1982, 77). Families, villages,
jobs, class, and acceptance are all intricately linked. Desirable jobs in the village
office, on the ferry, or with the DFO, are all "handed down" within families. For
most families it is the village-based histories, stories, memories, and traditions
that define how islanders think of themselves.

Annual Wood Island reunion

A brief look at Wood Island provides an entrée into the ties to village identities as well as to both the magnitude of change on Grand Manan and the
extent to which islanders are able to celebrate their past and adapt to change.
Wood Island has a special place in the history and traditions of Grand Manan,
memorialized each year by a group of Wood Islanders and friends who attend
a service at the church that celebrates their shared heritage. Until the late 1950s,
about 250 people had lived on Wood Island, serviced by a post office, school,
store, and church. The school closed in 1957, a year during which only five children were attending it. The last permanent residents, Ivan Green, Eugene Harvey, and Bernard Lorimer, finally left the island as old age made it increasingly
difficult for them to procure the food and heating required for basic subsis-
tence. Ivan moved to the main island around 1992, followed about six years
later by Bernard and Eugene. In 2004, Bernard was the last person to be buried
in the cemetery on the island. The main settlement had been largely abandoned in the early 1960s, with the widespread adoption of new boating technologies, including LORAN and, more recently, computer navigational aids.

When islanders talk about the people of Seal Cove, they often refer to Wood
Island because of the close connections due to both geographical proximity
and residential roots. When Wood Islanders moved to the main island, they

tended to stay in Seal Cove. Particular reference is made to their scholarship. Describing her experience of teaching on Wood Island in the 1950s, a retired teacher used the expression "as smart as a Wilcox" to talk about some of the better students she had encountered. Another former teacher referred to two Wood Island students she had had in the high school as being the smartest kids she had ever taught. Musing about why Wood Islanders were so smart, one woman speculated that perhaps "there wasn't much to do and they got to like learning." That the reputation for intellectual achievement had merit seemed to be substantiated in 2002, when a Wood Island descendant graduated from the high school with the highest marks anyone had seen. Reputations are an oft-cited issue for young people who complain that family histories can be either a stigma or a boost in their time at school. Reputations are linked in part to collective village identities and in part to family lineage. How these are attributed has changed even within the past decade, although the imprint of the village continues as a powerful perceptual model. Asked directly about differences between the villages, people will deny that there are any distinctive characteristics. But in conversations, the prevailing sympathies and critiques emerge unbidden. As a former Seal Cover said, without rancour, "We always thought North Headers were snooty." And a North Header would say, "Those bible thumpers from Seal Cove work too hard."

The legacy of hard work and religious faith continues to pervade lives in Seal Cove, with the descendents of Wood Island being at the heart of their traditions. In 2000, Lydia Wilcox Parker gave me a tour of the derelict sites that were so gracefully succumbing to the battering weather, seemingly being enfolded by encroaching trees, mosses, and bushes. She described the special places that had been her home for the first twenty years of her life. As we walked down Ghost Alley and scrambled through overgrown brush on Danger Hill, she pointed out the old post office and school, which were surrounded by clumps of wild rose and swaths of pink orchids. Sitting for lunch at the "watering place" next to a large patch of old rhubarb, we looked over at Outer Wood Island, past Oatmeal and Big weirs, to the remnants of another small school. She talked about the year her family had returned from Blacks Harbour, the year of the "big turnips, when you could get only four in a bag." One of twelve children, and now in her early seventies, her memories of the huge distances belied the relatively short walk down the island. But it was the size of the homes that she described in detail: "The rooms were so small, and all the girls were in one room, all the boys in the other. The real luxury was the three-holer, with three sizes for parents, teenagers and small children." Along Ghost Alley she

Historic home succumbing, Wood Island, 2003

pointed out the treasured horse chestnut "hugging tree," with its enormous spreading branches, perfect for climbing and children's games.

Even after fifty years she enjoyed recounting her memories of playing with her best friend, Delia (also one of ten children), as they slid down the hills or picked their way along the shore at Devil's Head. While church was an important part of their lives, it was not necessarily the practise of their faith that forms Lydia's happy memories. She related a story about a time when she and Delia had been sitting in the front pew, and the Sunday school teacher said they reminded her of a bird. Trying to guess what kind of bird she meant, they suggested all sorts of beautiful birds, such as goldfinches, tanagers, and cardinals. No, it was none of these: "You remind me of an owl. Every time the door opens you turn to look at who is coming in." One of her stories was of the postmaster, Uncle Roscoe, whom everybody loved except his son. His son, Thealand, had a girlfriend who lived away and who wrote letters to him. But being the postmaster, Uncle Roscoe was able to intercept the letters, and he never gave them to his son. So Theal died never knowing that the girl loved him. When Uncle Roscoe died, Thealand returned home to put great huge boulders on his grave. He is said to have exclaimed, "That'll keep the bugger down!"

Another former resident, Aurelie Tidd Inman, who lives on the mainland, returns every summer for a few weeks of renewal. Describing Wood Island as her "security blanket," she said that, although she left almost fifty years ago when she was fourteen, it is still her spiritual home and her annual visits are an important time of restoring balance and meaning to her life. She described some of her favourite spots, such as Black Duck Puddle and Pumpkin Island, and she pointed out the sites of the original barn and pig house as well as where their gardens had been. Her baby brother is buried in the cemetery, covered by a lovely, delicately carved headstone on which lies a tiny lamb. With parents whose lineage includes the Wilcox and Shepherd clans, Aurelie has direct claims to a long history on Grand Manan.

The last residents of the island had scarcely left in 2000, when new summer residents had begun to re-establish a presence on Wood Island. All descendents of Wood Islanders, these new residents have come under the watchful monitoring of a committee of Wood Island inheritors who hold title to the various properties. The island is not a Commons to be resettled at the whim of Grand Mananers or from-aways. The old school has been renovated as a summer home, and one of the old family homesteads is again being lived in by a former Wood Islander who now lives on the mainland. Another descendent has installed a Quonset hut with solar heating, allowing him extended seasonal visits. But it is the annual reunion and church service that really seems to reflect the sense of heritage and celebration that Wood Islanders feel for their home. Every summer, on the Sunday during the long weekend that has become known as "Rotary Weekend," anyone who wants to visit Wood Island is "taxied" over from the Seal Cove wharf by one of the fishers. Skiffs or dinghies ferry people from the large fishing boat onto the beach near Ivan Green's house, and they then walk to the Reformed Baptist Church for a service led by invited clergy. The elderly or infirm have ATVs provided to them should they find the walk too difficult. Past the well-tended cemetery, men, women, and children, mainly with Wood Island roots but also many who are curious or admiring visitors, they file towards the church.

The service usually attracts about fifty people and may last as long as two hours, depending on the sermon. But it is the singing that everyone most enjoys, with Brenda Tate playing the old pump organ and James and Audrey Ingalls playing violin and keyboard, brought especially for the day. As well, there is usually a solo sung by one or two of the congregation who may have rehearsed on the boat en route to the island. Even in its relaxed rhythm, the service exudes a sense of being an important ritual of affirmation. It is an expe-

Top: Lydia Parker and Brenda Tate: Practising hymns on the ferry to Wood Island
Bottom: "Junior" Ingalls and Fraser Shepherd: Conservators and Guardians of Wood Island

riential confirmation of people's heritage and of the central role of faith in their lives. In offering transportation on ATVs from the beach, and in maintaining the church through painting bees and ongoing donations, the Shepherd family is part of a significant history that Grand Mananers know about but do not always celebrate. Wood Island has special meanings for its descendents and continues to contribute to their personal and collective identities.

At the annual church service on Wood Island in 2000, the last resident, Eugene Harvey, came in as the proceedings were about to start, walked to the front of the church, and announced that he was the "mayor of Wood Island." Even as a few people cringed, most of the celebrants did not; rather, his sense of entitlement gave them a feeling of continuing life and significance in that the island was still inhabited and maintained important links with the past. Former Wood Island residents played the pump organ and sang hymns, such as "The Solid Rock" and "I Will Not Forget Thee." The clock, still working, registered the correct time, providing another affirmation of the continuity and ongoing meaning of this place. In his homily, Seal Cove resident Robin Wilcox recalled the services when he was a boy, when girls sat on the right of the church and he and the other boys were seated on the left. "There was a sense of security," he said, "that God was in our midst."

The Villages and Centralization

The clustered villages that are strung out along the eastern shore of Grand Manan all grew up around the wharves and sheds associated with the fishing industry. Home and work were geographically and psychically connected until the paving of the road in the late 1940s, which initiated a new sense of island life and hinted at a blurring of boundaries of identity. The villages of Grand Manan had been the incubators of personal and collective identities until an amalgam of major changes, described most succinctly as "centralization," began to encourage the movement of people across the island. In the discussion that follows, I describe the momentum of centralization, which began with the road and the schools and touched virtually every important gathering place that, for generations, had defined personal and collective identities. Schools, stores, the bank, the post office, garages, provincially mandated signage, a rural plan, and village halls were all affected by administrative and market-driven issues that had a direct impact upon how Grand Mananers saw themselves and how they experienced their daily activities.

Seal Cove, 2000

No single village better illustrates this changing reality than Seal Cove. Unquestionably the prettiest village on Grand Manan, Seal Cove was formerly named Benson Cove (Hydrographic Chart, original surveys 1855, 22nd ed., April 1925, US Hydrographic Office Publications). As one drives down the island from North Head, through Grand Harbour, up Spruce Hill, and then around the corner after Mark Hill (named circa 1917 when there was a "mark" placed on the crest that would be visible from sea) the twin spires of the Wesleyan and Baptist churches come into view as a focal point for the lovely valley of Seal Cove Brook. The small white, cream, and pale green houses spread out along its sides down to the wharf and breakwater sheds. Rounding the sharp corner to cross the bridge, one passes the old smoke sheds and the remnants of the last original island store, which closed in 2003. The rings of plastic that dominate the road along the breakwater and the empty fish plant at the wharf, with the missing letters in CONNORS, both testify to the changes that have overtaken the village and all of Grand Manan. There is a sadness about the emerging landscape of Seal Cove that reflects a common perception on the island: "Seal Cove is dying." Its calendar-quality beauty notwithstanding, the loss of stores, smoke sheds, fish plants, and tourist facilities have all meant that fewer islanders have any reason to drive down the island. While newcomers may "see the landscape primarily visually, as scenery; local people may perceive not only a visual but also a social landscape, with social ties coursing over the land" (Fitchen 1991, 261). Islanders perceive Seal Cove in very distinctive ways that are only now beginning to change.

Each year as tourist season begins people hope that McLaughlin's Wharf Inn will open again. Without it, the last restaurant will be gone. A map of island services clearly shows two important trends of the past twenty years. The first is the increasing centralization in Grand Harbour of services that cater to Grand Mananers. Indeed, an islander complained that the village council "is obsessed" with having all services located in the middle of the island.[1] The second is the growing importance of North Head as the tourism centre for the island. Apart from Peter Wilcox's puffin- and whale-watching tours, which leave from the Seal Cove wharf, the only tourist-oriented facilities at the south end of the island are two bed-and-breakfasts and, further down, several cottages that are for rent.[2] That residents recognize the disappearing landscapes was made clear in a 2005 essay in the *Grand Manan Historical Gazette*, which described the history of High Tide Grocery. Built originally as a poolroom in 1937, the building had been used for a variety of purposes, including as a tea room and a twine shed. Around 1949, it was opened as a store and, under several different owners, it served the village for the next fifty years. The last owner, Johnny Brown, enjoyed the daily visits by fishers, women, and children, many of whom shared their pains and problems at this informal "social centre." When "Johnny's store" closed in 2003, the business was moved to the former school house, which, because it is located away from the road, does not encourage community gatherings or idle chit-chat. Although the other villages are also losing their gathering places (stores, garages, and post offices), Seal Cove, having only two churches and one store, seems the most isolated.

In 2006, a compilation of renovations throughout the village was published as a report in the *Grand Manan Historical Gazette*. Noting that the Smoked Herring Sheds National Historic Site (created in 2001) would be a continuing focus for the Grand Manan Historical Society, the report pointed out that many other renovations in Seal Cove were aimed at ensuring a revitalized village and historic site. Reflecting concerns about the decline of the village infrastructure, the report wove together private and public accomplishments that

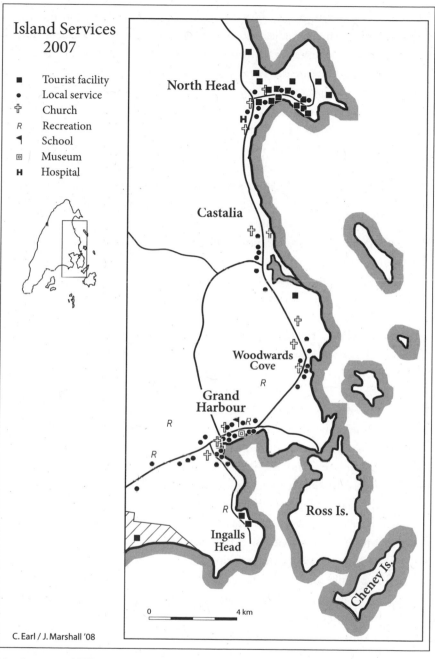

Island services, 2007

could contribute to the preservation of the village community. In drawing attention to improvements in Seal Cove, the *Gazette* hoped to encourage more initiatives as well as fundraising to protect the historic smoke sheds against further deterioration. Nonetheless, the islanders' level of concern for rejuvenating the Seal Cove landscape could be described as, at best, lukewarm. For most, it is not even a consideration.

One of the most noticeable changes in the island landscape concerns the distribution of grocery stores, which was affected by the paving of the roads, growing vehicle ownership, and the increasing dominance of the large supermarket chains. As a focus of village relations as well as food supplies, the stores provided important sites of interaction and community building until the early 1980s. Across the island, less than a decade after the mid-1970s, when a new national chain store arrived in Woodwards Cove, the number of stores had declined from about forty to fewer than six. Apart from the convenience factor, the stores had constituted a space for the men to gather after their evening meal and day's fishing. As one woman said cryptically: "They used to get home from fishing, and after supper they would head out to see their buddies at the store. They'd walk maybe two miles just to spend the evening out."

Traditionally, after a day's fishing, the men would walk to one of several nearby stores, gather around the stove, and share gossip and perhaps some games of cribbage. With the paving of the road in 1949, and the gradual increase in the number of vehicles on the island, transportation between the villages became easier. For most residents over fifty years old, their lives on the island had been village-centred, and they had had very little contact with families from other villages. Said one Seal Cover, "The only reason we ever went to North Head was for the dentist." Someone else described the annual cattle drive down the island, which provided families with their winter meat. The cattle that had been raised over the summer on the tender grasses of Nantucket Island were brought to the main island at low tide and herded down the twelve miles to Seal Cove, leaving a cow at various houses en route. Having extensive gardens, fish, usually some chickens, a pig, and a cow, islanders were essentially self-sufficient until the 1960s. Every village had its smoke sheds, drying grounds, and wharf facilities; schools, churches, and post office; and, of course, several stores that sold drygoods, hardware, and household products. The last smoke shed closed in 1997, and it is hard to find even traces of their former elegance. Today, wharf facilities are being upgraded because of the needs of the aquaculture industry.

Visitors' descriptions of Grand Manan provide some sense of the spirit of village life, although these accounts, which are from different time periods, also suggest differing perspectives on social life. In her memories of her fifty years on the island, Ina Small recounted the active social life of the 1920s and 1930s, which included "parties held in homes and halls, lively affairs with boisterous games and rousing singsongs. The Masons, Rebeccas, and Knights of Pythias were all going concerns ... every order held wonderful socials, each trying to outdo one another's clam or fish chowders or lobster stews" (Small 1989, 19). By the 1960s, when Katherine Scherman (1971, 23) visited Grand Manan, she experienced a much more subdued social scene: "Grand Manan attracts few visitors and has little frivolity, organized or unorganized." But both Scherman and Small agreed that festivities were never accompanied, at least publicly, by liquor (which was a real but hidden part of social lives). For Scherman it was the churches that provided the main venue for social activities: "Religion became almost their only entertainment, and remains so" (33).

The story of centralization really begins with the schools and moves on to the stores. And, although stories such as this have no "end," it would seem that the final chapter will be told through the churches. Throughout the period of centralization, the reconfiguration of spatial patterns has been inextricably linked to new social relations and village identities. An awareness of these links was evident in the comments and complaints of Grand Mananers. Even in 1995, almost twenty years after it happened, the bank's move from North Head to Grand Harbour was a constant irritant for many people, a reminder of the changing social and spatial rhythms of the island. In conversations people would grumble about the bank's move as a precursor to the gradual centralization of other services in Grand Harbour. The paving of the road in 1949 led first to all the high schools being centralized in Grand Harbour the same year. Then, in 1978, all the elementary schools, with the exception of the one at White Head, were also centralized in Grand Harbour. About the same time, many stores began to close. The closure of the village schools marked a major change in the degree of social integration among the village children. For the first time young children were all gathering in the same building on a daily basis. Those same children today are in their mid-fifties – adults who still feel the imprint of their village-based childhoods. While the high schools were integrated in 1949, at the time the road was paved, for young children the move to the "big school" in Grand Harbour was a memorable event. For administrators and teachers the improved communication between villages and the lower costs associated with maintaining a single facility were welcome benefits. But, as one astute observer noted, the amalgamation of the elementary schools had a far greater impact on integration than did the amalgamation of the high schools because so many children dropped out before reaching the latter.

Twenty years later, in March 1996, further efforts to rationalize the educational structures involved a much more controversial change. The provincial government centralized administration on the mainland, disbanding the board on Grand Manan completely. A former principal described the change: "The school board represented us; there was a sense of ownership. It was supported. But now there's no local autonomy. When local action went, pride, morale and expectations went." She was very discouraged, despite the fact that she had retired from the system. She felt that "big city values are being imposed" and that "we teach the curriculum, not the children." She said the academic level had declined because the "changes have been so fast."

Students who begin their education on White Head and then have to move to the main island in grade 7 experience real culture shock. In 1999, only six children were registered in White Head's small elementary school, which goes from kindergarten to grade 6. For these young people, the move to the high school was a traumatic event. Without exception, all of those with whom I spoke (both as adults and as teenagers) experienced moving into the high school on the main island as, at the very least, an intimidating event. Growing up in a small village of about 250 people, with few connections to the main island, the distinctive culture of White Head emphasizes family values and religious faith, with little apparent resistance from teenagers. In this small community, linked to the main island only by a ferry that runs until 4:30 PM (except by special request), there is a sense of relative isolation. This isolation has protected families from some of the magnetic temptations of the drug culture associated with youth on the main island. Asked about his experience as a White Head teenager moving into the high school, one young adult recalled his feelings clearly. He had found it very difficult, and he described his sense of being stigmatized. He said that even the teachers in the main island school would make jokes about White Headers. Today he is a university graduate and has a career on the mainland as a church minister.

The opening of the first national grocery chain in 1974 in Woodwards Cove initiated the rapid demise of between five and eight stores in each of the five main villages on the main island. By 1985, there were only single outlets in each village. North Head retained some grocery and general retail space, with the pharmacy being located at its corner store; Castalia had a small convenience store with its post office; in Grand Harbour, Newton's continued to provide a full range of grocery items; and in Seal Cove, Johnny Brown had a small store overlooking the "Crick," where men continued to gather, especially during inclement weather. Fifteen years later, only Johnny's was holding on, literally and figuratively, to its shoreline site, and it, too, closed in 2003, to be reopened in 2004 on top of the hill, away from the sightlines of passing traffic. The stores were important social centres for the men after fishing and for the women and children through the day. Virtually everyone over sixty years of age enjoyed talking about the stores, which had been such an important part of their social life until the early 1980s. Even today, directions often include references such as "where Cronk's store used to be" or "near Gaskill's store," both of which, by 2006, had been closed for over twenty years.

From a business perspective, the closures and eventual dominance of a national chain reflected the experience of every community in North America. While the entrepreneurial spirit of islanders was evidenced yet again in several forays into competing stores (e.g., the hardware store in Grand Harbour), in fact, the opening in January 2005 of a large new Save Easy in Woodwards Cove seemed to represent the final word in the new pattern of grocery shopping and store location. Moreover, the pattern of having a single centralized store is being replicated with regard to the gas stations.

The 1997 closure of the Irving station in Grand Harbour meant that only two gas stations served the island: one at North Head and one at Castalia. Like the stores, these businesses had been important gathering places for the men, beginning as early as 5:30 AM on their way to their boats. For the older men, especially those retired from fishing who want to keep up with the news of catches and maintain their social contacts, the garages offered coffee, doughnuts, and even hot dogs at midday, together with the all-important cribbage boards in a space usually reserved in a backroom. As has been mentioned, women were not welcome. When someone from off-island took over the North Head gas business in 2002, the gatherings disappeared and mainly focused on "Kenny's" in Castalia. However, ill-health eventually forced Kenny to try to sell his business. After a couple of years, he finally entered into negotiations with Irving, which resulted in the closure of both his garage and the one at North Head and the opening of a new gas bar, without any machine shop or extra room, that was attached to the Save Easy in Woodwards Cove. Its opening on the showery afternoon of 6 June 2006 seemed to mark the end of yet another gathering place for men.

In 2006, both the absence of any full-service garages (only a self-serve gas stop plus a mechanic shop at Cary Tire) and reliance upon a single chain supermarket represented a radical departure from the village-centred life of two decades earlier. The closure of the post offices removed a third important public space that had encouraged community sharing. Crib games had begun moving to the private fish sheds along Ingalls Head Road, still very much male territory, as they moved away from more inclusive public spaces such as the garages and stores. One wonders how much longer village identities will continue to define people's sense of place. At a small gathering of men in one of the sheds in 2004, the concentration on the crib game was occasionally interrupted by a cryptic comment. At the end of one game, several of them breathed a sigh of relief as Lester had finally lost a game: "The guy from North Head can't expect to win all the time!" In 2006, for adults over "a certain age," village identities were never far from personal identities. Moreover, whereas in the past the stores and garages represented public spaces that were inclusive of all men, the private sheds fostered and solidified divisions and political differences that were reflected in the debates within the GMFA, the municipal council, and even in the churches. Like-minded men, supporting (or not) the GMFA or village council, tended to gather in particular sheds, creating and/or confirming lines of division in the political life of Grand Manan.

For many islanders, the closing of the stores and gas stations and the relocation of the bank were all linked to increased mobility throughout the island due to the paved roads and greater number of vehicles. And, for the most part, they accepted the inevitability of these changes. However, they were not so sanguine about government-imposed changes that further eroded their sense of belonging to particular village communities. The provincially mandated village amalgamation that took effect in January 1996 seemed to represent a deterritorialization of island identities. Not only did the forced amalgamation require a single municipal council rather than five elected councils, but the government also demanded that street names and numbers be assigned and clearly posted at every civic address. Suddenly, large blue numbers jumped out of the landscape, and names on road signs identified "Whale Cove Road" or "Shore Road." Instead of islanders having to describe their house location in terms of "two doors down from the fire station," they were now able to provide a number. It took about two years, but soon everyone was using the numbers and road names to identify their locations. The transition did not happen easily, and there were some heated negotiations with the council over appropriate names or whether to be named at all. The most common objection came from residents who discovered that the naming seemed to be an invitation for tourists to drive down every little gravel pathway. While most of the names originated in designations that had long been used on the island, sometimes new names were decidedly atypical of the Grand Manan landscape and culture. "Bayview Heights" sounded more like a road in Toronto, while "Poodle Alley" and "Tea Kettle Lane" were both rooted in the history of the island. Gradually, more urban-oriented names began to permeate businesses. The opening of Eden Oasis Inn and Spa in 2004 and the building of the Rose Garden Estate condominiums a year later were further indications of changing island sensibilities.

The municipal amalgamation led inevitably to a government demand for the development of a "rural plan." For many islanders, this was adding insult to injury. For a culture defined by independence and isolation from mainland institutions, the idea of any restrictions on what they could do with their land or their homes was anathema. One spirited elderly woman who had obtained a copy of the plan prior to its public presentation in 1999 exclaimed: "This little island doesn't need this! They want to extend the marsh, and people are going to lose their land. There's a young fellow who just bought a piece for $10,000 and now he can't build, and they're not going to pay him for it! We were just fine with our three mayors! They weren't paid much. And with the two service districts, Woodwards Cove and Castalia, we've never been so good!"

Following on the municipal amalgamation in 1996 was the 1999 reorganization of the post office. Closing village post offices, which had been regular meeting places, was only part of the problem for islanders. That their addresses would now be officially "Grand Manan," deleting any reference to "Seal Cove" or "Grand Harbour," meant for many a further loss of identity. Furthermore, an exaggerated number of different postal codes was provided to this small island of 2,700 people. Sixty-three different postal codes were allocated, and the post office threatened to return any mail that did not use the required civic address as of September 1999. Nonetheless, one feisty North Head fisher declared that he would never use only "Grand Manan" on his return address: he lived in North Head and that defined who he was. Not only the Grand Manan designation but also civic numbers, which had been on houses since 1996, had by 1999 been legislated so that "emergency vehicles would know where to go." A conversation that autumn illuminated the gradual change in how people identified their houses. Describing where her house was, one woman mentioned "the police barracks" and "the sharp corner" then said: "Oh, yes, it's actually 1105; I keep forgetting we have numbers." The specificity of place was being eroded through both commercial interests (stores and banks) and institutional interests (schools and post offices).

Another loss for many people involved the use of the village halls for council meetings. Now that centralization meant that only one meeting place was needed, it was decided not to circulate among the various halls. In 2000, the council decided to test the popularity of having the meetings broadcast over the community channel, thereby necessitating a meeting at the Boys and Girls Club, where the equipment was all in place. The live broadcasts proved to be popular, but residents complained that if the halls were not used it would be one more blow to their local identities. Most of the village halls had originally been elementary schools, which had closed after 1978 and had continued as important village ecumenical gathering places. The prediction that the moving of village council meetings to a single location would result in the eventual closure of the halls was prescient.

Indeed, barely three years later, the final denouement came in 2003, when the council announced it wanted to sell the village halls. While the Woodwards Cove hall had already been quietly sold a couple of years earlier, the suggestion was to sell both the Seal Cove and North Head halls, which were no longer being used for council meetings. In North Head a citizen's group was galvanized to circulate a petition against the proposal. The intense reaction was summarized in the words of one North Header who complained that the sale of the hall would effectively erase the history of women's work and volunteerism. His mother, he pointed out, had been a member of the Glass Eye Society, and she had spent years knitting tea cozies and sewing aprons for church sales in order to raise money for the hall. "They worked, with pennies! Sale of the hall would be a travesty to their memory", he argued. "What do they think the money will be for? My taxes won't go down!" In the end, the North Head hall was saved.

In Seal Cove, the residents had other concerns about the imminent loss of their hall. For them the sale represented the solution to a problem. With the threatened closure of Johnny's store (High Tide Grocery), the prospects for the sale of the old building, which was barely clinging to the bank of a tidal creek, were negligible, especially given its poor state of repair. While there was an interested buyer, the lack of an affordable and accessible site was the main deterrent to selling the grocery business. The old school building, which dominated the centre of the village at the top of the hill, seemed to be an ideal location for a new store. As the council negotiated, in Seal Cove a petition was circulated that supported the sale of the hall to Farren Frost so that he could take over the grocery business. For residents it seemed to be an attractive proposition because they were afraid of losing the last store at the southern end of the island. In the end, the municipal council agreed to the sale, despite some murmurs of dissent from islanders living in other villages who did not want to see another hall disappear as a community centre.

In many other ways, Grand Mananers have felt the impact of centralization and have only reluctantly begun to accept that community events are island-wide rather than identified with particular villages. One of the first signs that activities were moving away from village identification was the demise of the competitive baseball games. Islanders still talk with pride about some of the men who were known as the best pitchers or runners on the different teams. The weekly inter-village games drew large crowds and symbolized an important part of belonging to one village or another. These games became fewer in the early 1990s, although there were teams that played during the summer (albeit without strong village affiliations). For youth, the village rivalries ended when the coaches of their teams decided to field a Grand Manan team that could compete on the mainland, strengthening the Grand Manan prospects through giving them access to a larger field of talent. For some parents this was a significant loss as the poorer players were no longer encouraged to play. But for a few families, the change provided an opportunity for high-calibre play that led several boys into national competition. Indeed, a front page-colour photo in the *Telegraph Journal* of 8 June 2006 told of the success story of one Grand Manan boy who had left the island with his family in 2002, partly to take advantage of opportunities for advancement in professional sport. Four years later, at nineteen years of age, he was drafted to an American team in the major league baseball draft.[3]

Also affected by centralization were all the village-centred events associated with Canada Day and the bank holiday in August. Traditionally, the 1 July celebrations had featured Seal Cove activities such as rope pulls across the Crick and skeet shooting at Red Point. Despite their popularity, they were discontinued in favour of island-wide events, although some of them were reportedly being reinstated in 2006. Similarly, during the Rotary weekend in August, the changed location of the parade (from North Head to the middle of the island) resulted in complaints from many North Headers. Despite denials when questioned about it directly, many islanders have a perception that North Head represents the "ruling class." And North Headers themselves occasionally act on this sense of entitlement. For example, it became known that at Christmas the village council received a letter of complaint from a woman who felt that if the island was going to have only one official Christmas tree it should be located in North Head, not Grand Harbour. Nevertheless, North Headers did manage to keep one of their traditional events: the setting off of the fireworks at the Castalia ball diamond during Rotary weekend. All of these events, eagerly anticipated and attracting half the island population, were important signifiers of the changing social relations and the increasing centralization of all island services.

By 2006, the village council was actively endorsing the trend to centralization, exerting its own pressure to ensure that all new recreational activities were located near Grand Harbour. Perhaps the most blatant example of this was the proposal to incorporate the village office into the new Boys and Girls Club plans, which were presented in early 2007. Notwithstanding the existing village office on Ingalls Head Road in Grand Harbour, there was an arguably well-suited facility with excellent space in North Head at the vacated business centre. Owned by the village, the business centre was a low-cost expanded facility that did not have other obvious prospects for tenancy. But it was summarily rejected because it was "not in the centre of the island." With the sudden support in September 2006 of two levels of government for a new Boys and Girls Club (see chapter 10), the village council agreed to be a third-party contributor, the proviso being that it be housed in the same facility.

Another indicator of the loss of village distinctiveness is the spread of newly built homes into the margins or in-between spaces between villages, onto cliff edges and hilltops, and into meadows overlooking the sea. No longer located in tight clusters of small houses and cottages, the newly built homes of young couples are large, often commanding views of the water. Their chosen road names reflect a new sensibility. "Cedar Ridge" and "Bayview Heights" suggest the urban areas of mainland Canada rather than the seacoast working landscapes of fishing villages. Returning for a family visit, some former islanders referred to these new houses as "urban sprawl" as they all merged into one another: "The island is now all one big village".

Interestingly, one the most distinctive differences between the villages is natural and uncontrollable: the weather. During my first visits to the island I would hear comments about the "fog in Seal Cove" or the "sun in North Head." I thought it was all a bit exaggerated. Soon I learned otherwise. On this island, which is only sixteen miles from north to south, there are distinct weather zones. The wild lupines bloom two weeks earlier and finish two weeks earlier in North Head than they do in Seal Cove. Vegetable gardens in Seal Cove are invariably two weeks behind those up the island. And, during the foggy month of June, tourists are advised to seek out their hiking trails on the north end of the island rather than hoping for any clearing at Southern Head. Weather divides the island, at Mark Hill just north of Seal Cove, at Thorough-fare Road as one leaves Grand Harbour, at Woodwards Cove, and at Castalia Marsh. The trophy for most foggy days always goes to Red Point Road, where the fog inevitably sits in against the hill at the top of Seal Cove Sound for days, while the rest of the island enjoys the sun.

Despite the apparent commercial and institutional centralization at Grand Harbour and, to a lesser degree, Woodwards Cove, the social centralization of the island is much more gradual. Boundaries between Seal Cove and North Head, between Ingalls Head and Castalia, continue in surreptitious ways. One woman who had been raised in Woodwards Cove described the divisions in terms of both class and travel patterns on the island. "It seems like the church-es in North Head are more elite churches," she said. "They're rougher in Seal Cove. We don't have any friends from Seal Cove. It seems that people from Seal Cove come up the island, we don't go down there." She said that she was never really outside of Woodwards Cove until after she graduated from high school in the 1970s. This experience was mirrored by a visiting former islander who had stayed in one of the tourist cottages in Deep Cove and, upon leaving, wrote the following in the Visitor's Book: "Having grown up in North Head I'd never have believed that I would enjoy the other 'end' of the island so much" (2005). One man who had grown up in Seal Cove at first denied any difference between the villages, but he then acknowledged that people did seem to attend church more regularly in Seal Cove. They also seemed to drink at lot more. "Maybe it's the repression of the church causes the other extreme," he suggest-ed. He recalled "us in our bib-jeans, they had their shirts and ties," and "even at baseball games we never talked to them. We just wanted to beat them!"

The notion of class differences permeates many assumptions about social distinctions – in school, among church leaders, and within the social work net-work. Despite the Seal Cove tradition of having hard-working, religious resi-dents, talk of troubled youth and problems with parental care seemed to focus on Seal Cove families. One Seal Cove resident thought that the people in North Head liked to socialize more than did the people in her village. She and her husband had more opportunity for pot luck suppers with their North Head friends than they did with their neighbours in Seal Cove. Similarly, she said that the men in Seal Cove drive around in their trucks with their dogs in the back, but they don't talk to each other as much as do the men in North Head, who always seem to be out on the wharves. Another young man who grew up in Seal Cove agreed that the men in North Head seemed to share information more than did the men in Seal Cove, probably because the former gathered in the garages or on the wharves in the morning, while the latter had no common meeting place.

Part of the perception of there being distinctive social differences between villages is undoubtedly related to their relative isolation over generations as well as to the historic competitions between them. George Ingersoll described growing up in Seal Cove, where he used to know every inch of the woods but only as far as Dark Harbour. There were clear boundaries beyond which he had no knowledge. "North and South, we were like enemies," he said. "Like North and South Ireland!" He related a story from over twenty-five years earlier, when he had been about forty years old. Someone was lost in the woods, and they called George to help in the search. The problem was the territory to be searched included the area to the north of Dark Harbour, where he felt totally lost himself. Eventually, the lost man emerged from the woods the next morning. But, as George acknowledged, for many islanders detailed knowledge of the woods is directly linked to their village ties. Asked about changes in village identity, one woman exclaimed: "Oh that hasn't changed at all. Amalgamation was in name only! In Seal Cove they even talk differently. And they all work, whether they need to or not, in the fish plant. It's a way of life; they would cry if you took it away. But really, all the villages stick together."

Thinking about Grand Harbour, she was not sure how she would describe the people who live there. But North Headers, "they stick together." As for the residents of White Head Island, she felt that they were very distinctive: "They're all on their own. They're strong. They have a strong sense of identity." Corroborating this perception, two fishers talked about buying their fuel and equipment from White Head because they liked dealing with the people there: "On White Head people help each other. Everyone stands around the store talking. It's a really strong community, like it used to be here [on the main island]."

Someone who has always lived on White Head has a sense of distinctiveness that is related to strong, faith-based family values. While residents reacted with mixed feelings to the proposal for a bridge connecting them to the main island, those with teenagers agreed that it would make their social lives much easier. For the most part, though, White Headers are happy to have the ferry, feeling that a bridge would only open up their small community to more problems from the main island. It is the businesses that have argued for the advantages of a bridge – an idea that seems to gain momentum at every election. White Headers feel that their strong family values and religious affiliation provide important supports for youth, functioning to prevent incursions of the drug culture that seems (to them) to be endemic on the main island. The island village of about 220 residents is not part of the Grand Manan municipality; rather, it stands alone as an independent, unaffiliated village with no elected representatives to the village council. It is not unusual for Grand Mananers who have grown up on the main island never to have been to White Head. Those who do visit usually do so in order to pick berries in the late summer, when bakeapple berries and blueberries are in abundance.

Another village institution is the restaurant. Clearly, the impact of tourism has meant that the focus has been on North Head, where a series of restaurants opened and closed between 1995 and 2005. For summer-only facilities, the continuing success of Laura Buckley's Whale Cove Inn Restaurant, and the restaurants in the Compass Rose and Shorecrest Inn (all located in North Head), were exceptions to a pattern of openings and closings that seemed to bedevil the attempts of local entrepreneurs to establish profitable enterprises. In Grand Harbour and Seal Cove there were also attempts to establish new year-round eateries that could withstand the spare months during the winter when owners had to rely upon local business. Harry's Bar (1999–2004), the Island Perc (2000–02), Uncles (2002–04), Water's Edge (1996–2001), the Sand Dollar (1994–99), McLaughlin's Take-Out (2000–03), and several fish-and-chip stands all vied for a place in the habits and hearts of both islanders and tourists. Only "Vern's" in Harbour Gifts managed to survive throughout the decade. The most recent entry, Keyser's Kafe in late December 2005, seems to offer a winning combination of excellent food at reasonable prices as well as a wine list. Unfortunately it, too, closed early in 2008. Meantime, Gallaway's and Area 38,[4] established in 2004 and 2003, respectively, continued to offer pre-prepared food with menus that never changed. Area 38 closed in late 2006, reopening a few months later, while Gallaway's closed temporarily in April 2008 for renovations that were to make it a "more family type restaurant [by] moving the bar to the back."[5] The climate of growing apprehension that jobs are not guaranteed and that the wild fishery will probably never be the same is having a profound impact upon all local businesses. Behind most of these restaurant enterprises were women who had dreamed of opening their own businesses and whose energies could not be sustained without the support of other family or community members. For those who were forced to give up, the sense of failure was assuaged only by the possibility of trying something else.

The centralization of tourism facilities in North Head has created a particular ambience in the village that is especially apparent during the summer months. While the fisherman's wharf and the several boat launches at the Whistle and at Whale Cove all provide important working areas for the fishers, it is the many tourist facilities that provide North Head with its identity and lively ambience. For all islanders and tourists, it is the ferry terminal, of course, that most defines the village. It is at North Head where the *Grand Manan V* docks four times a day, and the older, smaller *Grand Manan* (the "black boat") three times a day from late June until the end of September. As well, during the summer months the activity of North Head focuses on such tourist-related facilities as the Island Arts souvenir shop, several home-based jewellery and art shops, the bakery, the kayak rental shop, the whale research station, and the many inns and restaurants. Of note is that most of these are owned by people from-away, most of whom have moved to the island as retired couples, although several spend only the summer months there. The notable absence of native islanders in tourism is related to a number of factors, not the least of which is the industry's capital requirements and seasonality, which precludes an adequate annual income.

Within the tourism industry it is women who have made the greatest contribution, a situation that I discussed at greater length in chapter 10. For the men not in the fishery the entrepreneurial spirit tends to have focused on mechanical or material goods, including a Dollar Store, the Island Truck Repair, a grocery store, Home Hardware, and, notably, through many incarnations, the "L&P." The L&P enterprise, located in the in-between landscape between Grand Harbour and Seal Cove, began as a carwash in 1995, moved into lawn maintenance a year later, then became a convenience store, later a Chinese take-out, and finally a convenience store combined with a small terrace and take-out food service. While restaurant ventures tend to fare better in North Head, where tourism traffic can support them during the summer months, services for the island naturally gravitate to a more central location in Woodwards Cove or Grand Harbour. This centralization, which began with the road paving in 1949 but gained momentum in the 1990s, has been a significant force with regard to the dissolution of clear village identities over the past decade.

While Grand Manan is usually referred to as one place, with a single identity, it is clear that its long history of isolated villages and the independent spirits of hard-working fishers have created conditions for distinctive identities that continue today. As mentioned, despite denials of difference when asked directly, there is a consensus among islanders that accents vary, that working habits and social lives differ, that villages have distinctive religious and social values, and that youthful friendships tend to be village-centred. Although centralization may have erased institutional cleavages, and business efficiencies are dictating a spatial reconfiguration of distribution, villages nevertheless retain a patina of cultural identity that will fade only with generational change.

Dark Harbour

Dark Harbour epitomizes the shorefront zones of transition, a fascinating physical, social, and economic milieu that gives an "edge" to the island and offers photographic treasures to visitors. The name "Dark Harbour" evokes a place of mystery, adventure, and strangeness. All of these are accurate descriptors. But there are other elements too: gorgeous sunsets, curling wisps of fog climbing the edge of the cliffs, and the debris of past traditions touching the modern plastic cages of the salmon industry. Probably more than any other place on Grand Manan, Dark Harbour combines a dramatic physical landscape with a social reality that challenges normative island rules while also epitomizing the culture of camps and male spaces. It truly represents the shoreline transition between land and sea, its raison d'être being the harvest of dulse but its evening hours being filled with the sounds of clinking beer bottles and laughter as dulsers and their friends relax. At the Grand Manan Museum the photographs of the "Dark Harbour hermits," Darby and Lucy, suggest some of the interesting stories that locals have to tell about this place. While it is the unusual physical landscape that attracts first-time visitors and keeps photographers returning during all combinations of light and tidal conditions, it is the clues to the social lives that fascinate.

William and Lewis Green were two dulsers who lived out at Dark Harbour during the 1920s and 1930s and who, by all accounts, loved to entertain tourists and anyone else who might venture out to their camp on the cliffside. These two dulsers were known as Darby and Lucy, the Dark Harbour hermits. Their eccentricities were displayed in their unusual costumes and some of their crafts, which are today on display in the museum. They claimed they could commune with the spirits, and when they were not dulsing they would carve boxes and toys and small figures that would be built into constructions with peepholes and magnifying lenses.

Formed by the backward erosion of a deeply incised stream that plunges down a three-hundred-foot cliff, the harbour itself is totally enclosed by a natural seawall of rounded boulders and stones tossed up by the sea. About twenty-five feet high even at high tide, it is an impressive barrier that protects the brackish water it encloses. Almost totally cut off from the sea during the famous Saxby gale of 1870, its main entrance was on the eastern facing slope until the late 1950s, when a storm filled it in and created a new exit/entrance at the northern end, where it remains today. One is constantly confronted by the incessant and marvellous change of physical landscapes on the island, responding to storms, tides, and winds on a daily, weekly, seasonal, and annual basis. Cliffs erode (to the intense danger of unaware or careless hikers), shorelines retreat, and sand or gravel bars fill across sea coves to transform the coastline in an unending rhythm of change. Here in Dark Harbour these physical wonders are seemingly magnified or, at least, more obviously in motion.

In 2006, salmon cages shared space in the pond with an old weir that had not been "built" since about 1993 and with an experimental plastic "ring" that had been constructed as a floating weir for herring. These represent three technologies and two traditions, neither of which guarantees the future of the island. Among the relics that could be seen were a small wharf at the end of the road that wound its way down a steep hill, the remnants of a tower at the far eastern end of the pond, and, part way down the beach, a fascinating collection of swings and tires and hoops, obviously a home-designed playground for children and adults alike. But dominating the entire scene, if one is there near high tide, are the eighteen to twenty dories lightly floating at the edge of the tidal ripples like so many jewels hanging from a necklace. Mainly mustard yellow, a few of the dories are painted non-traditional colours such as blue, orange, and cream. Reportedly, the tower had been built to provide a bright light that would attract herring to the pond. (The new owners of the Swallowtail bed-and-breakfast in North Head used this bit of information to argue their case when angry weir owners complained that the lights from their establishment would prevent herring coming into Pettes Cove. So much for theories about why fish behave as they do!)

As recently as 1991 the tradition of "Dark Harbour Daze" continued to offer islanders a weekend of boozy fun and shenanigans. At that time, too, there was still a "mayor" of Dark Harbour, the longest reigning one probably being George Gaskill. But sometime in the early 1990s it seemed that the parties were getting too wild, or perhaps islanders were escaping more often to their camps in the woods on the weekends. No one is quite sure why, but the (in)famous Dark Harbour Daze disappeared from Grand Manan traditions in all but memory. Nevertheless, on summer weekends, when the tide is down so that vehicles have easy access to the camps along the cliffside, the small camps are filled with partiers. For several years the RCMP would station themselves at the T-junction coming out of Dark Harbour, and anyone who had had too much alcohol simply had to stay there overnight. Today, even this practice seems to have waned. The demise of Dark Harbour as "party central" probably relates to the proliferation of camps throughout the island's woods, which occurred in the early 1990s.

Meantime, in 2005 and 2006 the restoration of two weirs on the seaward side of the seawall has rejuvenated another of the Dark Harbour traditions. Connors' building of the Seawall weir in 2005 (renamed the SeaDream weir) and three young islanders' building of the North Air weir a year later were encouraging signs for many islanders who were becoming concerned about the demise of the herring industry. The coming together of dulsing, herring weirs, and salmon aquaculture in this one small pond seemed to encapsulate the changes that would contribute to the evolving cultural identity of Grand Manan.

Faith, Ritual, and Resistance

Plain living and high thinking are no more:
The homely beauty of the good old cause
Is gone; our peace, our fearful innocence,
And pure religion breathing household laws.
~Wordsworth, "London, 1802"

Written during a time of enormous change as the Industrial Revolution and scientific advances threatened the basic beliefs and structures of religious institutions, Wordsworth's poem still resonates in the lives of Grand Mananers. While it was the Anglican Church that first arrived on the island at the beginning of the nineteenth century, it was the arrival of the Baptists from the United States in the 1850s that solidified a strong religious tradition within the community. The ties to the United States, which were established by early migrations of United Empire Loyalists and later by incursions of American fishers, extended to and were strengthened by the spread of several religious sects and denominations that reflected an independent colonialist heritage more akin to American than to British sensibilities. As pointed out earlier, the New England imaginary was reflected in paintings that had strong religious overtones. American artists' depictions of Grand Manan show soaring cliffs reminiscent of cathedrals, with the ocean being related to the "biblical meaning of the wilderness-as-void" (Small 1997, 34). In the writing of Beecher Stowe, we also see evidence of a perception that island landscapes and boundaries could serve to protect religious values and the sanctity of Puritan households. In her novels, she extolled the virtues of holding to the Sabbath, which is something that was apparent in coastal communities. The churches that arrived on Grand Manan from the United States did not rely upon hundreds of years of rituals and textual interpretation; instead, they relied upon the needs of a new society for moral guidance and strategies that would protect family structures. The long absences of men while fishing, and the inherent strengths of women who sustained the family and community, were underpinned by the edicts of the church, which prohibited drinking and mandated men as heads of household.

Edmund Burke's description provides a window into this particular historical and cultural milieu: "The religion most prevalent in our northern colonies is a refinement on the principles of resistance: it is the dissidence of dissent, and the protestantism of the Protestant religion" (Burke 1780, cited in Bartlett 1980, 372). Indeed, the links between religion and society were explored and articulated in the classic treatise written by Alexis de Tocqueville in the 1830s.

> There is hardly any human action, however particular it may be, that does not originate in some very general idea men have conceived of the Deity, of his relation to mankind, of the nature of their own souls, and of their duties to their fellow creatures. Fixed ideas about God and human nature are indispensable to the daily practice of men's lives. (de Tocqueville 1990, 2:20)

In another view of the role of religion, Karl Mannheim (1968), acknowledging its essential ambiguity, described two distinct ways in which religion might function within society. He differentiated between ideological and utopian roles, arguing that different social groups have different life experiences that affect their "mind sets." He showed that people's ideas evolve as the group to which they belong undergoes significant changes. In other words, ideas are grounded in social reality, in the life experiences and historical meanings of different groups. Ideological religion legitimates the existing social order, defends the dominant values, enhances the authority of the dominant class, and creates an imagination suggesting that society is stable and enduring. Utopian religion, on the other hand, challenges the existing order and offers the possibility of creative change.

The nature of the religion-society relationship has changed in the twentieth century. While nineteenth-century interpretations were grounded in assumptions of the undeniable significance of religion, twentieth-century scholars have asked questions about its relevance and about links between religion, modernity, and secularization. Notably difficult to define precisely, "secularization" refers to a process of disengagement between religious adherence and social norms and practices. It refers to a decline in religious participation and a marginalization of religion as a source of reference for social values. While secularization tends to be characterized by universal processes, as David Martin (1978) has argued, there is no valid reason why this "must happen." Other scholars have also argued that secularization is not a linear process, despite its apparently common roots in industrialization and modernization. It is tempting to make such a connection when television and advertising beamed ideas into living rooms on a daily basis. But the reality is much more complex. Secularization is "not only a change occurring 'in' society, it is also a change 'of' society in its basic organization" (Wilson 1982, 148). It occurs in diverse ways and contexts, and it relates to a diminution in the social significance of religion.

Dissenting from a generalized perspective on secularization, Paul Tillich and Mircea Eliade both point to the sustained significance of symbols and rituals, which continue to give meaning to people's lives. Eliade (1957) shows that the significance of the church concerns not only a belief in God and Christ but also a belief in historical memories and events, rites, and rituals shared by family and community. There is an intimate link between the local and the global that provides both rootedness and a universal sense of belonging. The place-centredness of religious faith is a reflection of the significance of symbols of continuity and rootedness, which affirm identity. Tillich argues against the compartmentalization of religion, using the notion of a "boundary-situation" to describe the confrontation experienced as a support or threat, and suggesting that freedom comes from a support that can be the core of all creativity. Like Eliade, he denies the inevitable evolution of society towards a secular outcome, emphasizing the changing role of religion as a "turning towards reality ... penetrating in an intimate sense into what happens day-by-day" (Tillich, cited in Adams 1965, 11).

In Canada, the publication of Pierre Berton's *Comfortable Pew* in 1963 challenged the relevance of the church and provoked a debate that brought the church under broad public scrutiny for the first time. However, on Grand Manan the church has continued to be a pervasive presence. Understanding and describing the role and meaning of religion, faith, and the churches for the community of Grand Manan is especially difficult because of the complex and contradictory attitudes found in the community and in personal lives. Contradictions, paradoxes, resistance, and unquestioning support all describe the many relationships that Grand Mananers have with their faith and spiritual lives. Intertwined with active participation in their churches are norms and codes of conduct related to working on Sunday, gender relations, and alcohol. The change that has overwhelmed many island institutions is affecting the practise of faith in ways that are uneven (according to age and gender) and divisive with regard to specific community crises. According to social workers and psychologists who arrived on the island during the period of this study, the strong connections between religious affiliations and daily lives have affected their ability to offer solutions to family, marriage, and parenting problems.

As has been described in several earlier chapters, religion is woven into the fabric of Grand Manan lives in multiple ways, not only in the values and norms that guide daily behaviours but also in the conduct of the fishery itself. Examples of the latter include the tension-filled 9 July 2002 meeting over the Grey Zone issue, at which one man tried to calm the waters by reminding people they were in "the house of God." Other examples include the choice of the Baptist pastor to do the draw for the sea urchin licences in 1996 and the decision to make Tuesdays rather than Mondays the setting day for the lobster season so as to avoid working on Sundays. One of the most moving connections between the churches and the fishery may be seen in the annual service, on the Sunday before lobster setting day in November, when, with nets, traps, and model boats decorating the windows and aisles, special prayers and hymns acknowledge the dangers upon which the crews are about to embark.

In the gradual questioning that is beginning to emerge within Grand Manan conversations, there is no sense of a fundamental shift that could be described as "secularization." In the 2006 census, for example, fewer than 12 percent of the population said they had "no religious affiliation." Indeed, three days of revival meetings led by a team of visiting Baptists in June 2006 confirmed the continuing importance of religion on the island. All six sessions attracted hundreds of people who shared stories, sang, and went for healing. A week later, the North Head Baptist Church had a fundraising dinner for Harvest House, a faith-based drug addiction treatment organization begun in the Maritimes in 1997. Almost two hundred people attended, with many turned away at the door because there was simply no space. Led by the minister, Ron Ford, the dinner was followed by a moving testimonial by a young man who had been in rehabilitation for over two months, following eight years "lost" to drug abuse. While he described his conversion on Easter Sunday earlier that year, despite his prior rejection of "that Christian stuff," it was his obvious desire to change his life, to move away from the awful dependence that he had experienced, that was so powerfully obvious. At precisely the same time that he had been in treatment (spring 2006), eleven other Grand Mananers had also been in Harvest House facilities in New Brunswick and Nova Scotia.

The importance of the church in offering faith-based treatment cannot be minimized. For a community in which religion is part of the collective identity, and in which the church offers a supportive community upon return to the island, there is no question that an institution such as Harvest House can have a significant role in the process of treatment and social reintegration. For Grand Manan, the ultimate problem in assuring long-term rehabilitation is that, after treatment, a person's ability to find friends who are not taking drugs is limited. The awareness of the challenges was certainly corroborated by the number of people at the dinner that evening. Substantial efforts to provide an on-island rehabilitation facility were being made in the winter of 2006–07 by a group of volunteers that was refurbishing a historic hall in Castalia. From school house to funeral home to facility for the women's auxiliary of the Canadian Legion, this hall has been a central part of islanders' living and dying.

Nonetheless, the relationship to religion is more ambiguous than it was a decade ago, with evidence of some challenges to the authority of the church, resistance to particular edicts related to patriarchal structures, and even critiques related to the lack of leadership around issues of family and substance abuse. While there is little apparent decline in belief, in 2006, compared to even a decade earlier, there are fissures and edges of transition, manifested most clearly in declining Sunday school attendance. Regardless of the numbers, the prevailing ethos of being on Grand Manan continues to be defined by religious values and affiliation.

The long history of religious values linked to social mores has been described in other narratives, some of which are widely disparaged by islanders.

One of the most frequently quoted and quickly dismissed books about the island was written in 1971 by an American visitor. One can understand islander reactions in reading passages such as the following:

> There is also an extraordinary number of plain white painted churches, mostly Baptist and well patronized. Some years ago a Baptist revivalist team arrived on the island full of missionary zeal. The inhabitants had been chiefly Anglican, but the then minister was not popular. The Baptists held a few hearty prayer meetings with exhortations, hymns and public confessions of sin, and won the wholehearted support of the islanders. Religion became almost their only entertainment, and remains so ... The sale of liquor was forbidden until recently. It used to travel abundantly on the ferry from the mainland, depriving Grand Manan of a good source of income ... Now Grand Manan has, against strong protests by the churches, a package store. ... Tobacco is still hard to come by. If one asks for cigarettes there is chilly denial. We were introduced to one sinner who sold cigarettes. He is well patronized. I asked in a general store for lipstick. "He don't stock lipstick," said the clerk, sounding as if I had asked for heroin. However, somehow, people smoke and paint their faces. Grand Manan is not a monastery, however much the Baptists would like to have it so. (Scherman 1971, 32–3)

There are fifteen active churches on Grand Manan (including White Head), representing nine different denominations. The most popular of the denominations are the Baptist and Wesleyan Baptists, each with three active churches: one in North Head, one in Grand Harbour, and one in Seal Cove. Sunday services attract between forty and seventy faithful, while additional services on Wednesday evenings are also attended by ten to fifteen families. However, one committed Baptist complained that there "used to be 150 children in the Sunday school; now we don't even have young couples at church." In contrast, the two Anglican churches (in North Head and Grand Harbour, respectively) together do not serve more than about fifty people on any given Sunday. Despite the low Anglican numbers, both churches have very committed and active women's groups that serve suppers throughout the year. These are well attended by all religious affiliations.

The various denominations have different relationships not only with their own parishioners but also with the community of Grand Manan. They play distinctive roles within community structures and have different levels of power in terms of social meaning. While they have participated in an ecumenical "Ministerial" (see below) for many years, the Anglican clergy tend to maintain a greater distance from the community than do other denominations on the island, which combine social and spiritual activities in a more integrative way. For example, the Pentecostals hosted a Mother's Day special to honour one of the island women in May 2005. The Wesleyans have been active participants in community events, partly because of the energy and commitment of the popular pastor at the Central Wesleyan Church. He has been the unofficial host of many non-religious events: he was the announcer at a holiday baseball tournament, the emcee for the Rotary Club Variety concert, and the announcer for the provincial basketball tournaments. Similarly, it was the Wesleyans who held a prayer vigil for those who had been killed in vehicular accidents related to drugs and alcohol. On the other hand, community suppers that attract all denominations have traditionally been the purview of the Anglicans, including suppers for Remembrance Day, the Rotary Club, Shrove (Pancake) Tuesday during Lent, and various teas (such as the Strawberry Tea and Blueberry Tea during the summer months). Enjoyed by over 150 people, these events were among the few occasions when outsiders (i.e., summer residents and tourists) mixed with the islanders. So while the social functions hosted by the Baptists and Wesleyans tended to draw lively family groups and to be integrated with island secular events, the social functions hosted by the Anglicans tended to be more ecumenical, being widely advertised and open to outsiders and visitors.

In addition to the Anglicans, Roman Catholics, Baptists, Wesleyans, and Pentecostals, all with dedicated facilities, are the United Church members, Adventists, Jehovah's Witnesses, Ba'hai, Jews, and a sect that has no name but that has been nicknamed the "Go Ye's." While the United Church has been able to rent a building in Castalia, the others churches do not have their own facilities. The Jehovah's Witnesses did have a building until 2000, which was sold to a group of Grand Manan parents for a new Christian school. Said one observer, "They [the Christian parents] see it as a victory over the devil, to have bought their Kingdom Hall from them." It closed in 2007. Another group that was represented for a while on Grand Manan was the Zion Full Gospel Fellowship, which was disbanded in 1992. Their building was sold to the Grand Manan Historical Society for one dollar, eventually becoming the Grand Manan Art Gallery in 2005. Like most of the other denominations on the island, the Zion Full Gospel originated in the United States, and even in 2005 many of those who attended the funeral of a former member of that congregation were Americans.

In the past, even mainstream denominations such as the Roman Catholics have experienced difficulties in becoming accepted. When the Roman Catholic Church first arrived in the 1950s, several islanders said they would leave. Arson destroyed the first building, and for a long time after it was replaced people were afraid to attend it. Today, a small committed congregation of ten to twenty people support the small church in Woodwards Cove. According to several islanders, the Anglicans have always been viewed as vaguely related to the devil because of their rather liberal attitudes. However, with the establishment of an ecumenical group of clergy called "the Ministerial" in the early 1990s, attitudes towards all denominations gradually became much more relaxed. Nevertheless, identities, even of outsiders, continue to be defined in relation to religious affiliation. When, in 2003, a large North Head house built by Americans was sold, the islanders described it as "the house belonging to the Jehovah's Witness." One's faith and church membership may not define status unequivocally, but it continues to be one element that determines how people are described and how they identify themselves.

The establishment of a Christian school began with discussions in 1999, initiated by one of the Baptist pastors because of concerns about lack of discipline at the school. As rumours began to circulate in the community, speculation focused on who might be leading the concerned parents and where they would locate. At first no one knew who the main committee members were, although there seemed to be a consensus that the organizers were "respected members of the community," even as many doubts were expressed about the possibility that the idea would succeed. As one community leader observed, "Grand Manan people are hard to bring together." When they wanted something to change, two or three people would lobby the authorities, and when they were unsuccessful, "they would go off on their own." The question of a Christian school was only partly linked to the strong religious traditions on the island. Also important was the gradual centralization of the administrative power of the school boards and their increasing distance, both spatial and psychological, from local institutional structures. Just as the municipalities were being categorically restructured through the provincial government in Fredericton, the school boards were being centralized to such an extent that Grand Manan volunteers and teachers felt powerless. As rumours of the plans for a new school began to circulate in the community there was mounting concern that energies should be focused on the existing school rather than on a new venture.

But for many parents the frustrations of inappropriate curriculum development and inadequate discipline at the school were not being addressed, and their voices were not being heard. While the government sent touring bureaucrats to rural communities around the province, there was cynicism about the eventual outcome. The prevailing opinion was as follows: "The last time they sent someone was for the Equal Opportunity Programme; this time it is Social Policy Renewal. But there is no trust. People don't believe they'll be listened to." Another woman who had worked on various school committees for many years was upset with the complaints about powerlessness. "Why don't they just get together and go to Fredericton?" she asked. But the debates about education were not only focused on the administration and curriculum. A new addition to the school had engendered widespread dissension related to who would pay and whether or not a library was really needed. Even as the sod was about to be broken in 1999, the community was debating the value of a library (which they had never had) and was questioning the ability to pay for a new gymnasium (which was widely accepted as necessary). As quickly became apparent, however, the new library, which opened two years later, was probably the most significant new addition to Grand Manan in many years, perhaps since the paving of the road in 1949. The issue of a Christian school was only one part of a multidimensional debate around centralization, loss of local control, and lack of discipline among young people. At the same time, the church was seeking to assert leadership and to maintain its gradually loosening grip on the social mores of families and community groups.

Alcohol defined a whole gamut of changing attitudes and behaviours during the decade after 1995. Perhaps the most significant focus of social change, certainly the most obvious, on Grand Manan during this period, alcohol was a nexus of changing values, meanings, and economic structures, all evolving within the context of church dominance and patriarchal norms. As a visitor to the island several times between 1989 to 1993, I had not encountered the strict social mores that guided behaviours. But in the first week of 1995, as I began to define the parameters of my study, I learned some lessons in the hazards of ethnographic research. I had been staying at a bed-and-breakfast, depending on a traditional "snowball" technique of gradually meeting people and having a few interviews to provide a basic focus for my research. At the end of the week, I was invited to have dinner with the host couple, Alan and Hazel, both in their sixties and retired from a variety of fishing-related and Co-op store activities. I was touched by their kindness and, unthinkingly, immediately offered to "bring the wine." Hazel nodded and said: "That's fine; I won't have any but Alan would like it." Indeed, it was a lovely dinner, and Alan and I enjoyed our wine. It was almost a year later that I discovered that there had never been any liquor in the house, although Alan was known to have had "his share" in other locations. Moreover, their "house rules" were the norm for Grand Manan, with women mainly not drinking at all and men drinking mainly outside the home (in their trucks or at the camps but rarely on the boats). Slowly I began to realize the extent to which alcohol was at the centre of several important community structures. It defined family relations, gendered spaces, religious edicts, youth activities, relationships with outsiders, and social problems. Over the years, as Hazel became a good friend, I recounted the story to her, offered an apology, and expressed my appreciation for the generosity of spirit displayed by her response at the time. I include this mea culpa in my teaching of research methods at McGill University as a warning about the hidden dangers of unacknowledged assumptions.

News articles from 1975 reveal the persistence of several issues that, in 2006, continue to preoccupy the community, including schedules and charges for the ferry, government subsidies for island businesses (tourism facilities), and recreational facilities for young people. But perhaps the most contentious issue during the late 1970s was the question of a liquor licence. Until 1972, liquor could be ordered through a local freight agent and picked up the next day when the ferry had arrived. There was, however, also a well-organized after hours smuggling route from the United States, linked to the regular boat traffic provided by the fishers in those years. The establishment of a provincial liquor store on the island in 1972 effectively curtailed that trade, although the wry stories and chuckles regarding various episodes were still part of island lore twenty-five years later. Then, in 1975, an outsider from Toronto who had bought the Marathon Inn applied for a liquor licence. Community-wide opposition was intense and was led by religious leaders. As a columnist described the hearings for the application, the church involvement was significant: "A friend tells me he went to hear the application of Grand Manan's Marathon Hotel for a beer and wine licence. As he was about to leave the elevator he ran into a large number of people being led in prayer by a clergyman." Later, when the provincial government had granted the licence, the local MLA, James Tucker, was quoted as saying: "It's an atrocity. I can't understand why the board would turn the application down five times and grant it the sixth. I'm only taking the side of ninety-five percent of the island opposition." He pointed to the Island Ministerial Association, which had been able to raise about $10,000 to fund the court appeal. "There's nothing wrong with wanting to preserve a way of life", he said.

The granting of the licence represented a major challenge to island normative practices that officially prohibited drinking but acknowledged a serious problem with binge alcoholism away from home. The women, historically the guardians of family and community values and institutions, acted as gatekeepers against alcohol at home; even into the 1990s, women were never seen buying liquor at the provincial outlet. Drinking was a strongly male activity, not condoned for public social gatherings, although women were known to imbibe at private mixed parties. A woman in her sixties described the mixed parties she attended as being quite "liquid," although "after a while I stopped because it just got too much." Asked about the gender difference in drinking habits, one observer suggested: "Well, maybe because when the fishermen were away for long stretches, Sunday to Saturday, the women had to look after it all, the responsibility." She said they used to drink out at Dark Harbour at sunset, parked alongside the road. But "they'd never bother you. They weren't rowdy." In describing the role of the church in enforcing these community norms, one woman talked about the power of the churches to confer status as well as to reorient people's behaviour. She described one of the young men in his early forties who had a reputation for binge drinking and abusive behaviour towards his wife: "She had a hard time. But he's in the church now and has calmed down; there's less drinking." Another fisher was described as having been transformed when he "received Jesus into his life" and no longer engaged in binge drinking.

The Marathon Inn's liquor licence was revoked a year after it was granted but was reinstated in 1982 under the official provision that the inn was to serve only hotel guests, not the general island community. This restrictive situation continued until 1998, when another perceived outsider, this time with some "legitimate" island family relations, was able to obtain a licence for a pub. Nonetheless, their tenuous connections (as returnees after a long absence) tended to render them outsiders when their actions did not conform to island norms. The pub started as a video rental store in 1995, became a pizza parlour in 1996, and then a pool hall in 1997. The original liquor licence, granted in late 1997, had been for a dining room, "which was a mistake" according to one of the owners because, under provincial laws, dining rooms could not have pool tables. Even without the liquor permit a year earlier, the pool tables had been a draw for fifty to sixty people after 5:00 PM. But the new liquor licence proved to be a difficult adjustment for the owners. Asked about island opposition, one of the owners acknowledged that there had not been any overt hostility but that the village council had stalled on granting a permit.

The opposition was not from the churches, this owner said, but seemed rather to be the product of a general community resentment towards a new business started by people from-away. As well, she noted the strong sense of a prevailing morality regarding how people should behave. Indeed, one islander had ventured the opinion that the pub owners had "come here and tried to impose themselves instead of going with island ways." She admitted that there had been some rowdiness and that neighbours had complained, but she felt that, gradually, the pub might become accepted. While community response was certainly less vociferous than it had been fifteen years earlier, it continued to be characterized by the strongly negative opinions of community leaders. Although the churches were not involved in opposing the licence, a councillor voiced the opinion that people simply "won't go".– a sort of unspoken boycott resulting from community values, which deride public drinking. As the news spread around the community in the spring of 1998, others voiced similar opinions. "Who would dare go?" asked one islander. "The RCMP will be watching it like a hawk! The people here don't want change." But, in fact, change was occurring.

When a second pub and licensed restaurant opened in North Head in 2003 there was not a murmur of dissent, and even community leaders were seen to be patronizing the restaurant, if not the bar itself. Interestingly, had anyone predicted which village might be the first to accept a bar, the vote would probably have opted for North Head, which has always been perceived as the most sociable and least "god-fearing" village. Changes were also noted in the attitudes towards women buying liquor. Whereas in 1995 most island women had never been in a liquor store, a decade later it was not uncommon to see them there. Nonetheless, even in 2003 I was told a story that illuminates the tension that still existed for many people. A woman visiting relatives in Hamilton, Ontario, said that when she went into a Hamilton liquor outlet she looked around self-consciously, hoping that no one would recognize her. She told this story about herself, recognizing just how ingrained were Grand Manan behaviours and expectations.

As noted earlier, religion touches every aspect of daily life on Grand Manan. Another area affected by religious values is the fishery itself, although the "rules" have been more relaxed in some villages than others. In 1995, a Seal Cove lobster fisher noted that his village adhered to a strict "no-Sunday" rule. When I asked what would happen if a lobster fisher went out to check traps on a Sunday, the reply was, "Probably he would find them cut." While no such restrictions apply generally in North Head, at least one North Head resident fisher avoids work on Sunday. Describing the different religious values in the two villages, a fisher from North Head pointed out the practical problems associated with the no-Sunday rule. The weir fishers were expected to harvest fish on Sunday in order to ensure that the Connors' packing plants had fish on Monday morning. While Sunday harvests have been the norm in North Head since about 1990, in Seal Cove in 2000 they continued to observe the "Lord's Day," thus being unable to supply the fish plants on Mondays.

Despite a general assumption that churches promote community solidarity, the experience of some islanders has been the opposite. For some people, the churches of Seal Cove have been a negative factor, inhibiting social cohesion and community spirit. The two beautiful churches that dominate the skyline and village centre of Seal Cove are the centres for cliques, said one resident: "They split the community." According to her, not only did these cliques preclude exchanges between the two churches and within the community but they also attempted to exert control over people's behaviour. The example cited was a newly opened restaurant that was boycotted by some of the villagers because it was owned by a homosexual couple. "The church tries to control too much," she said indignantly, describing how someone had phoned her several years earlier to complain that she should not be having a picnic outdoors with her children on a Sunday. While she admitted that there seemed to be somewhat less rigidity in 2000, there continued to be a lack of cooperation between the churches, even with regard to youth groups. A couple of years later, this same woman was critical of the school for allowing a dance on Maundy Thursday. "Things are changing," she said. "There's less interest in and respect for the church." Another observer felt that the church was partly responsible for women's decision not to find paid employment when the family needed money. The church emphasizes a traditional role for women, she said. Nonetheless, while acknowledging the historical predominance of this attitude, another woman said that it was much less apparent in 2006.

Occasionally, divisions in the community became apparent in the way people ascribed reasons for specific events. One example of the tensions associated with strong religious affiliations and the loosening of rules and behaviours came in the aftermath of a camp in the woods that had been burned to the ground. Built by several teenagers with great pride, and used as a gathering place for their parties, its apparently intentional destruction caused the parents to place blame on a "religious group" that they knew thought of the camp "as evil." Rarely acknowledged but occasionally surfacing in conversations are religious divisions that can result in acts of revenge and hostility.

The responses to direct questions about the role of religion on the island reflected people's strong religious convictions as well as their reticence to openly challenge a prevailing ethos that extends back over one hundred years, even though there is a growing sense that some aspects of church power need to be resisted. Typically, young people felt that the church was too interventionist. When asked what change he would most like to see on the island, one male youth exclaimed: "Have the church butt out!" Paradoxically, when asked to rank various aspects of his life (education, family, church, friends, etc.), this same young man ranked church as number two, right behind family. A seventeen-year-old elaborated: "When something happens that's good, like the pool hall, the church puts its nose in it. We need places to go. It's okay to be Christian, but not with recreational activities!" Women generally defend the church's social prerogative to provide leadership on family issues (such as male dominance). Said one divorced woman: "Jesus said men were the head; and that's how it is." Then she went on to explain that the only way she could develop personally was to leave her marriage. The churches feel an ongoing sense of entitlement to protect social norms, as is illuminated by the sign outside the Grand Harbour Baptist Church, which, in July 2004, read: OBEDIENCE IS THE PATHWAY TO JOY.

For outsiders arriving on the island as new spouses, or, in a few cases, as young married couples, the churches offer one of the most effective ways of integrating into the community. According to one young family, their children especially benefited through their participation in youth groups at the Baptist Church, even though neither of the parents had been members prior to their arrival on the island. They were very positive about the role of the church in the lives of their two children, who, in 2004, were both entering high school. For the large group of Newfoundlanders who migrated to Grand Manan after 1996, joining one of the churches provided an important way of meeting islanders and gaining acceptance. While those Newfoundlanders who came only for the summers and lived in the trailer park, located a mile above Seal Cove, did not tend to participate in any of the churches, the approximately ten families that chose to move permanently to the island eventually joined one of them. Usually they attended one near their home, and within a year or two they might move to another church. In the case of one family, which settled on the Pentecostal Church, the choice was based on the welcome they received: "When we got to the island, we weren't Christians. The kids started going to Sunday School, and he [her husband] was going to a men's group. After I got saved, it changed my life totally. I had no group of friends, but after I was saved there were mentors for me. People would talk to me. I started to fit in, and to know the kids. So we could stay on Grand Manan." For this family the church provided an important community: "The church activities are our life now. We take baskets to seniors."

In another case, a Newfoundland family felt that Grand Manan services were too short, that people expected to be "in and out in an hour" and did not seem to enjoy staying to chat. They felt that there was a lack of real community feeling, unlike in their Newfoundland home where people would stay on for a couple of hours after the evening service. Nonetheless, the churches were aware of the struggles faced by the many young families that had migrated from Newfoundland to the island. During the first couple of summers, several groups collected food baskets as a way of easing the impoverishment of a number of the new arrivals. And, in 2004, the Seal Cove Baptist Church had a picnic for the Newfoundlanders, attracting about sixty people. As one of the Newfoundlanders commented: "It was nice. Some of the Grand Manan women talked with us. The kids had fun."

As well as integration into the community, the church provided credibility. During the autumn of 2001 there was a by-election for mayor following the abrupt resignation of the man who had been elected a year earlier. As possible candidates were being discussed in the community, two men seemed to be the most probable contenders. One of the men, several years earlier, had been mayor of one of the villages prior to amalgamation in 1996, and he was also a leader in his church. When I asked the other possible candidate about his plans, he responded hesitantly: "If Peter runs, I won't run. He's a God-fearing man." In many ways the church leaders are seen as the informal leaders of the community, and they tend to be the ones involved in organizations such as the Rotary Club and fundraising for the Boys and Girls Club. Nonetheless, the consensus about "leaders" and role models on Grand Manan is remarkably weak. It is as though the history of fishers as individualists and, until recently, the separateness of the villages have together inhibited the emergence of an effective leadership structure. While the church serves as a crucial institutional basis for the development of leadership, the distinctiveness of its many denominations serves to undermine the emergence of a cohesive structure.

For the members of many young families it is their religious faith that underpins their resolve to improve island choices for youth. For example, for one island family, the church had provided a crucial social and spiritual foundation with regard to the home schooling of their four children. I had interviewed both parents in 1996 at the beginning of my study, and I returned to them in 2005 to explore the changes they had experienced over the intervening years. Asked about what they saw as the most significant change on the island over the ten years, the husband, with his wife nodding in agreement, said that it was the deepening of their faith and the increasing role of the church in their lives. Both parents were involved in bible study groups.

The active intervention of the churches in the social values and norms of the island occasionally met with resistance. When several parents active in their churches expressed concern about the prevalence of drug use on aquaculture sites, they decided to do something. Together they sent a letter to each of the companies requesting that they adopt of a policy of zero-tolerance towards drug use. To their dismay, the company response was: "Impossible to enforce; it won't work." Less obviously, the church has often affected the work of social workers and addiction counsellors. While many church activists have supported the goals of volunteer organizations such as Community Reach, some, according to social work professionals, have been unwilling to adopt treatment strategies such as visualization. Some islanders saw such techniques as inappropriate and dangerous, describing them as the intrusion of the devil. In attempting to solve problems of substance abuse and childhood traumas, counsellors frequently felt constrained by the islanders' deeply embedded religious values.

Weddings and funerals are key markers of life transitions, reflecting personal and collective spiritual values. Until about 2003, weddings, to which the entire community was traditionally invited, were always performed in a church. In July 1998, a friend, whose husband had died suddenly three years earlier, was about to be married for the second time. Over the phone, just days before I was to arrive on the island and a couple of weeks before the planned wedding, she mentioned that she would be expecting me there. I demurred, noting that I had not received an invitation. Shocked, she quickly informed me that "no one on the island ever sends out invitations" as it is understood that everyone is invited. Planning for the catering, including the contributions of many friends and relatives, was a challenge in flexibility and good guesswork. Their decision to have the service in a garden because of their mixed affiliations (Baptist and Anglican, respectively) represented a break in island norms. Given the long-established traditions of church weddings and community-wide invitations, the changes that began to occur after 2000 were quite startling. Young couples dared to have their wedding vows exchanged not in a church but in a private home or garden. Then, in February 2004, one well-known couple was married with no prior announcement and in the presence of only a few friends and relatives at their new home. Their breaking with social conventions was the talk of the island for months afterwards. When one of their friends announced his engagement the following year, most people were not surprised when they learned that he would be married in his garden. Perhaps the most startling break with long-established wedding traditions was the at-home wedding, on Halloween of October 2005, when wedding guests were requested to attend in costume. It seemed the dam had burst and that everyone wanted the most innovative ceremony they could dream up.

Funerals too involved very specific norms of participation. As Pocius (1991) described for Calvert, Newfoundland, the importance of rituals related to death and dying are significant markers of belonging and affirmation of community. Within the memory of most Grand Mananers over forty, it had been common practice for the body to be laid out in the home parlour prior to burial. More recently, funerals have been organized by a commercial company, a family island business in which parents, son, and daughter are all trained morticians and are capable of dealing with the many complex arrangements that might be required. They carefully orchestrate every detail, from visitation to service to burial. As an outsider who happened to be on the island at the time of the death of a Grand Mananer, the changes in daily movements and activities across the island were immediately obvious: lines of cars parked outside the funeral home; conversations in the stores; vehicles coming and going from the house of the deceased; preparations at the church, with speaker systems at adjacent halls; and, on the day of the funeral, excursions planned to avoid passing the church during the service. Even the seating in the church is choreographed, with sections set aside for the various organizations in which the individual had been active, such as the Legion, the Rotary Club, or the church women. Unlike the trends in the rest of the country, where cremation is increasingly popular, on Grand Manan over 98 percent opt for traditional burial. When there was a cremation in the spring of 2005, people commented on it but expressed no surprise since the woman's son, who had predeceased her a decade earlier, had also been cremated, something that everyone remembered. An open casket was the norm (a closed casket was a disappointment), often incorporating characteristic touches that reflected the individual's personality. The detailed description I received of one funeral included the fact that there were over 450 people there and that the deceased "had on his baseball cap."

Even traditional funerals may offer possibilities for the creative solutions that characterize Grand Manan enterprises. In the late 1990s, the death of a well-known island character occurred at an inopportune time for the funeral home. As someone who loved to tell stories and had created many humorous lawn creations, such as painted sculptures and carefully constructed witches flying from telephone poles, it seemed fitting that his departure from this earth should have been marked by a similar sense of whimsy. According to the story, when he died very suddenly, the contract he had signed with the funeral home could not be honoured immediately. The solution devised by one of the employees and the local taxi driver was simple: the latter arrived with his car to pick up the body, placing Albert in the back seat with the air conditioning turned full on. Albert was kept comfortable until the funeral director was available to proceed with the funeral arrangements.

For denominations regarded with some suspicion, such as the Jehovah's Witnesses, funerals were regarded with a mixture of curiosity and dismissiveness. On the other hand, they could be an opportunity for improved communication and awareness. In the case of the death in 2003 of a man who had been a member of the "Go Ye's," a small sect with approximately five members, the reaction of islanders who attended the funeral could be described collectively as one of "amazed respect." Held in the high school gymnasium, the funeral, according to reports, was attended by over two hundred people, mainly from off-island. There were "lots of children, who all sat in the front as quiet as you can imagine," said one admiring Baptist observer. Similarly, the beautiful Baha'i service in October 2007 for Helen Charters, a well-known and respected artist who had moved to Grand Manan from Toronto in the mid-1970s, was attended by about sixty islanders who seemed moved by the outdoor location of the graveside service. The special relationship of the deceased artist with her adopted island had helped to blur the boundaries between "us" and "them" in ways that were beginning to be widely acknowledged throughout the community. In many ways, she represented new modes of belonging and the gradual change in community identities.

Following a death, particularly if the individual has had a colourful history on the island, the stories told in the kitchens and on the wharves can be most revealing. In the case of one man, whose public face had always been as a bon vivant, welcoming and conversational, his death revealed the reality of his domestic violence – a contradiction that came as a shock. Several people said that his wife had had "a very hard time." Describing the exploits of one woman who had been at the heart of church organizations and who was a leader in school, Legion, and Rotary activities, several people chuckled as they mentioned her "Wednesday man." Apparently it had been common knowledge, accepted without social censure, that on the Wednesdays when her husband met with his Rotary buddies, she would meet with a neighbour. Sick in hospital, so the stories elaborated, she was visited by both her husband and her neighbour, who sat on either side of her hospital bed, holding her hands. "These are my two favourite people in the world," she is reported to have said to one of the nurses, with great affection in her voice.

Another story was told by an eighty-one-year-old shut-in widow, whose contacts kept her well informed about all the weddings and funerals on the island. In the summer of 2003, a well known and colourful fisher who was in his eighties died, leaving a legacy of mixed memories. Apparently he had died on a Thursday, and even before the news had spread through the community, on Friday he had come to life again. It seems that he had a pacemaker and that it "just started up again! You've gotta be careful. Imagine. I told them if someone dies they should take the pacemaker out. It could start up again after you're buried! It's scary." A year later, the death of a young man of forty-one from cancer galvanized the entire island for several days. As one person observed, "It's such a big family, and almost everyone here has some connection to them. It'll be a long service. The visitation has gone on for two days." Funerals invariably cause islanders to review family histories and to comment on relationships. In trying to figure out someone's lineage, a conversation might recount a whole series of events that place the person, establish a maiden name, and finally conclude with: "Oh, that's who she is!"

Despite the continuing relevance of the churches on Grand Manan, there is evidence that their power is gradually declining. This is seen in both blunt comments and in decreasing attendance at regular services. A teacher whose own involvement in the church had been somewhat irregular felt that church influence was waning. In his grade 10 class in 2000, only two students were regular attendees, and, according to him, most "do not know the Bible. Most young people do not attend with their families after about fourteen years of age." This assessment is supported by other active members who have begun to explore the notion of merging churches of the same denomination. In response to declining attendance at regular Sunday services, Baptists and Anglicans are openly discussing the possibility of merging their respective multiple-point parish churches by closing one of their buildings.

Finally, religious faith permeates not only social relationships and community norms but also perceptions and interpretations of the natural environment. A poem attributed to Annie Johnson Flint and published in the *Atlantic Advocate* in 1959 draws together the strong bond between religion and landscape in the hearts of Grand Mananers:

"The Crag of the Cross"
Beside the bleak coast of the northland,
Where winds with the northland keep tryst,
Amid a wild welter of waters,
An Island looms out of the mist;

Forever the high tides of Fundy
Sweep past with a rush and a roar,
Forever the gulls cry their warning
When fog wreathes the desolate shore;

Above the grey billows the cliffs frown,
Above the grim cliffs bends the sky,
But clear against cliffside and heaven,
The Crag of the Cross rises high.

(Man) spendeth his life as a shadow,
And only his passing is sure;
But through all the ages unchanging,
The Cross and its Glory endure!
(cited in Ingersoll 1963, 70)

This description of the island's Southern Cross at Southwest Head did not predict this famous landmark's eventual erosion. Today it is only a remnant among the crags and pinnacles of Southwest Head.

Small spaces of identity, place, and community

To describe Grand Mananers' identities as evolving only in terms of villages and churches would be to misconstrue the social lives of those who rarely see the inside of a church or describe themselves as being from a particular village. Aside from the men's camps, which have been discussed at length, another marker of island identity is the adoption of nicknames, which apply to and, for the most

Multi-talented "Smiles" Green
Bottom photo: Nicole Wolfe

part, are used solely by the men. Interestingly, it seems that the adoption of nick-names is not being passed on as it is invariably only the older men (over forty) who have them. Another tradition may be passing as well. The multi-talented Smiles (Oliver) Green comes from a very large family, all of whom had nick-names such as "Snooks" and "Link." Other nicknames well known to islanders include "Cowboy," "Snap," "Bud," and "Rusty." In fact, so embedded were these names that, for some islanders, the problem was to find each other in the tele-phone book when they did not know each other's "real" names.

For people from-away and summer residents there are much more limited options for gathering places where casual encounters allow for exchange of news, invitations, and a sense of shared belonging. Although the tennis courts provide some of these benefits, the two most important venues are the North Head Bakery and the North Head Saturday morning market. It is significant that both of the seasonal markets, catering as much to tourists as to local resi-dents, are located in North Head. Both of them are open for limited seasons and therefore cannot fulfill the integrative role of a truly "public space" for most island residents. However, at the Saturday markets from 1 July to Labour Day

The Whistle at sunset

summer residents can mix with islanders (and tourists, although they do not really count) and feel that they are participating in island activities and enjoying some of the local crafts and garden produce. But it is the bakery, open from about mid-May to mid-October, that probably offers the greatest opportunity for casual encounters and is an important drop-in place for morning coffee and croissants.

One of the most interesting public spaces, and perhaps the only one where everyone can meet equitably throughout most of the year, is the Whistle, just beyond the Long Eddy lighthouse at the northeastern tip of the island. It is a space that is significantly demarcated by a long bench built by an enterprising and convivial man of many skills and by a fence that protects people from falling down a steep stream fissure. At the end of a road and atop a three-hundred-foot cliff, the outlook towards the American shore, which also overlooks the Gully weir, offers spectacular views of sunsets. A steep road continues down to the rock beach, providing access for dulse harvesters and weir fishers to launch dories that are tethered to large boulders. Moreover, because of its location at the tip of the island, tidal currents converge here, with a mesmerizing confluence of ripples, waves, feeding gulls, osprey, porpoises, and whales. It is an enthralling lookout for fishers to check on weather patterns, herring runs, the activities of other fishers (at their weirs or tending lobster traps), and for generally hanging out. Male and female, young and old, islander and tourist can meet, chat, and exchange news and views in an informal way not available anywhere else on the island. At noon tourists arrive with their lunches, bird books, and binoculars, hoping to see whales or birds that they can add to their lists. For the men who arrive in their trucks, often with cold beer in a cooler in the back, the apparently thoughtless parking strategies of cars from Maine or Ontario may provoke comment. But usually they enjoy the innocent questions about whales and fishing, and may even engage in some wicked exaggerations that the unsuspecting tourists take as truth.

But it is during the evening hours, after supper until the sun sets, that the locals gather (mainly a group of twenty- to thirty-years-olds) and the Whistle is at its most lively. Older men (mainly) entertain with stories of past exploits, women flirt, tourists mostly listen, and someone always has a camera to capture the last rays of the red sunset as it disappears along the Eastport shore. The only sporadic intervention may be the unwelcome checks of the RCMP, who may drop by to ensure that drinking is subdued if not entirely curtailed, depending on the officer in charge and recent events on the island. Until the opening of the pub on Ingalls Head Road in 1998, there had been no other public space for sharing a beer, and the Whistle was for many years a controversial space, beloved by some locals and tourists but feared by parents of teenagers and vilified by many in the churches. The bench was built by a local beer-drinking, socializing islander who, over at least twenty years, has continued to repair it, keep the grass mowed, and the garbage bin emptied – a voluntary contribution to a sense of community and a welcome for visitors.

As discussed earlier, for women, the loss of the smoke sheds represented not only a loss of employment but also the loss of a place for social interaction that nurtured identities and a sense of community. For the young women who arrived as spouses in the late 1970s and early 1980s there were opportunities to integrate into the community through the informally organized herring stands, where they could string or bone herring in between family duties. They could be seen to be participating in the daily rhythms of the island in ways that islanders would appreciate. Today, young spouses are more likely to be teachers or nurses, with their contacts in the community defined by their jobs, which are not seen as intrinsically "of Grand Manan." For them, the challenges of integrating and being accepted as "islanders" are ongoing. The lack of spaces for informal socializing inhibit these women's smooth entry into new cultures of being, the topic of the next chapter.

9

Habitus and Cultural Change

The geographic and historic isolation of Grand Manan has been alluded to throughout the preceding pages. It is now time to examine the meaning of this isolation with regard to socio-cultural formation, which ultimately informs the possibility of resilience and the capacity for adaptation. Central to any discussion of isolation is the obverse notion of connectivity and networks of evolving relationships. Grand Manan has had an uneasy relationship with the mainland and with new arrivals on the island. Crucially, the island boundary is not only a physical, geographic reality but also a socio-cultural one. The history of very low rates of in-migration has contributed to a distinctive culture that is resistant to change, effectively inhibiting the acceptance of new arrivals. Discussion to this point has focused on changes of landscape, economic activities, and religious adherence, all of which suggest long-term cultural change. Contacts between cultures are also important when thinking about communities in the midst of change, and the concept of "habitus" helps us to encapsulate the fluidity and complexity of cultural formation.

Habitus is a concept that mediates between objective and subjective realities in that it "enables an intelligible and necessary relationship to be established between practices and a situation, the meaning of which is produced ... through categories of perception and appreciation that are themselves produced by an observable social condition" (Bourdieu 1984, 101). One of the critiques of concepts such as "society" and "culture" is that we objectify them in order to isolate and understand their characteristics. Moreover, as Bottomley (1992, 12) points out, "the truth of social interaction is never entirely in the interaction as observed," and we need to recognize the structural constraints on perceptions. Bourdieu's concept of habitus represents a mediating stance between social positions and practices in which the fluidity of meanings can be seen as social products. The processes of culture are understood as objective relations of power that are inherently political and that arise out of structures that constitute a particular social environment.

Habitus is the embodiment of history exemplified in a sense of place that is "literally embodied, i.e. written on the body, in language and in particular ways of being-in-the-world" (Bottomley 1992, 13). In considering the in-migration of Newfoundlanders, especially after 1999, for example, habitus offers us a way of understanding the merging of two distinct cultures. The importance of habitus to an understanding of the role of migration in transforming a community relates to the way in which it illuminates the relations of contact, producing new positions in social and historical space and incorporating new categories of perception and appreciation rather than separating the subjective and objective experiences of socio-cultural change. Implying a dialectical relationship between structured circumstances and people's actions and perceptions, habitus demands that we go beyond our revered dualisms. As Bottomley points out in her discussion of the concept, habitus "manifests itself in practice, in action and movement, in the way one orients oneself in relation to specific social fields" (123). In the context of recent debates in Canada about the meaning of multiculturalism and the need to accommodate distinctive religious and ethnic realities, habitus can be seen as the negotiated confluence of cultural traits, ideas, and values that are neither merged nor distinct but always fluid. Whether its value can be appreciated in this controversial and immediate context is still open for consideration.

Directly linked to identity and self, habitus reflects the *longue durée* approach to history, which emphasizes long-term historical structure over events. In situations of in-migration, even when the objective criteria of employment, income, educational, and environmental histories might suggest comparable worldviews, the reality of experience and sense-of-place, and the relation between practices and situation (habitus), may produce very different meanings and categories of perceptions and understandings. New migrants may generate processes of change through which the habitus of communities becomes altered in ways that are biased by pre-existing conditions. The ways in which new relationships evolve between social structure and human agency is "different from place to place and depends crucially on the particular arena of encounter" (Livingstone 1992, 357).

New Arrivals: "From-Aways" and Summer Residents

For Grand Manan, isolated by space and time from the mainland and from evolving Canadian values and social conventions, new arrivals have always, at the very least, provoked discussion and, occasionally, debate and controversy. When a sect arrived in the mid-1980s seeking to establish a religious retreat and small business by starting up a lobster pound, the local population was aghast. After barely three years the sect quietly left, but not without a trail of stories that islanders continue to relate. While for the most part Grand Manan has woven those who have stayed, and whose children have attended local schools, into its fabric, the divide between "insiders" and "outsiders" in island social relations has been a fundamental characteristic of social networks, institutional cleavages, and participation in activities on the island. Even after decades, spouses who have married onto the island often retain a sense of being "Other." The arrival of a new spouse engenders both curiosity and, often, suspicion, especially if she is an attractive young woman. For wives who are newcomers, the stresses of trying to integrate into island life have been largely unappreciated by their island husbands. The women who have arrived from-away never feel quite accepted, despite their working in the fish sheds and picking dulse. The many unacknowledged ways in which the insider/outsider divide underpins social relations can extend even to perceptions of illegal behaviour. Several people suggested that a factor in the altercations over drug use in July 2006, when an alleged crack house was burned (see below), could have been related to the house owner's being from-away. As one observer who had lived on the island for twenty years remarked, "There's islanders and then there's the rest," and the targeted house was owned by a man from-away (*La Presse*, 30 July 2006). His comment, reported in a Montreal newspaper, reflected the way in which attitudes towards from-aways permeates all relationships on the island.

Certainly an important issue for people who migrate to the island is that they have not had the lived experiences that create the "nebulous threads" of island culture. It is not possible for them to fully understand the "subterranean level of meaning" that would allow them to truly belong (Cohen 1982, 11). But it is also important to recognize that migrants are not a homogeneous group. Until the early 1990s, there had been three main groups of incomers who defined the patterns of in-migration and who had distinctive relationships with the island: single women, retired couples, and spouses who married onto the island. A fourth group of from-aways who consider themselves islanders are the summer residents, who come from as far away as California and who have houses on the island to which generations of their families have returned each summer for decades. For some of these people, many of whom are from the United States, their sense of being a "native" is expressed in both subtle and not-so-subtle ways. For example, when one summer resident returned to the island and went to church the next day, she announced her return by standing at the church door as people arrived, acknowledging all their welcomes with a huge smile. In other cases, people affirm their sense of belonging by participating in the Grand Manan Historical Society or by hosting an annual occasion (such as a barn dance). Some, like Peter Cunningham, a professional photographer who spends about a month on the island each year, have tried to lay explicit claim to their islander status. In establishing his claim to a lineage stretching back seventy years, he wrote an article in the *Island Times* entitled "What Does the Name Peter Cunningham Mean to You?" He pointed out that his connection "is a bond forged nearly seventy years ago when the islands within the archipelago became the stomping ground to Peter Cunningham's 16 year old father Robert ... The Cunningham family remain loyal to their seasonal trips back to Grand Manan, a place that has shaped them as people, given direction to their careers, and continues to provide a haven from the fast-paced lives their family has led off-island" (*Island Times*, 1 July 2006). Generally however, most native islanders do not know summer residents (except those who happen to live close by).

In a column in a local newspaper, a slightly different way of categorizing islanders was proposed, tongue-in-cheek, by a from-away resident and former journalist. Her categories included, in order of status: native islanders, "choosies," summer residents, former islanders who have returned, and tourists (Joan Barberis, "Grand Manan Who's Who," *Telegraph Journal*, 30 July 1994). In her humorous analysis, she argued that people from-away who have chosen the island as their home should be ranked second because, unlike the "disloyal" islanders who left and returned later, they have proven their commitment to the island.

Some have quietly lived on the island from May through to September, aware of the various challenges confronting local residents but never openly participating in their organizations or politics. Nevertheless, significant financial donations to the Boys and Girls Club and to provide computers for the school are two of the ways that summer residents have found to contribute to their adopted community. These gifts would never be known about in the wider Grand Manan community. For the most part, their participation is through tennis, clearing trails, church activities, and supporting the museum and historical society. There is, however, one significant area in which they have contributed to developments on Grand Manan: tourism. As I describe later in this section, most of the inns and many of the stores, restaurants, and whale-watching businesses aimed at tourists have been the initiatives of people from-away.

In the 1990s, three changes occurred in the typical patterns of in-migration, and these involved three separate groups. The first group consisted of young teachers who came for a year or two and then decided to stay. The second group, which was more important in terms of numbers, consisted of families from Newfoundland. Prior to 1995, a few Newfoundlanders moved to Grand Manan, encouraged by an active recruitment drive on the part of the owner of a fish brokerage company. After 1995, and especially after 1999, there was a significant migration from Newfoundland of both seasonal labourers and those who wanted to stay, and this created a new dimension that was to have long-term impacts on the social structure of the island. The third group consisted of Aboriginal people (the Maliseet) from three inland reserves in New Brunswick. Then, after 2000, and clearly associated with rapidly escalating real estate prices on the mainland, a fourth group was added to the changing pattern of in-migration. This group consisted of increasing numbers of retired couples, from both the United States and urban Canada, who bought homes and renovated them or built large new homes in prime locations. Most of the couples moving to the island after 2000 were in their sixties. Several had artistic backgrounds and almost immediately opened studios out of which they sold their work, including glass-bead jewellery, quilts, and paintings. These initiatives mirrored the earlier trends of incoming couples, who invested in tourist-related ventures such as small inns or cottages.

Prior to the 1990s, the first two groups – single women and retired couples, described as the "choosies" – had developed an uneasy alliance with native islanders. Even after many years as permanent residents, many of them tended not to mix with native islanders except through formal activities within the church, Rotary Club, or curling. These from-away groups often experienced tensions associated with volunteer activities, especially when they tried to introduce new ideas or to assume leadership positions. For most of them an adjustment period involved learning not to put themselves forward but, instead, to adopt a supportive role. For the single women and retired couples from the mainland, most of whom were in the same age group (late fifties and sixties), there were many shared interests across a broad spectrum of social activities (e.g., bird watching, gardening, music, and book clubs). "Somehow we just seem to gravitate to each other," one explained to me. As Bourdieu (1984, 170) describes the conditions of existence, there is a distinctive habitus that derives from "not only a structuring structure, which organizes practices and perceptions of practices, but also a structured structure: the principle of division into logical classes which organizes the perception of the social world is itself the product of internalization of the division into social classes."

Among these incomers have been some very interesting individuals whose quiet contributions to the island have provided significant support to evolving structures. While islanders tend to retain the roles as president or chair of various organizations, people from-away have served on boards of the museum, the Rotary Club, the Chamber of Commerce and Tourism Association, the Boys and Girls Club, and Community Reach. Their past experiences as professionals (e.g., teachers, business executives, or architects) or as volunteers have enabled them to contribute important skills to various organizations. While often unheralded and sometimes resisted, they have nevertheless been important for the ongoing success of many island organizations that have benefited from their reliable commitment and expertise. Their integration has also been encouraged through informal activities such as tennis and the maintenance of hiking trails. While it has been people from-away who have organized trail maintenance (including providing signs) and have written an excellent guide book (including maps), increasing numbers of Grand Manan residents have begun to take an interest in the trails. The Chamber of Commerce and Tourism Association provides another important vehicle through which people from-away can contribute to, and be seen to be enjoying, the island without interfering in ways that islanders find offensive.

As has been mentioned, perhaps the most famous from-away summer resident was Willa Cather, who spent all her summers on Grand Manan between 1922 and 1942, staying at what is today Whale Cove Cottages. For her, the island "seemed the only foothold left on earth" (Lewis, cited in Stich 1989, 153). Her influence has bestowed upon the cottages and the island a sense of peace and connection to nature that has continued to attract visitors from across North America. Cather's fame as an author, and the many women who accompanied her, created a kind of mystique that lingers to this day. It would be a mistake to ignore the importance of Whale Cove Cottages and other historic tourist accommodations, which attract seasonal visitors who would return year after year. At the time, the decades-long association of Cather and her entourage with the island offered virtually the only ongoing, sustained connection that islanders had with people from-away. The cottages where they stayed, along with the Marathon Inn, the Shorecrest Inn, and, later, the Compass Rose Inn and Aristotle's Lantern, were all owned and operated by people from-away. And their clientele consisted of summer visitors.

For some of the owners of these inns the island was a summer retreat or vacation spot that ultimately captivated them. They looked for an opportunity to stay by investing in a tourist business. In contrast to the bed-and-breakfasts on Grand Manan, almost all of which have been operated by island women, the inns require substantial financial outlays as well as management skills and dining-room expertise, which have not been easily developed within Grand Manan culture. The hard work and creative talents of entrepreneurial women such as Laura Buckley, Linda L'Aventure, Cecelia Bowden, and Helen Charters have resulted in significant tourist facilities that were crucial to encouraging the development of such activities as whale watching and a kayaking business. As well, by attracting bird watchers and artists, and encouraging art classes, the inn operators have indirectly supported other year-round businesses (e.g., variety stores and restaurants). The history of the Marathon Inn, the stories of its various owners and operators, and the variety of ways in which it has contributed to the island is worthy of a book on its own. Photographs of the inn that date back to the turn of the twentieth century, when tourism was just beginning, hang in the museum and show its dominant location in North Head, overlooking the harbour. From school classes to summer camps to Elderhostel, the Marathon Inn has hosted visits by many educational groups. For many visitors this first introduction to Grand Manan is what brings them back as individuals and families in later years. The from-away owners and operators of the Marathon Inn have offered a key entrée to Grand Manan that helps sustain a vibrant, if relatively small, tourist industry.

Others who moved to the island did so as permanent residents. They were usually in their sixties and were recent retirees. Despite their low profile among islanders, these people from-away have impressive and interesting backgrounds. When Joan Barberis decided to move to Grand Manan from Ontario in the mid-1980s, she was a recently retired CBC journalist. Upon the death of her mother in Toronto, and with no other close family in Canada, she decided to move to Grand Manan, where her "pension would go much further." Joan had been a WREN during the Second World War and had been stationed in Nova Scotia as a radio communications officer. In 1961, she took a year's sabbatical from her civilian job as a journalist and travelled alone to Hong Kong, Singapore, Mainland China, and across to Moscow on the Trans-Siberian Railroad. Given her obvious independence and willingness to take risks, it was no surprise that she eventually chose to move to an island where she knew nobody. At the same time, however, she was a canny realist. In order to ease her integration into the community, Joan volunteered, over a period of two years, to pick up garbage for the elderly and shut-ins at a time when garbage disposal was still the responsibility of individuals. As the "garbage lady," she quickly became known by many islanders. Her renowned sense of humour always served her well. Following an operation for cataracts that had forced her to drive more than three hundred miles to Halifax with a friend, she went to the doctor's office for a follow-up. Upon walking into the office and seeing about ten people ranged around the room, all with the same eye bandages, Joan laughed and exclaimed: "Oh my, it seems I've come to the meeting of some secret society!" She was often the grateful recipient of gifts of lobsters and scallops from several fishers, and she was well known in North Head for her daily walks to the bakery for her coffee, croissant, and morning chat with whoever happened to come by. Living alone and gleefully supervising the splitting and stacking of her winter's wood supply each fall, she tended her wood stove and looked after a small vegetable garden until her stroke in March 2007. Mercifully, she died quietly a few days later.

Another from-away island resident, Helen, who had early attachments to Grand Manan through her mother's summer tourism business during the 1940s and 1950s, came to the island from Toronto as a divorcee, with her four children. They established their home on Grand Manan, and even four decades later, in 2005, each of them had stories to tell of their common struggles for acceptance, although the details of their experiences varied according to their various activities. For Helen, who had arrived in New Brunswick as a single mother without a high school diploma, the challenges to provide for her children were daunting to say the least. But she eventually obtained a university degree, taught school, and then began a tourist business as the owner-operator of Aristotle's Lantern. In the meantime, she had studied art and had become a well-known painter and print maker. Very much her own person, not relying upon islander affirmation or the company of people from-away, in many ways she exemplified the indomitable and independent spirit that characterized these incomers to Grand Manan. Using the old inn as her home and studio, she took in a few guests and offered tea every afternoon during the summer, while developing her artistic abilities with oils, watercolours, and prints as she depicted the world around her. Whales and fishers, smokehouses and lighthouses: these were the subjects of her work. Her artwork can be found in collections throughout North America and Europe. In describing the problems of having to integrate into Grand Manan life, she said that her greatest concern was for her children: "Learning fishing when it had never been in the family was hard. But Ben was lucky. There were some wonderful men on White Head who looked after him and essentially taught him how to be a fisherman." A measure of how broadly she had been accepted by islanders could be seen in their large and respectful attendance at her graveside funeral service in October 2007. Today her son Ben, who is in his fifties, is a successful lobster fisher. Debbie, his from-away wife, is a full-time nurse assistant to the local doctor. Although both Ben and Debbie are from-away, their quiet commitment to working within traditional island work structures has helped them to feel like, and to be seen as, Grand Mananers. Nevertheless, cultural and social dissonances are always evident. For their children, the challenges have been somewhat more hazardous (see chapter 10).

The opening of other art studios and galleries by from-away migrants has occurred in recent years all around the island but is notably concentrated towards the north end, which is more accessible to visitors. Openings include galleries in Grand Harbour (2000–06), Castalia (2006), North Head (2006), and Seal Cove (2006). Then the opening of the Grand Manan Historical Society's art gallery (2006) in Castalia, by islanders themselves, seemed to suggest a new sensibility towards cultural expression might be evolving on the island.

Those who have moved to the island as retirees may not be seen as "islanders," and they will never deny the challenges of being accepted on Grand Manan, but their commitment to island life has provided for a mutually interesting relationship between them and the natives. Among many who have arrived as permanent residents are two architects, a filmmaker, a teacher, a school principal, a glass artist, an English professor, a former naval officer (who was awarded the French Legion of Honour for his D-Day landing on the beaches of Normandy), and an Australian who had worked for the Canadian Embassy in Washington, DC. Their backgrounds are diverse, but all of these people arrived knowing that they would always be seen as "Other." Talking to one couple from-away about their plans as they age and recognize their increasing dependence on the health system, I had the sense that they had contemplated moving back to Ontario to be closer to their grown children. But in the end they acknowledged that, despite knowing they would never be seen as islanders, their twenty-five years on the island had given them a sense of belonging that would be difficult to achieve in Ottawa or Hamilton. "We're content here," they said. In one sense, the collective impact of the retirees who have moved to Grand Manan has been relatively minor. And yet, when one looks closely at their involvement, which is often hidden from view, there are many areas in which their talents proved significant. As treasurer of the museum, initiator of the business centre, members of the Rotary Club or the Boys and Girls Club, they were able to make a small contribution to island initiatives. As pointed out earlier, some positions would never go to those from-away (such as chair of fundraising for a major project). When asked to accept such roles, they usually understood that to do so would in fact jeopardize the success of the project. Islanders acknowledge that they are happy to welcome these newcomers but add that they "shouldn't try to change us."

Although the couples who have moved to Grand Manan as permanent residents were mainly retirees, there were also a few young couples. But the latter's resolve to settle on the island was sorely tested by daily struggles to balance the many issues related to schooling and the relative isolation of island life. For them, the demands of raising children and assuring steady incomes were difficult because they had to be involved with the school and to adapt to challenging youth cultures. Then the events that occurred between 1999 and 2001 (five drug-related deaths, discussed in chapter 10) threatened their ability to provide their children with secure childhoods. One of the from-away couples left the island within a year of those events, and a second couple moved away from the island for the first school year (2004–05) and then had their son stay with friends on the mainland for the second school year. The problems and choices that confronted these families are discussed in chapter 10.

For those who have married onto the island, experience, social relations, and perceptions are quite different. In the vast majority of cases, it was the men who returned with off-island spouses, although in more recent years these "mixed marriages" are taking place between male islanders and female teachers or nurses who had come to the island for a brief period of work and then decided to marry and stay. Grand Manan women who go away to college or university tend to marry away and not to return to the island, except for family reunions and special holidays. Lineage and family history are crucial factors for one's ability to integrate into the community, even after decades of residence. Even after twenty years of living on the island, one forty-five-year-old woman continued to feel the effects of being from-away. Recalling vividly the experience of her arrival, she said: "I took to it like a duck to water; but then there was a complete turnaround ... I sensed a betrayal by the people we were socializing with. They were drawing conclusions thinking they knew my business." The insularity of island culture, naturally protected by time and space, is a significant characteristic of its identity.

One young woman, a marine biologist, who married a fisher and stayed on the island contributed to the possibility of establishing a robust tourist economy. Her arrival in the mid-1980s coincided with a Toronto couple's recent purchase of the Marathon Inn. Together they conceived of the Whale Research Station, which today supports the research of several universities that are studying seabirds and marine mammals, and established a whale-watching boat. It seemed that there was a growing interest in whale watching. As mentioned in chapter 6, in the early 1980s, a lobster fisher from Seal Cove established the first whale-watching tour business on Grand Manan, taking advantage of the closure of lobster season at the end of June, his boat, and his knowledge of the ocean. Preston Wilcox established Sea Watch out of Seal Cove, a business that he has passed on to his son Peter, who takes tours out to the puffin colony at Machias Seal Island through to mid-July and then moves into whale watching until the end of August.

Over the twelve years of this study, there seems to have been a marked change in the reception offered to incomers. Those who have arrived since 1995 have experienced a somewhat more welcoming community, and this is associated with the multiple new connections provided through the growing aquaculture industry and links to mainland institutions. For younger couples in particular, the transition seems to be easier due to their participation in the working life of the island. While most of the in-migration, even after 1995, has consisted of retired couples, a few younger couples moved to the island beginning in the early 1990s. As has been mentioned, their decision to stay after the events that occurred between 1999 and 2001 (see chapter 10) depended to a large extent on whether or not they had children (nine or ten years old) about to enter the middle school. The significant exodus of young professionals in 2001 and 2002 marked a momentous and unfortunate loss to the island. In fact, while most of these people tended to leave quietly without providing reasons, the community recognized the losses that were being incurred, even if only because they were losing members of their own families.

The 2006 Census reveals the significance of this out-migration. The population of Grand Manan had decreased from 2,610 in 2001 to fewer than 2,500 in 2006 – a decline of 5.7 percent. This marked the first decline in population since the turn of the century and is suggestive of a new reality. One of the major factors in island changes, as described in chapter 6, has been the transformation from the structure of the traditional fishery to the structure of the global agro-food industry. When combined with the escalating costs of licences and the degradation of the marine Commons, this implies major changes in the nature of available work, with the result that young people may not have the option of entering the multi-skilled, entrepreneurial enterprises of their fathers. Unless new entrepreneurial initiatives can be developed, their choices will be severely constrained.

On the other hand, there have been some hopeful signs since 2002, notably the increased number of young teachers from-away who have energized the school and brought new ideas for activities and recreational programs. Introducing new activities, with the possibility of redirecting the goals of young people and gradually transforming the cultural habitus, could be a long-term endeavour for everyone.

The two major influxes of newcomers, which occurred in the decade between 1995 and 2005, were of an entirely different magnitude than had ever been experienced on Grand Manan. When Ron Benson first encouraged two Newfoundland families to move to the island in 1991 just prior to the declaration of the Newfoundland cod moratorium, little did he appreciate the extent to which his initiative would affect hundreds of lives. For both Newfoundlanders who eventually came to the island, especially after 1998, and for islanders themselves, the meeting of two cultures was both difficult and life-changing. The second influx, much smaller and less integrated into island structures than the first, consisted of Aboriginal people from three mainland Maliseet reserves in 2000. This second group migration had very different implications than did the first, and these were mainly associated with confrontation and outsider status (see chapter 4). Rather than actively contributing to a changing habitus, the Aboriginal presence opened up debate about access to the Commons and government involvement in resource allocation and regulation.

The Newfoundlanders

In the period between 1991 and 1996, approximately three Newfoundland families had migrated to the island, all related by marriage, all from the same region of Newfoundland (centred in Comfort Cove and extending to the Twillingate area), and usually undergoing a series of trial periods of residency prior to making a final commitment. In conversations with islanders it was apparent, however, that even these few families were a new phenomenon for Grand Manan. Whereas most in-migration to Grand Manan had historically been in the form of individuals, the Newfoundlanders were arriving as young families. Nevertheless, despite three Newfoundland children being enrolled in the school and despite the Newfoundlanders' involvement in a variety of island activities, their impact upon the community was generally muted prior to 1999. While the discussion that follows focuses on Grand Manan, it is important to recognize that this is only part of a much larger story of the out-migration of Newfoundlanders. Having major implications for Canadian policy at all levels, the trajectory of Newfoundland migrations and the experiences of Newfoundlanders need to be explored and communicated. Census data have confirmed that the trend towards an aging population in Newfoundland and Labrador is related to the out-migration of its young and middle-aged families. However, data alone do not acknowledge the terrible toll on individual families that these moves involve. The narratives that follow offer glimpses into the struggles and courage of a few families who moved to Grand Manan in search of work and better lives for their children.

In 1999, there was a dramatic change in the number of Newfoundland arrivals on Grand Manan. As a result of a recruiting trip by the manager of Connors fish plant to the Comfort Cove area of Newfoundland, approximately fifteen families moved to the island. His success in attracting so many families allowed him to introduce a second shift at the Seal Cove plant. Of those who came during this initial recruitment period some stayed through the winter, although most returned to their homes in Newfoundland when the plant closed in late October. In 2000, the Connors manager David Green again went to recruit more workers, and this resulted in the arrival of approximately thirty families between May and July of that year. This phase of the migration represented a new pattern insofar as most of the migrants planned to return to Newfoundland, whereas in the period between 1995 and 1998, the four families who moved planned to stay as permanent residents. With Connors' active recruitment in 1999, the pattern of migration became more complex, with most families electing to return to Newfoundland each winter, usually after the plant closed in late October. Some, however, stayed, and for those with children the adjustment proved to be particularly difficult. This is reflected in the decision of about ten of these families, who had stayed throughout 2001, to return to Newfoundland in the autumn of that year. A few families would send children home in September with relatives, while the parents stayed until the Connors plant closed; and a couple of others stayed for the first two weeks of lobster season (which opens during the second week in November) in order to make the money to pay for the trip home. In other words, there were variable patterns of migration behaviours, largely determined by family structure and the ages of children.

The school records of children registered for classes at the local school indicate that approximately eight families elected to stay through the winter of 1999–2000. While some of these considered themselves as permanent and expressly described their situation as involving finding a new home, others continued to vacillate. As one woman said, "We're trying it this year, but if the school doesn't work out we'll leave in October next fall." Their sense of feeling that they were living between communities, and that they were struggling with choosing between economic security and social networks, was palpable in almost all the interviews, particularly when children and school were involved. For most, incomes and job experience had been tied to the fishery, though a few had worked in the woods for Abitibi-Consolidated. For both men and women, the common problem was an assured income, at least during the summer months, that would allow them enough Employment Insurance to keep them off welfare.

With regard to histories of the fisheries, there were important differences between the experiences of Newfoundlanders and Grand Mananers. Whereas the migrating families had all been affected by the closure of the cod fishery, and had only limited reprieves in the more recent crab fishery, fishers on Grand Manan have been largely protected by the diversity of fish species and capitalization. In contrast to the devastation of the groundfish stocks around Newfoundland, some groundfish is still being caught around Grand Manan, albeit in much smaller amounts than formerly. More important is the variety of other fisheries on Grand Manan – notably lobster and scallop – both of which are lucrative and, apparently, in reasonably good health. Grand Mananers have quite simply not had to contemplate the prospect of having no income or the possibility of having to migrate in order to find work.

While most of the Newfoundland arrivals worked in the Connors plant, and couples would split their times so that there was always someone at the trailer with the children, others worked on aquaculture sites or in the tourism industry cleaning cottages. The work at the plant was the most desirable because the pay and working conditions were most favourable. On the scissors line (which was piece-work pay) women could earn between twelve dollars and eighteen dollars per hour, depending upon their adeptness, whereas cleaning cottages paid only nine dollars per hour, or about $270 per week. But, according to one woman, it was enough to get her stamps. People back in Newfoundland asked them what they did with all their money. Her husband laughed, exclaiming that he "wouldn't be here if I was rich!" Many Newfoundlanders took advantage of the niche activities to supplement their incomes. While only a few harvested dulse because "Grand Mananers don't like it much when we do," many picked periwinkles. A bucket could hold fifty pounds, easily picked on a tide. Selling for between seventy-nine cents and eighty-nine cents per pound, periwinkles enabled the women to make about twenty dollars per hour, or sixty dollars on a tide.

Arriving on the island, Newfoundlanders found much that was familiar; but there were also significant differences. Both the process of adapting to the new community and the responses of islanders to the new arrivals created a reflexive relationship that ultimately changed the meaning of community on Grand Manan. This relationship was informed by (1) the differences in historical experiences, as tied to the fishery and other resource-based activities; (2) the similarities in fishery experiences and educational and income levels; and (3) the strong sense of place and history rooted in family lineage that defines Grand Manan culture.

The community of Grand Manan's response to the Newfoundland arrivals was significantly different from that accorded to most new arrivals. Grand Mananers felt a degree of kinship with the Newfoundlanders in that they had both experienced the vagaries and struggles associated with the wild fishery. Whereas an islander might complain that people from-away are "always trying to tell us how to do it," their comments about the Newfoundlanders were more likely to focus on their work habits: "Oh, they're mighty workers!" However, given their own transition to an economy that was increasingly dependent on aquaculture and hourly paid labour, and the changing distribution of wealth, islanders had their own difficulties with which to contend. The traditional strengths of the nuclear family and strong religious affiliations were being undermined, creating tensions throughout island social structures.

For Newfoundlanders confronting this turbulent situation, there were serious doubts about the validity of shared worldviews. Even though Grand Mananers have generally been welcoming to Newfoundland migrants, they soon realized that their individual and collective identities, arising out of the long-term historical structures of history (*longue durée*), a strong sense of family, the central importance of religion, and common experiences in the fishery, were not necessarily enough to assure a stable basis for constructing new communities of belonging. As well, there were distinctions within the Newfoundland migrants that created their own basis for different degrees of integration.

Three factors defined these differences: the first was the decision to rent a home on the island rather than to stay in the trailer park; the second was the decision to buy a home, with plans to stay permanently on Grand Manan; the third was family status (i.e., whether or not there were children). For those who migrated seasonally the work on Grand Manan represented a purely economic choice that was meant to provide a better livelihood for their family. The Comfort Cove Trailer Park, named after the village from which most of the original migrants had come, had been developed by the manager of the Connors plant as a direct response to the need for accommodation. But its location presented particular problems. Situated near the southern end of the island, twelve miles from the ferry terminal and a mile out of Seal Cove (where the fish plant was located), it was relatively isolated from important centres of activity. The location of the trailer park was a crucial element in Newfoundlanders' experience of the island. Because of its spatial segregation from the community, the trailer park represented a social separation that affected islander perceptions as well as the ways in which Newfoundlanders were able to integrate into the community for the five months of their stay. As one of the trailer park residents described it, on Saturday nights "we're the local tourist attraction. They drive through just to see us." Not only were the Newfoundlanders spatially isolated from stores and other services but they were also far from recreational facilities such as the swimming pool and the Boys and Girls Club, both of which are in Grand Harbour, which could have provided activities for their young children.

For the Newfoundlanders who found other rental accommodation (e.g., in Grand Harbour or Ingalls Head), the sense of isolation was less pronounced. This was especially important for families whose children enrolled in the elementary school for the month of June and for the two months at the beginning of term just prior to their return to Newfoundland. There is no question that the issue of children was a crucial factor in both the need to find work on Grand Manan and in the eventual decisions of families not to return to the island. Their capacity for work and their determination to provide for their families were strong motivators that drew them to the island, but the sacrifices made with respect to their children's education were substantial. One parent was convinced that her son suffered a setback when they moved to the island because the Grand Manan school curriculum was behind Newfoundland's. She was also concerned that the children were allowed to get away with bad behaviour: "It's lax here compared to at home. The kids don't take it seriously here. They're

Newfoundlanders at Comfort Cove Trailer Park, 2003

always misbehaving. I don't know how they get away with that!" She described the taunts her son had experienced: "They look at Newfoundlanders as being stupid." That some Grand Mananers viewed the Newfoundlanders as "tough" was corroborated in several interviews with island youth, who described them as part of the marginalized "skunks" group. One Grand Manan youth acknowledged that "they hear jokes [about themselves]. It can't make them feel good." Despite these experiences, some children were welcomed. Little Mayah, who was in grade 2, had a good year in 2003, even given the short time she was on the island. When she left in early November, she proudly showed me a good-bye card from her class that everyone had signed: "They all like her," said her mother happily, "she's done really well." And she showed me Mayah's exercise books, which contained many congratulatory stickers.

For other parents it was the struggle to ensure continuity in their children's education that concerned them, and this was made especially difficult by the dates of the annual migration season, when children were shuttled between schools at both the start and end of the school year. Leaving the island late in 2003 (7 November) because the fish plant had not closed, one couple explained that they might get the afternoon boat so that Mayah could finish her day at school. But if the plant was open one more day, then they would take the 7:30 boat in the evening. After five years of seasonal moves so that both parents could work – one in the daytime, the other at night – they were feeling the strain. As Austin rushed around to winterize their camper, sealing windows, flushing toilets, putting in anti-freeze, Dale admitted that, as the children got older, it got harder. The long daily trip along a dirt, potholed road to pick up their daughter from the school bus (a few hundred yards) each day was just one of their ongoing frustrations. Another stress involved not knowing exactly when the plant would close. They were committed to be there until closing, and yet, during some weeks at the end of the season, there would be only six hours of work. The Newfoundlanders – a courageous, persevering, and determined group – found their patience severely tested during the final two weeks of each year. The depth of their anxiety was obvious to a Grand Manan fish plant worker when, on a Friday, she heard plant manager David Green begin the expected announcement that the plant would be closing with the words, "If we open on Monday …" "You should have seen their faces," said Bonnie. "It was a joke, but it wasn't nice of him. He really scared them!"

With regard to another family who had come to Grand Manan over a period of three summers, in 2002 the husband decided he preferred to work in the woods. So he went to northern Ontario while his wife Helen and his daughter went to Grand Manan. The family was feeling the strains of the annual migration and questioned how much longer they could be sustained. He returned from Ontario to Newfoundland at the end of October 2003, while she returned from Grand Manan with their daughter when the plant closed two week later.

Describing their experience, Helen said that the worst part was the twice-a-year moves, packing up and wondering if all the ferries would be running. As well, the return trip cost $1,500, which was a major expense. But she was no less unhappy about trailer life itself: "The trailers are so uncomfortable, cold in the winter, too hot in the summer." She pointed to one end of the trailer, where her husband would sleep, saying that there was a steady cold draft. She also said that "last year, when the water lines froze, David Green arrived with a hair dryer to try to unfreeze them. The electricals are a constant problem, like we can't plug in the kettle when the toaster is in because there's no 220 wiring." They paid $200 per month for the season for the trailer space, which included electricity, water, and sewage. The laundry building was built in 1999, when only fourteen trailers used the park. By 2004, when there were thirty-eight trailers in the park, there were only three showers for the men, which of course meant long waiting lines. Laundry days had to be carefully planned. The couple felt that it would be preferable to be able to rent a small apartment.

Even after four or five years of seasonal work on Grand Manan, the couples who lived in the trailer park said they did not know the other Newfoundlanders who lived permanently on the island. Both the ghetto-like physical separation of the park and their temporary status tended to isolate them from those who had moved permanently in the late 1990s. The division within the Newfoundland community was defined not only by migration history and residential location on Grand Manan but also by their own home areas back in Newfoundland. At the park in the early years, from 1999 to 2002, the inhabitants were virtually all from Comfort Cove or nearby villages. But by 2003 that had changed as word of mouth brought a young couple from Marysville on the other side of Newfoundland; and another large group (eleven trailers) arrived from Fogo Island to participate in the rockweed harvest. Underlining the non-homogeneous nature of migration histories, even Newfoundlanders were quick to point out that Fogo Islanders were "different" and that they tended to stay together. Their trailers were in a distinct area of the trailer park, arranged at right angles to all the others. Their part of the park was seen as "where the rockweeders stay, like as if they're a separate group." Spatial relations reflected social relations and tended to reinforce them.

Although they could not know of the devastating announcement that Connors would make a year later, by 2003, many Newfoundlanders who had been coming to Grand Manan for four to five years were finding the annual moves very difficult. Upon their return in the summer of 2004, Mayah's parents, Dale and Austin, found that there were fewer double shifts at the plant and that the amount of work had fallen off significantly. They questioned whether or not they would return in 2005. Indeed, the decision was made for them that same year with the closure of the plant in December 2004. By February 2005, their trailer was for sale, and Austin was in western Canada looking for work.

Children's education was a major problem for many families. Two couples who had tried to integrate into the community by moving out of the trailer park and finding rental accommodation on other parts of the island, and who had tried staying over the winter for two years, finally decided that they would move back to Newfoundland. When their children reached grades 8 and 9, respectively, they decided to remain in Newfoundland despite the uncertainty of job prospects because they felt their children needed a secure educational environment. According to her parents, during the summer of 2003, twelve-year-old Nicki had real problems trying to cope with the social world of Grand Manan, and, near the end of that summer, her grandparents had driven the almost a thousand miles from Newfoundland to pick her up in Saint John. When the plant closed in November that year, her parents decided not to return for 2004.

For both Grand Mananers and the large group of Newfoundlanders who migrated to the island between 1999 and 2004, the closure of the Connors plant in late 2004 and the restructuring of the aquaculture industry (begun in early 2005) marked the end of an important chapter. Most of the Newfoundlanders who had lived in the trailer park and in rented accommodation did not return to the island in 2006. The school records for 2005-06 indicate only a few children who had originally come from Newfoundland, and none who was there only for the months at the start and end of the school year. The trailer park had been sold. Nonetheless, there remained on the island about fourteen households originally from Newfoundland,[1] some of whom had bought homes, had children graduate from the high school in 2001, married islanders (2004), and had children of their own (2005). Nonetheless, according to those Newfoundlanders who stayed, the beginning of integration between the two cultures was not without ongoing strains.

On the day of the first wedding between the two communities there was friction during the reception, which marred the event. One of the Newfoundlanders who had been among the first to arrive in the early 1990s expressed her frustration: "It was the first time in a long time I'd heard comments. Why can't they just let it go? They feel threatened." After more than a decade they were beginning to feel at home, and, rather than returning to Newfoundland for their vacation, they had taken their holidays in Florida. "Little by little", she said, "the ties are being broken." Someone had called from Newfoundland to tell her about a piece of waterfront land that was for sale at $4,000. "Here on Grand Manan it would be $100,000! I thought about it, but finally said no. But wasn't it hard!" She had bought a small business that delivered soft drinks on the island and she also worked two days a week at one of the fish companies. She said that, by 2004, most of the businesses on the island had at least one Newfoundlander working for them full time. Indeed, a new habitus of cultural meaning was evolving out of this important period. After a decade, the Newfoundlanders who stayed were beginning to feel a new sense of home and a stronger sense of belonging. The boundaries between Newfoundlanders as outsiders and Grand Mananers as insiders were becoming blurred.

In the nexus between two cultures, the details of which define variations on habitus, there are significant tensions that eventually transform both communities. To the extent that school is a prime arena for the working out of values, norms, and expectations, conflicts may be defined as being institutional in nature, whereas in fact they are rooted in different cultural milieux. The habitus of Grand Manan, which has been sustained over generations, is linked both to island isolation and to family lineage but is underpinned by a diverse fishery that created a sense of security and generated very different kinds of networks, modes of familial responsibilities, and expectations than those found in the migrating Newfoundlanders. Grand Mananers and Newfoundlanders may be seen as two objectivities, but they must be understood as evolving in the context of particular "categories of perception and appreciation" (Bourdieu 1984). In expressing concerns about the differences in religious practices and school behaviours on Grand Manan Newfoundlanders are expressing more than a sense of being viewed as "Other." They are articulating a different habitus as it was played out in the daily minutiae of particular situations, and they are experiencing the reality of the mediating constructs that defined their migration experience. The issues raised by the coming together of these two communities are important both for their impacts upon individual lives and for the way in which they were replicated in many other communities in Canada. "Between belonging" is a difficult world, and it is being negotiated by large groups of migrant and migrating workers and their receiving communities everywhere. And by far the most affected are the indomitable Newfoundlanders, whose diaspora across the country is marked by special stores devoted to Newfoundland products.

The media are powerful perpetrators of both the bucolic myths of rural life and the critiques of governments that "inexcusably" reinforce them. Contrary to the arguments of John Ibbitson in the *Globe and Mail* (22 July 2005, A10; 12 July 2006 A4; 14 March 2007, A6), the solution to the rural-urban divide and the increasing regional inequities of lifestyles in Canada must not be defined in terms of migration. Indeed, he argues that the failure of wealth to "trickle down" from large cities to their rural hinterlands occurs because "Canada's labour force is lamentably immobile" (*Globe and Mail*, 12 July 2006, A5), a perception not supported by the experiences of intrepid Newfoundlanders. Apart from the basic moving expense involved (which no government has offered to pay), and the fact that urban areas would not be able to cope with such radical influxes of population, there are fundamental issues of morality. To adopt policies that would encourage migration as a solution to underemployment would be to deny and to negate our historical roots as a country as well as the importance of belonging and community relations. Furthermore, as *Tides of Change* makes clear, such policies would inevitably fall on the backs of the people who can least afford them.

Ibbitson's argument that politicians have catered to rural voters by siphoning "scarce resources to prop up the hinterlands" (*Globe and Mail*, 12 July 2006, A5) is both exaggerated and reflective of a simplistic analysis of the meanings of lives lived in a rural context. My descriptions can hardly do justice to what is experienced by individual families who have had to leave parents in Newfoundland, or take children out of school, or struggle to adapt to new forms of community hostility. Their perseverance and determination must not be accepted as a norm. It would be immoral to enact policies that assume that thousands of people should endure the sacrifices that I describe in this book. In one of his columns in the summer of 2006, Roy MacGregor described the reaction of one Ottawa Valley town to rural struggles for survival. Entitled "Expressing Rural Rage with a Musical Comedy," he described the creative energies of Killaloe, which had produced a musical comedy called "Here to Stay," as representing "an act of defiance" in its "look at the new Two Solitudes of Canada, rural and urban" (*Globe and Mail*, 28 July 2006, A2). In an interview with MacGregor, the songwriter Ish Theilheimer, referring to farming in the area, argued: "We believe there's a strength in rural communities and something worthwhile to be said in the ability of a nation to feed itself." The contradictions, struggles, and fears of rural dwellers can be found across the country in the bemused columns of newspaper writers and frustrated letters-to-the-editor. For example, referring to issues of gun control and RCMP killings in Saskatchewan, writer Patricia Robertson wondered whether "Harper can be tough on rural crime too?" (Patricia Robertson, *Globe and Mail*, 29 July 2006, F2). "The myth of bucolic rural splendour that we peddle to cloistered Central Canadians is now in jeopardy," she suggested. Perhaps Canadians are beginning to recognize the problems associated with the "rural plight" defined by their "shrinking and aging population, lack of basic services, dismal economic prospects" (Letters-to-the-Editor, *Globe and Mail*, 13 November 2007, A18). As I discuss in chapter 10, these questions are complex and demand attention from policy makers at the highest levels, but this attention must take the form of a dialogue with local communities. As the preceding letter-to-the-editor argues, "We need a national debate on the future of rural Canada and on the role of the federal government in ensuring that future is brighter than it looks now."

10

Resilience: "Between History and Tomorrow"

For some young people the best they can hope for themselves is rather narrow, and this is more punishing now than ever before because of how much more they have learned to want thanks to the reach of globalised cultural production. These issues, which are environmental in the broadest sense, have deep, and largely unexplored, effects upon young people's constructions of identity, in how they see their "place" in the world, and ultimately, in how they produce the world to come.
~ (Katz 1998, 142)

In the previous four chapters I have explored the changes on Grand Manan with respect to globalization (chapter 6); belonging (chapter 7); identity, place, and community (chapter 8); and habitus and culture (chapter 9). All of these are interrelated perspectives on the effects of socio-economic change associated with the forces of globalization. In looking forward towards the possibilities for adaptive change and resilience, in this chapter I focus on two crucial groups in the community: women and youth. They are, I believe, both the indicators of important change (the "canary in the mine") and the keys to encouraging constructive long-term resilience. On the one hand, youth represent the future; on the other hand, women represent both the history of Grand Manan (in their memories) and the future (through their strength and determination), offering hope for successful adaptive strategies.

The idea of resilience is important because it allows for adaptive change, which is what ultimately underpins strong community relations and cohesive structures that encourage new forms of activities and productive relations. The challenge is to mitigate the powerful threats from globalization and to find ways to strengthen the potential for creative change so as to ensure the long-term sustainability of the community. An understanding of the links between identity and community is intricately bound up with the nature of social and economic relationships, especially in periods of radical change. While some have argued that the forces of modernity and globalization have actually widened the scope of community (Hewitt 1989; Gusfield 1975), others have pointed to a decline in the significance of community life. On the one hand, there is an emphasis on the stretching of connections, and on the other, there is the decreasing importance of place-based relationships.

> The expansive possibilities of community contained in modernity repre-
> sent the source of a potentially fulfilled self, but a self always under stress
> of conflicted definitions and meanings. Weakened forms of organic com-
> munity survive in such institutions as the family and religion and are
> reproduced in such locally based structures as neighborhoods, schools,
> workplaces, and voluntary associations. Such forms offer a degree of
> stability and security in the face of rapid social change but can also incur
> major obligations on the part of the individual. (Dunn 1998, 61)

For rural areas, the impacts of economic restructuring and the intrusion of exogenous forces of change have particular importance because of the less complex nature of their social institutions. In terms of the ability to adapt to rapid change, the implications of complexity and rurality are not clear. On the one hand, it would be reasonable to expect that resilience will be more problematic for rural communities due to their lack of economic options, which may create social insecurities that manifest themselves as resistance and frustration among vulnerable segments of the population. On the other hand, some argue persuasively for a theory of social change that focuses on the role of inertia as a significant barrier that affects complex (core) societies due to their large investments in maintaining the system (Dodgshon 1998; Gregory 1990). Dodgshon suggests that the more stable and complex a society, the more likely it is that a greater amount of its available energy must be expended on maintaining received forms of knowledge. Therefore, he argues, it is peripheral societies, not core societies, that exhibit the potential for creative change. Crucially, however, his argument contains within it the dilemma that "those who gain most from a system are less likely to will change to it" (Dodgshon 1998, 183).

The capacity for resilience has been linked to trust within community settings insofar as residents are able to articulate and negotiate issues that require new strategies (Fukuyama 1995; Misztal 1996; Portes 1998; Woolcock and

Narayan 2001). In a community where relationships have been based upon individual entrepreneurship within a patriarchal culture, such that women's voices have not contributed to broad community structures, opportunities to nurture collaborative projects have been thwarted. The structures within which trust might be expected to evolve have not been developed. Trust, as a central mode of being, can only become embedded within networks of cooperative ventures that are inclusive and supportive. For fishing communities, highly supportive relationships emerge during times of crisis for both families and individuals. But when livelihoods are threatened throughout the community there is an equally strong competitive urge to survive, which can undermine trust and mutual support. Trust is not simply "desirable," it is essential. And without it the ability of communities to adapt will be impaired.

In the context of New England, a discussion of different situational characterizations of cooperation and trust examined the responses of fishers in a small American fishing community. It was pointed out that knowledge of what the competition is up to is one of the virtues of entrepreneurship: "In the fishing business, however, exactly where one builds a fence between fraternity and camaraderie, on the one hand, and prudent self-interest, on the other, is a matter for the kind of internal microanalysis that Brian sees underlying every gesture of hunting-and-gathering" (Carey 1999, 133). There is, in other words, a challenging dilemma for the community that affects the previous taken-for-granted boundaries of norms and values associated with the fishery. In the new world of globalized structures, these boundaries must be reassessed and renegotiated in the context of new relationships, both economic and social.

On Grand Manan, history has conspired to embed a profound sense of rootedness and lineage, and island geography has magnified a sense of isolation that has limited contacts, making mobility and interaction between the island and other regions very challenging. Both history and geography have created an entwined network of relationships that act not as threads through time but, rather, as chains that have anchored families and young people to the island in a culture that is tied to family lineage. Despite their less obvious presence within soco-economic structures, however, it is women who have shown the strength and determination to chart new courses of action.

The complex position of women and their contradictory relationships with men form an integral part of understanding Grand Manan culture in the context of both its history and its future. Both youth relationships and gender relationships are truly situated "between history and tomorrow."[1] The most poignant and valid way of illuminating these complex changing relationships is through the words of those directly involved, which reveal worlds of struggle and challenge. Their stories reflect major changes in the "becoming" of island lives and identities. On the one hand, as many of the interviews indicate, young people are acutely aware of the role that history has had with regard to who they are; on the other hand, many of them actively resist traditional choices, while being constrained by a limited education and a reliance upon the wild fishery. For both women and youth, as the following narrative makes clear, resistance, challenging traditional ways, and taking a different path are neither straightforward nor necessarily ultimately successful endeavours. Nonetheless, there are significant signs of hope. To an important degree, the stories that unfold here are tied to two crucial time periods: before 2001 and after 2001.

Gender

He rules the roost,
She rules him.
(Fridge magnet)

The truths, perceptions, and values contained in humorous birthday cards and magnets clinging to refrigerator doors are not trivial. The one quoted here, seen in a Grand Manan kitchen, reflects a reality within island culture that is characterized by separate social spaces for women and men. This reality encompasses the home, village, wharves, boats, and woods, extending to recreational and social spaces such as the curling rink and camps. Nevertheless, with economic restructuring, there is a shifting balance in the relative importance of various roles and relationships, representing a series of shifts away from cultural roles rooted in traditional belief systems towards new economic and social roles. New patterns of interaction and different views of the world are changing the ways in which islanders communicate, socialize, organize their institutions, and work together. Nowhere are the changes more apparent than in gender relations, which reflect the character of being "between history and tomorrow."

The story of a woman I shall call Carla reflects many of the issues that have defined the lives of women on Grand Manan over generations. When Carla, one of five children, was thirteen and in grade 8, she began dating a tall, good-looking boy who was sixteen and entering his senior year at school. A few years later, when she was seventeen and in grade 11, she became pregnant, leading immediately to her dropping out of school and getting married. That she sub-

sequently suffered a miscarriage changed very little for her. She did not return
to school, and, to this day, over fifty years later, admits that she feels a sense of
failure in never having completed her schooling. Her first child, who was born
when she was nineteen, was soon followed by four more, the last born in 1972
when Carla was thirty-three. In addition to her responsibilities looking after
the children and a large house, Carla supplemented their meagre income by
gathering dulse and stringing herring. While her husband was involved in all
the fisheries as crew or was harvesting dulse and periwinkles, her freedom to
move about the island was constrained by home duties. She soon learned to
cope with his binge drinking, which might occur every three to four weeks,
leaving him non-functioning for days at a time. While she had always felt the
urge to leave the island, her early marriage and large family were constraints
that she never overcame. With a wry laugh she said: "I was the one who most
wanted to leave, but it was all the rest of my family who left, and me who
stayed." But her sense of needing personal outlets for her talents was finally
realized in 1981, when she decided to enrol in a nursing course on the mainland,
which necessitated week-long sojourns for two winters. Recounting the experience of exerting her independence in those two winters, which involved never
knowing when her husband might indulge in another alcoholic binge, Carla's
emotions were obvious. Her stepping out was a rare challenge to island culture.
When her husband was away for a week or two at a time on a fishing boat she
had a sense of freedom that was severely restricted while he was at home, when
"I always felt bound by what I thought he wanted."

For the small community of Grand Manan, the patriarchal structures that
define gender relations function as an essential constraint on creative change.
Indicative of the link between identity and gender relations on Grand Manan is
the habit of naming women according to their husband's first name. Because of
the relatively few surnames on the island (and tens of any single surname, such
as Green, Brown, or Ingersoll), a woman might be known as "Marion David" or
"Ruth Charlie" rather than as Marion Brown or Ruth Green. This type of naming has provoked vigorous resistance from newcomers to the island. Said one of
the Newfoundlanders who had moved several years earlier: "It's archaic, naming
women by their husband's name; like they have no identity of their own!" She
did not think it odd that her surname was that of her husband. Even for mainland women who had moved onto the island many years earlier as spouses of
Grand Mananers these naming norms were offensive. After twenty years one
woman recalled her feelings around being known as "Alice Lester": "It brought
my hackles up!" In the past decade there is evidence of a move away from these

namings, although they continue to be used by the older generation. Another naming norm reflects the overlap of common names – for example, Alice Brown, who married another Brown, is referred to as a "double."

Gendered social norms have been a defining characteristic of Grand Manan culture. The experiences of women who moved to the island as young wives are particularly illuminating with respect to gendered expectations and values. Their sense of isolation living on an island was exacerbated by their separation from family support networks. Especially during pregnancies and in the first years of child bearing they felt that they were completely on their own, while at the same time experiencing the stresses of always feeling they had to fit in. Adjusting to new societal norms and values as young wives was especially difficult because their husbands, as islanders, had no understanding of how different their lives on the island were to their former lives. Talking about Grand Manan attitudes towards alcohol, the women referred to their initial attempts to resist island mores by ignoring the prevailing norms, which prohibited women from drinking. Eventually, however, they were defeated, feeling forced to comply by at least "beginning to tone it down." The first few years were "a real awakening" for new wives as they sought to redefine their identities within a strong island culture. It did not help that, to island women, the from-away women were the "natural enemy." The tragedy of one from-away wife who had married on to the island twenty years earlier, and whose husband had died, evoked not only sympathy but also musings as to whether or not she would leave. Thinking about their own situations, wives from-away were ambivalent, pointing out that their friends were now all islanders and that perhaps there really was no place to go.

For women from-away who have spent forty or fifty years of their married lives on Grand Manan, the choice to "stay or go" when they are in their sixties and are either widowed or have to cope with an ill husband is especially difficult. Having a chronic illness on Grand Manan can involve many full-day trips to the hospital in Saint John, and this brings with it the stresses of the ferry schedule and weather conditions. As a widow there are also concerns about support networks in the absence of an extended family. Talking with Joanne, whose two grown children were living on the mainland and whose husband had just recovered from a stroke, I learned that she had been trying to persuade him to move to the mainland. She took a lease on an apartment on the mainland and moved many of their belongings there one winter, hoping that his opposition might gradually wane. However, she found that he avoided spending time there, and so she had to return to the island at least once a month to ensure that bills were paid and to see that he was looking after himself. When I talked with her it was at the ferry terminal as we were both waiting for the ferry. The back of her car was filled to the roof with small furniture items, boxes of appliances, and kitchen paraphernalia. She seemed quite defeated. After her year of retreat from Grand Manan she was returning in order to be with her husband, who would not consider leaving. He had spent just ten days on the mainland over the year because he "couldn't stand the four walls." She explained that they "have no family on the island and you need people in case of problems. If I need help all my friends are the same age as me and either have sick husbands or can't do much themselves. So who is there to call?" Ironically, after pointing out that she had to return each month to pay the bills, in the next breath she said: "The problem is he is so independent."

At a birthday party for a feisty ninety-year-old, I talked with two women whose fisher husbands were about to retire. Both of them were looking forward to the possibility of new homes on the mainland, where they would have access to a wider variety of activities. One of them had advertised her house for sale but admitted that her husband was not being supportive of the idea. In the case of the other woman, Anne, she and her husband had bought a small home on the mainland, but he continued to return to Grand Manan for lobster fishing. Both of them seemed depressed about their struggles to redefine their future away from the island. For many women the challenge is to encourage their children to leave, even if only for a year or two. While the men tend not to acknowledge any value in time away, the women almost universally recognize the benefits and improved choices that being away can offer their children, and it is they who can make a difference by encouraging the boys to leave (the girls tend to leave of their own volition).

After several turbulent years with her two teenage boys, one mother who had married onto the island expressed profound relief when both of them decided to leave Grand Manan. As a direct result of the tragedies that occurred between 1999 and 2001 (see below), some youth realized that their futures depended upon making drastic changes in their lives. In 2003, this woman's twenty-two-year-old son joined the navy. Despite his almost backing out at the last minute, she said, "He is a changed boy! Good grades, trying hard so he'll get the posting he wants." She was delighted. "When he was at four funerals of friends in one year he was very depressed. We got him grief counselling." Her other son was in his fourth year at university, and was on a national baseball team, hoping he might get a scholarship to an American college. She described both sons as being motivated, having goals and enthusiasm that they had not had before.

For women whose family lineage goes back generations on Grand Manan, the changing structures and relationships have been equally challenging, but for somewhat different reasons. Societal change is necessarily contextual, and any theory must be "sensitive to the different ways in which society's capacities for change varies geographically across societies and between societies" (Dodgshon 1998, 50). Traditional values and historically rooted social relations continue to inform responses to new economic realities, patterns of interaction, and power structures. As described in earlier chapters, the centralization of island institutions, the decline of village-based networks, the changing markets for smoked herring, and the radical transformation in corporate structures associated with the closure of the Connors packing plant have combined to irreversibly alter the potential for sustained women's networks. While the villages used to provide the spatial milieu that enabled women to meet in the boning sheds, fish plants, or stores, today those jobs and stores are gone.

For generations, stringing and boning herring provided crucial social spaces that allowed women to share experiences, problems, and support in ways that encouraged a sense of strength and mutual survival, even in the context of the men's long absences. While, on the one hand, these activities created and perpetuated a gendered spatial and emotional separation, on the other hand, they also provided necessary support. Removing these activities has meant both opportunity for greater gender integration in daily lives and a loss of traditional female support systems. For those women who work, they now do so as store clerks, book keepers at the gas station, fish brokers, teachers, or nurses. With the exception of the nurses and those women who work in the nursing home (in the kitchen or laundry), island women have few opportunities to meet casually. Service jobs in stores are essentially isolating. Unlike the work in the herring sheds, the shift work in the hospital and nursing home requires structured hours. In today's working environment there are no mutually shared spaces where social networks can be nurtured. For women, the changes of the decade between 1995 and 2006 have effectively transformed employment choices, separating social and economic spheres of activity. Even in casual exchanges as men drive about the island in their trucks, one sees the raised finger of greeting as they pass on the road. Few women participate in this highly symbolic but non-verbal form of communication, which contributes to a sense of belonging among the men.

The churches (see chapter 8) provide some opportunities for socialization, but only on a less than monthly basis, when teas, bazaars, or suppers bring the women together to prepare the events. For women who do not work, social spaces are limited as well. Whereas retired men continue to enjoy contacts at the wharves and sheds, the women have only the formalized structures of curling, bingo, and the fitness centre. There are no drop-in areas that allow for impromptu, unplanned discussions. In other words, for women, networks are dependent to a significant degree on phone calls. While some women, especially those who work, do not miss these informal contacts, others (particularly young mothers) often feel isolated. On Grand Manan it is the men who have the gossip and the news first, and frequently it is not passed on to wives.

The economic changes over the decade since 1995 have altered the clearly demarcated territories and spaces of being that dominated lives for generations. Nonetheless, some traditionally female or male spaces continue to exert considerable constraints on evolving social relationships. Some of the most prominent of these traditional gendered spaces include the kitchens, stores, garages, churches, boats, wharves, trucks, and camps. In describing how these spaces have evolved as centres of relationships and community identities, the words and stories of individuals highlight their meaning for people.

One of the most striking aspects of the homes on Grand Manan (at least for an urban outsider) is the interior design, the layout of rooms. As in most rural areas, invariably one enters homes through the back door directly into the kitchen, where, almost always, one is invited to sit down. Most socializing takes place in the kitchen, certainly in households headed by people over forty. Even where there is a front door it is not used. Doorbells are largely absent, except on houses owned by people from-away or, increasingly, on newly built homes. People are expected to simply walk in – a habit that, for newcomers, takes a bit of getting used to. They walk into kitchens and call out if no one is there. Kitchens are both the centre of activity and the welcoming entrance hall. Through the kitchens the women direct family activities. These areas are control central. Even in the new homes being built by young couples the kitchen is designed as the focal point, with immediate access to the main door. These activity centres become the arena for haircuts, gossip, coffee klatches, shared stories, and problem solving. In the kitchen, women reigned supreme.

For the men the spaces of shared camaraderie and social networks have been the stores (until the late 1970s) and the garages (until 2006). From 5:30 AM onward, those locales found men playing crib, drinking coffee, chatting, arguing, and generally learning about the news of catches, failed businesses, births, and deaths. Grand Manan men have always maintained their social networks through places such as the garages and through casually driving about onto wharves and through the villages. Their trucks have provided them with mobility that, until very recently, women were largely denied. While women's integration into island spaces has occurred only in the past ten to fifteen years, men's trucks have allowed them the freedom to venture into every lane and forest road across the island. For women, therefore, the spatial knowledge of the island has been much more circumscribed by village boundaries than it has been for men. Furthermore, because of the historical constraints on their mobility, women have not acquired the habit of exploring new roads or new developments as the men do. A new house being built or a new business being renovated is examined and discussed by the men within days of its inception. Only on the weekends, usually Sunday afternoon, could it be expected that the men might venture out with their wives in the truck to see these new sites of change.

One of the first things I heard about the island was that its women were "strong." However, as a researcher several years later, I began to discover a much more complicated reality. On the island it seemed that women's strength was commonly hidden behind a veneer of self-deprecating humour or cynical comment. Especially among the older women, there was a feeling that they could not have anything interesting to tell me because: "We're nothing!" Still, some women recognized their contribution to sustaining community relations. In describing women's activism one woman said: "The men are busy, and they've given up; they think nothing can be done. The women have more time and tenacity." A common thread through all conversations concerned the long absences of the men. One woman described her husband's fishing trips of twenty years ago when the children were young: "I was working shift work, and we'd meet in the doorway as he was leaving. He would leave on Sunday afternoon and they would go until Saturday. Our first son was six weeks old before Gary saw him!" One multi-skilled man who had worked intermittently in the fishery thought that those men who were full-time fishers must "hate their wives." While this might be a non-typical point of view, the sense that marital relationships had been made more difficult by the demands of the traditional fishery permeated almost all conversations about social relations on the island. In considering how we conceptualize patriarchy, it is essential that we not fall prey to reductionism. Patriarchy has to be seen in terms of integrated processes concerning the "reproduction of individuals in their daily lives through

Men on the wharves

housework and mother work, and the production of gender identity and ideology" (Fox 1988, 164). Understanding gender inequality "requires analysis at the level of social structure and at the level of the individual."

Over the generations, the long absences of the men have had a profound impact on social relations, including on the nature of male-female communications, attitudes towards alcohol and religion, and the value given education. The women stayed home and were responsible for family and community and for ensuring that the finances were in order, the groceries bought, and the children clothed. Their reliance upon each other within their village settings was crucial to their survival. Virtually all of the women with whom I spoke described a past that was a struggle and a present that offered most of the comforts enjoyed by any middle-class Canadian. In describing the changes, they were unanimous in feeling that there "has been a huge change in the attitudes of young men. More Dads are coming to the clinics with their babies and children; and they're taking pre-natal classes, and are at the delivery of their babies." Historically, women had been the watchdogs of community moral codes, but the men dictated the family rules when they were at home. Wives "felt bound" by their husbands' wishes while they were home. For many of these women, their lives continue to be defined by past constraints and struggles.

In a lively conversation with an eighty-one-year-old survivor of abuse, who had courageously stepped out of her marriage in the late 1980s at a time when such a move was rare, I was warned, "Don't ever take up with a man who don't like animals. He won't use you good!" She referred several times to her cats, who loved her "more than any man." Despite evidence of abuse, many island women have shown a spirit of survival and creative enterprise. A younger woman who had stayed at home to raise four children while also tending to her husband's accounts and business interests, exemplified some of the real challenges with which many rural communities are faced. Although many islanders described her as a perpetual victim of abuse, supplying some horrific details, no one in the community was willing to make an official complaint that would call her husband to account. There is a cloak of secrecy that envelopes family lives, despite the very public awareness of particular situations. In another instance, a woman described the spanking she would receive from her husband if she was late returning home. When queried about tolerating such behaviour, she implied that she had deserved it since she had been late. In a small community there is a delicate balancing act between maintaining respect for privacy and ensuring the safety and protection of everyone. While there is less tolerance for child abuse on the part of non-family members, the possibility that date-rape and family abuse will go unreported continues to be a significant issue in the community. The perception that women are the chattel of men, and that men have the right to control their lives, is a major barrier to combating many of these problems.

Throughout the community, one of the barriers to disclosure is a prevailing sense of the importance of work. There is enormous respect for men who have established reputations as good workers, and this can undermine efforts to break cycles of abuse. One day on the wharf I was standing with a DFO officer watching the men loading the lobster traps prior to setting day. When the name of a young fisher was mentioned I was startled because, on numerous occasions, I had heard it negatively associated with several young women. The DFO officer noted my expression and simply said: "He's a really good fisherman!" Similarly, islanders who know of the plight of the woman described above also recognize that her husband has worked very hard and "come up from nothing." The most common expressions on Grand Manan are: "He's got ambition" and "He's a worker!" These explain an important element of the value system that informs behaviours and expectations and that leads to concessions in the cases of suspected spousal abuse.

For researchers these situations pose particular hazards. Listening to the stories from all ages of women, I began to wonder about the possibility of a drop-in centre for them. In the summer of 1998 I began to systematically explore with women the idea that they might enjoy having available to them a neutral (i.e., non-denominational) place (i.e., indoors) where they could drop in for casual conversation. Whereas at that time the men could enjoy crib, cards, and shared coffees in two garages and several other public areas around the island, from 5:30 AM until 9:00 PM the women had no casual places for meeting. I wondered, why not a drop-in centre? It soon became apparent that the idea was quite revolutionary. The names in the stories below have been changed.

Approaching a couple of older women, I was surprised to find that, while not hostile, they were quite lukewarm towards the idea. Indeed, one woman suggested: "Perhaps it would be a good idea to start with the younger women, and they might encourage us older folks to try it!" The older women simply doubted that a centre was needed or would be used – two quite different problems. Two interviews stand out in revealing the views of women and women regarding the idea of a "public space" for women that would be comparable to the men's garages. In both cases the comments of the men were uninvited, but they are probably very reflective of an island consensus around a "woman's place" being at home (unless she had a paying job outside).

The first interview was with Susan, a young woman with two children, a teacher from-away who has lived on the island for twelve years. We sat in the kitchen while she fed one child in a highchair and watched the other playing on the floor. Her husband, David, a native islander and a carpenter, sat in the living room within earshot. I asked her about the opportunities for women to get together informally and wondered if a centre of some sort might provide at least a facility to make it possible.

SUSAN: "They wouldn't have time. They're tied to the home."
DAVID, from the next room: "Why would they want that?"
SUSAN: "You can drive up the island anytime and stop at Vern's for coffee. Men don't have a commitment to stay at home." [Then she continued, thinking out loud:] "There might be a need. But I don't know how much women would use it. I looked for a group when Philip was born. I tried to get a support group going for new moms. But there wasn't much interest. Jennifer [the social worker] tried, but it was in her office, at night. It couldn't work, especially at night when moms are tired. To see

other women we have to call and make an appointment!" [She remarked
on a resource centre she knew of on the mainland that seemed to be
successful and wondered if islanders might learn from their experience.]
DAVID (still in the next room): "Women always want to organize things.
The first thing they would do is schedule who would make coffee and
cookies each day. We just take our chances and drop in."
INTERVIEWER: "When was the last time that a male friend dropped in
at your house?"
DAVID: "When Susan was away!" [They both laughed, knowing the truth
of this even as they seemed somewhat chagrined to have to admit it.]
SUSAN: "I'd be less likely to go if the old women are there, who I have
no interest in."
DAVID: "That's another difference compared to the men. We meet
everybody."

As the interview continued, it was apparent that David saw many obstacles to
the success of any initiative for a woman's centre. I was left wondering how
much of his negativity grew from his discomfort at the idea of women having
freedom in the public sphere.

The second interview is with Carol, a native islander in her mid-fifties
whose grown children live on the island with their own young families. Her
husband, Nick, sits with us, listening as he sips his tea.

CAROL: "Well, maybe people would like it [pause]. I think it would
be good for the younger women who need to get out. But I don't know
if someone like me would ever go there. We're sort of used to the way
things are."
NICK: "I don't think the men would like it!"
CAROL: "But it's not for them!"
NICK: "But why should women have their own place. They can stop
at the garages if they want!"

In another interview, thirty-seven-year-old Diane admitted that she liked the
idea but that the place would have to be "hidden" so that people could not see
their vehicles parked there. She thought that a hall such as the one in Woodwards Cove (sheltered behind trees and homes) might be a possible location. It
quickly became apparent that, despite the need for it or the wishes of women
themselves, men would not be supportive. My experience in raising the ques-

tion of a women's centre pointed out some important issues not only with respect to gender relations on the island but also with respect to my role as a researcher. Allowing my values to intrude, even with the best of intentions, suggested an activist role that could both threaten my research position on the island and upset social relations in a way that might ultimately be a deterrent to more gradual change.

The idea of a women's meeting place challenged deeply embedded cultural norms, and it is islanders themselves who must take the initiative to change or modify them. As a social worker acknowledged, there can be real dangers, in the form of violence and abuse at home, for women who step out. As the identities of men engaged in the traditional fishery are being threatened by declining fish stocks and disappearing markets, their insecurities may be played out at home. If, at the same time, women are seen to be gaining freedom from their traditional ties to the family, the situation may be exacerbated and women may be the scapegoats. Barb, who was about twenty-seven years old, described her difficulties in staying at home with her two-year-old daughter. She felt isolated because her husband took the truck and they could not afford a second vehicle. It was a thirty-five-minute walk to her sister's house. Barb worked during the summer at a small inn but had been unable to find permanent work during the winter. She had enrolled in the program entitled Nobody's Perfect (a Health Canada initiative that encourages new parents to meeting to exchange experiences and problems related to patenting) but was concerned that men did not participate and could not understand the problems women were encountering. She felt that her lack of mobility inhibited her connections within the community. A single program could not meet her long-term needs regarding sharing daily problems of childrearing. Barb liked the idea of a women's drop-in centre but wondered how she might get there without a vehicle.

One area in which the patriarchal structure on Grand Manan is illuminated is in the low participation of women on the municipal council. Like many other communities across the country, Grand Manan has not encouraged women to run for public office. In 2006, throughout New Brunswick only thirteen mayors were women. On Grand Manan, the only woman to have successfully run for office was Clare Foster, who was originally from-away but who has lived on the island for over forty years. She also held the record as the longest-serving councillor – twenty-seven years when she was defeated by two votes in 2001. Nonetheless, it reflected a major change when six women put their names forward in the municipal election in 2008. Equally significant, none was elected. Despite the patriarchal constraints experienced by women, they have developed many strategies of resistance and have found various creative solutions to ensure that they have fulfilling lives. Women on Grand Manan exemplify the entrepreneurial spirit of the island not only through the more obvious tourism-related businesses such as selling arts and crafts or opening a bed-and-breakfast but also through developing innovative ideas such as offering "wharf walks" to visitors. While some move from job to job at gas stations, groceries, and offices, others seek out ways to develop new businesses.

Women's flexibility is reflected in the working lives of many of them. Julie, for example, was a trained nurse who became head of the local hospital and then moved on to start her own gardening business. This meant investing in a small greenhouse and having to work within the constraints of a ferry schedule and mainland suppliers. Very successful for about seven years, she eventually sold the business and went on to be the accountant for the marine engines company on the island. Another successful initiative is Pam Cronk's "Whale Cove Knitters," which relies on about twelve women around the island to knit hats, sweaters, and socks that are sold mainly through a weekly open-air market during the summer months. The expansion of her business is indicated by the increasing demand, from about fifty sweaters in 1997 to ninety in 1998. As well as running this business from her home, she is also a community nurse, making house calls three days a week. Other women have opened restaurants or become known for their ability to make elaborate birthday cakes. Most working women have had at least five or six jobs, each of which often requires very different skills and knowledge.

In a sense, the very changes that removed the natural social networks (notably the closure of the smoke sheds and, finally, the 2004 closure of the Connors plant) became forces of empowerment as women found it necessary to find other sources of income. Women were becoming more self-confident and were creating the possibility for a more public acceptance of their strengths. When a young woman about to be married was asked about her future husband, she described him in glowing terms: "He is the greatest hunk on the island!" When it was suggested that she was so much smarter than her fiancé, she replied: "Oh, that's how it's supposed to be!" Reflecting this new sense of confidence, more women are contributing to the formal activities within the various clubs and associations in the community.

In trying to understand how gender relations affect the community, perhaps the most illuminating event during my twelve years on the island was the community meeting of 4 June 2001. A spontaneous response to the fourth tragic death of a young man in less than two years, the meeting attracted about three hundred people who packed the high school gymnasium. Anger, hurt, fear, sadness, frustration, and despair were all apparent on the faces and in the voices of those who spoke that evening. The meeting, facilitated by a social worker who had worked on the island for six years prior to her departure three years earlier, lasted over two hours and was explicitly designed to address a process related to four areas: (1) the current situation on the island; (2) what would constitute a desirable future; (3) what the barriers to this future were and how islanders defined the problems facing them; and (4) the steps and strategies they saw as solutions. Among those who were in attendance there was no one under about twenty years of age and only a few under thirty years of age. This was notable because the focus of the evening was drug and alcohol addictions among youth. But even more remarkable was the distinctive gendered nature of the discussion. Apart from men who represented specific roles on the island (mayor, emergency responder, pastor), only two men offered their comments or expressed views about the situation. Throughout the evening, virtually all of the input came from women: mothers, teachers, nurses, grandmothers, and sisters. It was the women who described the loneliness of youth, their desperate search for meaning in their lives, the need for love, the search for affirmation. The strong sense of not wanting to point fingers at neighbours permeated the narratives, underlining the difficult boundaries of discretion and honesty that small communities constantly negotiate. Only very gradually was there acknowledgment of denial as a deeply embedded obstacle to finding solutions. Introduced by the facilitator, the word "denial" began to emerge within the discourse as the women told their stories, drawing on personal experiences and family problems. It was an intense night that provided some cleansing and perhaps some clarity, but, as subsequent events were to reveal, solutions were a long way in the future. And it was apparent that it would be the women who would ultimately accept responsibility for change.

To some degree, gender relations can be described in terms of the now more public face of women's roles in the community. Whereas women have historically been responsible for nurturing and sustaining community relations, today their contributions are often island-wide rather than restricted to the villages, and they usually occur within the formal organization of established charities or projects. Therefore, in considering the possibility for adaptive change in the face of tensions rooted in economic restructuring, the role of volunteerism seems to offer women an avenue towards experience and empowerment. Furthermore, examining the structures and relationships within the volunteer sector provides another perspective on the central importance of gender relations.

Leadership and Volunteerism

Many volunteer groups on Grand Manan have had a long-standing problem with regard to finding leaders, people who will accept roles as chair or president of organizations. While many women may be actively involved in the church, school, or food bank, few have been willing to take on the leadership that all organizations need. As for the men, few of them participate regularly at any level. Of course there are notable exceptions, such as Reg Flagg, George Ingersoll, and Al Hobbs (among the older men), who were the core of the Rotary and the Legion activities. While a few younger men have begun to take on leadership in areas such as tourism and the skating rink, for the most part volunteerism is left to the women. As one observer said, the men "have traditionally been away fishing and now that's become their excuse."

As noted earlier, as the traditional fishery has diminished in importance the men have lost many traditional gathering places, including the stores, garages, and even the work sheds. Their increased dependence on private sheds has changed the social dynamics for men, with the constraints on their casual gatherings hampering the easy conviviality and community-building fostered by shared relationships. Unlike women, who have devoted energy to volunteer activities, the men continue to resist responsibilities in this area. When the board of the Boys and Girls Club decided on the need for major renovation and expansion, the search began for someone to head the fundraising committee. Focusing on the men, those whose names were mentioned were either "from-away," and therefore lacking in credibility, or, as one member said, "not exactly a role model in the community."

In exploring the ideas of leadership and volunteerism, I talked with a woman who explained that people represent different interests but that no one really "speaks for the island." She mentioned the Sonnenbergs, who run the GMFA, pointing out that they are articulate and able to conduct interviews to get particular views across regarding the fishing interests of the island. However, she also pointed out that their views were not always popular and that there really was no single individual who could represent island views. Even the council could not engender any unanimous or widespread support: "Grand Mananers never want anyone else to be 'better' than they are. They feel that anyone's profits must be on the back of the islanders." Those who have money tend to keep a low profile, although they are usually generous when called upon for donations to specific projects. Talking about volunteers in the community, she felt that "there's a lot of apathy. It's hard to find volunteers to organize things. They'll cook and drop things off, but not organize." One woman said her husband was trying to dissuade her from joining any board, despite acknowledging that "there's a real problem with leadership."

Understanding the intersection of gender and social structures through community organizations raises issues of unacknowledged attributions of "status" and "class," mainly related to fishing, wealth, and the church. If a family is seen to represent all three of these, then its acceptance at the top of the social hierarchy is affirmed. Others, not directly involved in the fishery but who have benefited from the rapid expansion of aquaculture (through trucking or machine maintenance, for example), have also been viewed as successful community leaders. The ambiguity that hovers when attempting to establish any hierarchy is linked to an island disposition that tries not to claim dominance even as their competitive instincts drive islanders to new ventures. Community status is blurred or even negated by a variety of factors, including church participation (or not), an intentionally low profile, not being directly involved in the fishery, and, for some, legacies of various family struggles and even rumoured illegal activities. With regard to community leadership that could provide creative ideas and support for change, one finds few options among the men. Certainly one key descriptor that permeates islanders' perceptions of leadership potential among men is the holding of multiple fishing licences and investments in brokerage operations, plus the active business support of their wives. This rooted fishery connection is still seen as crucial to their continued growing assets and to where they stand in the eyes of the community. However, with the exception of Maurice Green and Ron Benson, both of whom have run for positions on village council, these men have not accepted leadership roles in the community. Despite their potential as spokespersons for raising funds for a community centre or the renovation of the Boys and Girls Club, the library, or the museum, for example, they have maintained very low profiles. Their generous contributions to individual student scholarships notwithstanding, their role as leaders is untested and they do not have a public face as philanthropists. Whether this has to do with conscious efforts to maintain a perception of egalitarian status, or whether it is related to an island predisposition to not appear to be taking over or to be critical of others, is difficult to discern. Moreover, given that the aquaculture companies are based on the mainland, the increasingly dominant role of aquaculture in the economic structures of the island does not offer the potential for solving this leadership gap.

However, an inventory of the number of volunteer activities, and those who are directly participating in them, provides an impressive list of which any community would be proud. And in terms of dollars raised, there is probably no community in Canada that contributes more per capita to the Heart and Stroke Foundation, the Cancer Society, the children's hospital in Halifax, or to individuals in need within the community than Grand Manan. Collecting annually for the national organizations, women go from door to door in each village, obtaining donations. Or, in the case of special needs such as a house fire or special medical treatment, boxes for donations appear at each of the check-out counters on the island, collecting for the families in need. Volunteers run the curling rink, the Little League baseball program, Scouts, Guides, Meals-on-Wheels, the food bank, the historical society, the museum, and school committees. Probably the most visible and continuous volunteer activities are those centred in the churches. The women's groups organize teas, suppers, bazaars, and bake sales. For many women, these activities are significant opportunities for ongoing social contacts and networking, providing important outlets for shared stories, problems, and news. Other groups have been formed outside the formal institutions, having started as semi-social groups that would meet to sew or quilt to raise money for various charities. One of these, the Helping Hands Club, was started by Gladys Leighton. It met weekly for about twenty years, although it was disbanded in the early 1990s. The heart of most volunteer groups is the women. Describing the women who are the main organizers, one woman told me that the main criterion was their capacity to "reach out to everyone, not just stay with select groups." As she named about six key women in the community (from all villages), she also noted that "it's important not to have enemies." And, indeed, she had none. The committed engagement of Helen Bass in many community organzations, including her volunteer work in the Legion, Historical Society, the school board, Rotary Isle Estates (a twenty-unit senior citizens apartment building in North Head), the museum, and many others places, was celebrated on her sudden death in March 2007. But there were few to take up her spiritual torch. Although the curling rink and weekly bingos are run mainly by men, as in many other communities across the country it is the women who attend school meetings, participate in the Boys and Girls Club, and try to find solutions to community problems affecting youth.

The strong youth culture (see below) on Grand Manan and the problems associated with it have been the focus of several important volunteer initiatives.

One of these was the formation of Community Reach in 1997. Another is the Boys and Girls Club. With many overlapping board members, including the professionals on the island (the executive director of the club, the social worker, and several nurses), the boards of these two organizations sought to provide both activities that would enrich the lives of youth and programs that would help find solutions to problems associated with family breakdown and substance abuse. Obviously, given the issues they were trying to address within the context of limited resources and the island isolation (which constrained access to mainland options), there were real challenges.

For volunteers in the Boys and Girls Club one of the central problems was how to balance programs targeted at "youth-at-risk" as compared to achievers. Because there was a common perception on the island that activities such as Teen Night attracted only the "losers," the possibility for role models to have an impact on behaviours and expectations was very limited. For the professionals working with youth and families the need for effective role models to support youth-at-risk was a paramount objective. However, as off-islanders, these same professionals lost their persuasive power in the discussions with parent volunteers on the committees who felt that there should be more attention paid to those who were achieving at school and who would benefit by special programs (such as debating or film studies). In June 1999 a meeting called to brainstorm and create a "wish list" for new programs became a debate between two radically different points of view. For islander volunteers, the recent destruction of a pool table by the at-risk group meant that they did not deserve "more money and effort being invested in them." In the past the achievers, represented by the school's Keystone Club, had been very active. But since about 1995 the marginalized kids had taken over, with the result that fewer and fewer of the good students participated in club activities. The board felt that it would be impossible to encourage the mixing of the two groups, an opinion I subsequently corroborated with several teenagers.

For the two professionals who were arguing for more resources to be devoted to the at-risk group of teenagers, the meeting had been "brutal." They were discouraged not only by the attitudes of parents who wanted to further marginalize the problem youth but also by the lack of clear purpose in how they defined their goals. Both felt that without islanders taking a proactive leadership role there would be no possibility for a successful fundraising campaign, let alone the development of plans for renovation. Trying to find someone to head an expansion committee in 2002, the executive director said that he had several people who had agreed to be on the committee but that no one would commit to be the chair. The challenge, he said, was to find an islander who was articulate, motivated, reliable, able to communicate with heads of large businesses and government, and respected in the community. Reiterating a common refrain in the community, he said that "leadership and role models is a big problem." When the Parents Supporting Parents Program was inaugurated in 2002, the participants all expressed the same concern: "I need to feel I'm doing something. There has to be a change in the community mindset. So many things need changing!" Indeed, as we see in the next section, by 2004 the seeds of change were beginning to take root.

For the professionals involved (social workers, addiction counsellors, teachers) the solutions were complex; for the parents the solutions seemed to be directly linked to more activity options for their children. While there was a consensus that attitudes needed to change, most people also argued that the possibility for a variety of activities could be supported through the increased and long-term funding of both the Boys and Girls Club and specific projects such as the media project. As one instructor pointed out, youth are seeking support; they are eager to talk to adults who will offer their non-judgmental time. However, the available institutions have not yet provided the informal and formal networks of support they need. While some parents had worked hard to develop new structures of support, there was not the continuity and depth of commitment that were required. There was no belief among youth that they could rely upon the availability of long-term programs. And yet it was not an issue of not caring; there was no lack of community support for specific programs or activities. Any school event, for example, had packed halls and parents baking and bidding on auction foods as fundraisers, and the churches would easily raise several thousand dollars to send a group on a mission to build homes in the Caribbean. Leadership was, and remains, the key issue.

The problems are deep, and even the good will of many cannot quickly undo generations of independent spirits and isolation that have not only underpinned struggles for survival but that have also inhibited resilience and the ability to adapt to new realities. As the community evolves between "yesterday and tomorrow," it is the women and youth who most obviously bear the strains. But, as counsellors will testify, in a less public way the men too are beginning to acknowledge the tensions concomitant with change and to consider their options.

Youth

Many that are first shall be last; and the last shall be first. ~ (Matthew 19:30)

Biblical wisdom seems especially appropriate in thinking about how youth are key to any discussion of encouraging resilience and adaptive change on Grand Manan. They are more than simply an age category in the island community: they are its future. But they are also defined largely by the island's past, very much between history and tomorrow. In this book so far, it may not be obvious that youth are central actors; but, of course, such an impression would be erroneous. Youth reflect the struggles of the past and they represent the challenges of moving forward within new cultures of being and becoming. The boundaries for them are blurred zones of transition that may appear either as impermeable barriers or as surmountable hurdles. There have been no clear routes through these edges and spaces of meaning, despite the interventions of parents, churches, and counsellors. Successes and failures are never clearly defined, one turning into the other as young people seek the security and meaning that can sustain them through future decades.

For the discussion that follows I begin, in one sense, with the end, briefly describing the tragic events that took place between 1999 and 2002, in 2004, and in 2006. For islanders there is no question that they define both an end and a beginning, despite the eventual denouement in July 2006, which may have marked the climactic closure to a period of turbulence, alienation, and dashed hopes and dreams. I then go on to provide an overview of the milieu within which youth have grown, based on interviews and data collected over the period from 1995 to 2006, finally returning to consider the series of tragic events in light of these complex and often contradictory indicators of struggles and hopes for a secure future.

The Events

- 18 September 1999: Kristin Leonard (age twenty), a passenger in a car, is killed in what is believed to have been an alcohol-related accident late one evening. Some people complained that there were no consequences, no proper investigation following the accident.
- 30 June 2000: Tim Rideout (age twenty-one), a recently arrived Newfoundlander and father of a young child, is killed in the middle of the day by an allegedly drug-impaired driver whose vehicle was on the wrong side of the road. Again, there was a sense that it was unfortunate but that nothing could

be done, despite the eventual conviction of a young man for unsafe driving, which resulted in a short prison term.

- 16 September 2000: Ryan Cronk (age twenty-one) is killed in a truck with two other (uninjured) passengers while driving away from a party at the Whistle around midnight. The investigation never clearly established who had been the driver of the vehicle. The tow-truck driver reportedly did not smell any alcohol at the scene.
- 10 May 2001: Haydn Cronk (age nineteen) is killed around 3:00 AM walking along the road, apparently returning from a beach party at which it was alleged that copious drugs and alcohol had been available. There were multiple conflicting stories regarding the events that led up to his death, and the specific cause of death was never known (although at least one car was certainly involved).
- 22 July 2001: Alex Parker (age seventeen) is killed around 7:00 AM alone in his Pontiac Grand Prix, apparently driving too fast to get to his job on a salmon site after a late night out.
- 9 October 2004: Four young adults in their thirties are killed late one night, apparently driving too fast around a bend in the Ingalls Head Road.
- 22 July 2006: The home of an alleged drug dealer is burned to the ground during a riot involving about forty people. This event made national news, and a year later an article in the *New Yorker* told the story (Calvin Trillin, "The House across the way," July 2007).

Five young people killed in drug- or alcohol-related accidents within less than two years in a community of 2,700 people; and nine killed in almost five years. These are statistics that probably cannot be replicated anywhere else in Canada, with the possible exception of First Nations reserves. As the community struggled to understand the roots of the problem, many pointed the finger at inadequate policing; others blamed poor parenting. No one, however, could suggest a solution that might provide immediate respite. And, almost unanimously, the community voiced the opinion that, really, it was no different here than anywhere else. Thus, even as community meetings sought to bring people together to build a consensus towards adopting new initiatives, people remained frustrated, angry, and fearful. These emotions erupted on the night of 22 July 2006, when approximately forty people, wielding baseball bats, knives, and guns, burned the house of an alleged cocaine dealer. The story made national news. In 2008, the event is still referred to as simply "Cedar Street."

What was going on?

In the discussion that follows I rely upon a variety of sources, including high school yearbooks and personal interviews with youth and adults. The yearbooks proved to be a rich source of information. With help from a teacher, they provided me with a census of those who were still on the island as residents, dating back to the 1960s, data related to male/female participation in school sports and leadership activities, and information about the deaths of young people (through the "In Memoriam" tributes that appear at the front of each yearbook).

The process of interviewing young people and the nature of the results deserve special comment. While my use of a tape recorder when interviewing adults on the island (with the exception of the Newfoundlanders) was invariably vetoed, the young people, without exception, actually seemed pleased that I thought that their views and opinions were worthy of being recorded. The tapes reveal a fairly homogeneous view of island living, with notable distinctions between the worldviews of girls and boys. But, overall, they reveal some startling perceptions and values that were at odds with what their parents would have expected (at least according to the parents with whom I subsequently spoke). Most of the youth interviews were conducted during the summer of 1999 and the spring of 2000, and the ages ranged from twelve years to twenty-two years. I include in my discussion some comments by young people in their early twenties, even though they might not properly be defined as youth. As well, in the period between 2004 and 2006, with the help of school staff, the data were updated, showing encouraging evidence of turnarounds in school dropout rates and in the number of those continuing on to postsecondary education.

The taped interviews were usually about ninety minutes long and were based upon a common set of questions whose purpose was to gain some insight into how youth experience and perceive life on the island and the ways in which gender shapes their choices and options for the future. The questions explored values and expectations, religion, sports, leadership, drugs and alcohol, village affiliation, and leaving the island. Perhaps the most illuminating question related to how youth would rank particular areas of their lives in terms of what was most important for them, such as education, the church, friends, and fishing. There was a distinct gender differentiation in the responses to the entire range of questions, but this was especially the case in the ranking of areas of importance. There was also an interesting degree of significance attached to religion, which contradicted their behaviours. The behaviours, values, and experiences of youth often seemed to be at odds with those associated with a traditional, resource-based society dominated by patriarchal norms. Younger women are beginning, in small ways, to challenge their expected roles. As one older woman said to me regarding the growing number of young women who can be seen jogging along the road: "Ten years ago they would have been seen as looking to be picked up." Indeed, a woman in her late seventies who had been an avid walker until her friend and companion was no longer able to accompany her said to me, "I don't walk now because I'd feel like I stand out too much."

In the question related to ranking areas of importance, one of the most interesting responses was initiated by a young girl who said that her most important goal was to be her "own boss." As a result, that response was added to the choices offered, and it was invariably given a very high-ranking by the young women. For the boys, of course, it was not even an issue; they simply took it for granted that they would have control over their lives.

A second interesting pattern is the relatively high status given to religion. Whereas, according to many of my informants, including youths themselves, the attendance of teenagers at church has fallen off in recent years. But the fact is that many go to church camp in the summers and attend Bethany Bible College upon completing high school. In other words, they continue to strongly support religious institutions, including through participating in several church-run youth groups. What is significant here is not only the adherence to religious values but also the significance of gender relations in the context of church teachings. The continuing power of the church over young people will undoubtedly have a profound impact upon the patriarchal structures of tomorrow. This may be seen both through husbands who insist on their wives' attending church, often without themselves attending, and through strong normative statements from the pulpits. As one forty-year-old woman said to me, "The Bible says men have to be head of the family." Interestingly, even as she expressed her support of this idea, she described her efforts to be trained as a truck driver and showed me the manuals she was studying to learn how to use air brakes. As noted earlier, the ambiguous sign outside the Grand Harbour Baptist Church in 2004 that proclaimed OBEDIENCE IS THE PATHWAY TO JOY suggested a particular interpretation of family relationships.

However, the undeniable power of the church in the lives of islanders is being questioned by youth in tentative but significant ways. For example, one teenager ranked the church as of top importance for him; yet, when I asked what would be the single most important change he would like to see on the island, his response was: "Have the church butt out!" He noted that whenever some new activity was proposed for teenagers, the churches seemed to have a reason to be against it.

In exploring how younger women are dealing with the strong patriarchal relations on Grand Manan, I asked adults how young girls seemed to be responding. One articulate thirty-seven-year-old, recently divorced, said:

> The young women are starting to change. Very few are getting married young. There were some married in my class! When I graduated three were pregnant. The girls are more scrappy now ... one girl beat her boyfriend up outside in the schoolyard. They're more aggressive, more physical. They spit at each other. Some people say it's because they see it on television; but I don't think so. I think they see it at home. They've been hiding under the bed for five years, and now they're coming out.

Her observation seems to be corroborated in the following comments of a seventeen-year-old girl whom I interviewed 1997. She referred to her temper and said that she often gets mad at her boyfriend "and even attacks him." She "feels badly afterward" and doesn't know why she does it. Nonetheless, eight years later she was happily married with a young son.

Quite another youthful perspective on gender relations was given to me during an informal conversation with two young men, ages twenty-one and twenty-eight, respectively. One a fisher, the other a worker on the ferry, they both had new vehicles and cell phones, and one had his own house. Both had girlfriends, to whom they both referred as "the boss." While one may be tempted to dismiss the comment as a common sort of complaint, in fact they went on to discuss why they felt this was an accurate description. Both agreed that women make all the household decisions, control the money, and set the rules within a social context. Whereas the men dominate the exterior spaces – the woods, the wharves, and the boats – at home it is the women who determine the rhythms and details of life. That being said, men dictate whether or not women are allowed to engage in activities outside the home. They insist on the women taking the children to church, regardless of whether or not they too attend; they will not allow their wives to socialize in the evenings; and they may try to control where they work outside the home.

Gender relations are potent conveyors of culture, beginning with their formulation within families and youth relationships. The ways in which teenagers use space on Grand Manan reflect both the continuing significance of patriarchal structures and unequal gender relations on the island and the weakening boundaries between male/female spaces of meaning. Doreen Massey (1994, 1997) has explored the importance of spatial attributes of gender relations, showing how mobility over space reflects and preserves a balance of power that characterizes relations between men and women. She has also considered the meaning of social spaces for the nurturing and expression of youth cultures: "Social spaces are best thought of in terms of complicated nets of interactions in which each particular culture is differently located, but as networks that are certainly not tidily organisable into distinct 'scales'" (Massey 1998, 124). Massey asks questions about the importance of understanding youth cultures: "Was the youth culture of the Yucatan countryside the entry-point for external influences into (and maybe the eventual break-up of) this inherited Mayan culture which had lasted so long and endured so much?" (121). As I considered the behaviours and social activities of Grand Manan youth I began to wonder about this possibility – that youth were in fact both the carriers of the new elements of habitus and, at the same time, the mirrors of resistance. If we were able to take advantage of this dynamic potential, it could provide the community with the vision and focus to develop effective programs and activities.

For teenagers on Grand Manan, the access to vehicles is the single most important requirement for status and power among their peers. The differences between boys and girls in this regard are stark. It is the boys who drive the trucks, have the ATVs, and whose mobility ensures their claim over wide island spaces. In the same ways that their elders acknowledge each other with the raised finger salute as they pass in their trucks, male youth are out networking and solidifying relationships within public spaces that have been largely denied to women. There are distinctive spaces of activity not only with respect to dating but also, and even more significantly, with respect to informal activities and times spent "hanging out." For the boys, hanging out involves groups, throwing "hoops" at the Boys and Girls Club basketball court, or driving around playing "car tag." For the girls, in contrast, hanging out always occurs in the homes of friends, usually with only one or two others. This distinction between home versus public space, and large group activities versus small group activities, as it relates to teenage boys and girls is a direct reflection of adult patterns, and it seems to be perpetuating the traditional separation of both male/female activity patterns and the spatial relations that define them.

Not surprisingly, one area in which gender relations are played out is sports. Two sports that dominate recreational and school activities are basketball and baseball. Baseball has a long tradition on the island, with a fierce rivalry between villages that essentially ensured that community identities were male-defined. Basketball, on the other hand, is totally connected to the school. Between 1990 and 2005 Grand Manan, with a population of only 2,700, won the senior boys provincial championship four times. Between 2003 and 2005 the boys' team won three years in a row and, as a result, was moved into a higher league. Even at the higher level their strength took them to the provincial championship again in 2006. The girls' teams were in the championship finals each of the years between 2004 and 2006. Then, on Friday night 24 February 2006, the girls won their first provincial championship in over twenty years. Interviews, photographs, and media coverage followed. For these weekend tournaments, always in Fredericton, about five hundred people leave the island to support their teams. The diversity of strategies to ensure a place on the ferry reflected the importance attached to getting to the games. The cheers could almost be heard on Grand Manan. The popular expression on those weekends is: "The last to leave should turn off the lights." The island was proud indeed. A year later, in 2007, both senior teams, girls and boys, won their provincial championships. Hardly a soul remained on the island that weekend.

The importance of the championship weekend was underlined in 2006 by the request made to the provincial minister of transport to allow a change in the ferry schedule for the Saturday return trip. Announced at the previous weekend games, MLA Eric Allaby thanked the minister for "the delay to permit the many fans of the Grand Manan Breakers to return home on Saturday" (*Saint Croix Courier*, 21 February 2006, B2). The newspaper headline read: "Ferry Delayed Saturday for Final." So important is basketball as a source of status among peers and community both, that when a new principal at the high school (in 2000) insisted on enforcing the rule of a pass in all courses if students were to be allowed to play on the team, parents threatened to sue when their sons were excluded due to low marks. However, this high esteem and deference towards basketball is restricted to the senior boys team. When it comes to girls, there is a much lower profile – and smaller crowds.

One seventeen-year-old complained that the girls' team not only received less support in attendance but also fewer tangible resources from the student council. When one of the star players on the boys' team was asked about this, his comment was: "The girls are hobby players. They're less willing to learn." The consensus seems to be that the girls don't count. At the 4 June 2001 community meeting described above, one woman talked about her experience playing volleyball and reaching the provincial championships. "Only three people came to those games," she said, "But at the boys' basketball tournaments half the island was there." She used the example to underline the need to support and nurture boys and girls equally, and to point out the existing discrepancies that permeated island culture. The effort of coaches to mitigate the inherent discrimination was apparent in the March 2000 edition of the *Island Times*, which carried a full-page story about the successes of the girls' basketball team that year, written by the wife of their coach. Several years later the girls won the provincial championship.

The interviews with young people also revealed a deep sense of difference between the roles of males and the roles of females in relation to decision making. The interviews all included the following question: "Tell me about who you think are the leaders in your class or social group." A twelve-year-old boy rhymed off a series of names, all of which were those of boys. I asked him whether there were any girls that might be included. "Well," he said, "maybe. They sort of are but really not." This same boy, when asked what he might like to do when he was older, said: "I wouldn't mind fishing. Or maybe a vet; they make lots of money. Or maybe I'll be a plumber or electrician. There are lots of single women around, who are divorced, and they need someone to fix things." Interestingly, the confidence and sense of power articulated in naming himself and his friends as the "popular leaders" was carried over into his rankings: he did not have to rank "doing what I like" highly because he knew he would be able to do exactly what he liked. Not one of the interviews with girls indicated this level of self-confidence.

Another distinction in youth behaviours relates to residency in late teens and early twenties. Whereas girls almost invariably live at home until they become attached to a male partner, as soon as boys reach twenty or twenty-one they have made plans, or have actually begun, to build a home. Large homes for men in their early twenties (who do not necessarily have steady girlfriends) is not an unusual situation. And these homes are often built by the young men themselves. Because the families often have large amounts of land or have access to woodlots, the men begin cutting the trees and preparing the lumber for their homes over a period of a couple of years.

Two interesting patterns seem to refute the contention that girls have low levels of self-esteem. The first is the historically higher level of educational achievement reached by girls, which is not unusual in resource-based communities; and the second is the significantly higher proportion of young women as opposed to young men who leave the island following graduation. For the entire population over twenty-five years of age, in 1996 only 54 percent had a high school certificate, compared to 61 percent for the Province of New Brunswick (StatsCan 1996). Ten years later, if the older age cohorts are subtracted, we see that in 2006 the Census reported that over 70 percent of the population between ages twenty and thirty-four had a high school certificate, compared to 90 percent for New Brunswick (StatsCan 2006). Clearly, the average educational levels were rising. But, when one compares those who have completed high school, there is a striking discrepancy in island educational patterns between females and males. Forty-seven percent of the men compared to 62 percent of the women have high school certificates – a disparity of 15 percent. The main reason for this, of course, is that boys have historically dropped out to go fishing with their fathers. Now that there are strict provincial laws in place assuring that youth must stay in school until age sixteen, this is changing. Nonetheless, school data suggest that the change may be related to more than legal requirements. Concerned that Grand Manan had the highest dropout rate in District 10, a school committee was established to explore various options that might offer solutions. Its figures, shown in table 10.1, show a high dropout rate until 2004, when there was a significant turnaround. The sharp increase in dropouts in 2002 reflects the departures of young families following the events that occurred between 1999 and 2002.

Table 10.1
Dropout rate, Grand Manan High School, 1992–2005: Grades 7–12

Year	Enrolment	No. dropouts	Dropout rate
1992	236	13	5.5
1993	244	10	4.1
1994	248	9	3.6
1995	232	10	4.3
1996	230	10	4.3
1997	221	15	6.8
1998	206	7	3.4
1999	210	16	7.6
2000	211	16	7.6
2001	198	5	2.5
2002	196	16*	8.2*
2003	204	9	4.4
2004	214	4	1.9
2005	231	5	2.2

Source: Prepared by the Grand Manan High School administration.
* Reflects families moving away.

One explanation for this is directly attributable to the efforts of the high school to improve its retention rates. According to teachers in the school, several initiatives after 2001 contributed to the improvement, which was seen by everyone as a tentative sign of hope for future graduates, with implications for island culture. Before elaborating upon these initiatives, further discussion of the history of migration patterns and the context of peer groups is helpful in understanding the extent to which relative isolation has been important to the intransigence of cultural change.

Migration patterns and outside contacts have been an often neglected aspect of cultural change, especially in the context of small communities. For Grand Manan, the pattern of net migration has been especially important because it relates to expectations, goals, and the future direction of the island. It is also significant for its gendered distinctiveness, with a predominance of female departures throughout the period. As the accompanying graph shows, increasing numbers and percentages of young people chose to stay on the island.[2] There are three distinct periods of migration history: (1) the early 1950s, when over 65 percent of islanders left the island permanently; (2) between 1956 and 1970, when slightly more than 50 percent left the island; and (3) after 1970, when fewer than 30 percent left the island permanently. Figure 10.1, representing the period from 1948–49 to 1996, shows the number who have stayed (H) versus the number who are living away (A). The line above the bar graph delineates the percentage of all graduates who stayed away. The reference point of 50 percent serves as a basis for comparing the earlier period from 1950 to 1970 with the later years after 1970, when a higher proportion of graduates stayed on the island. This later period clearly reflects two factors. One is the difficulty of finding a job on the mainland during the recession of 1972 to 1974 and the other, more important perhaps, is the new fisheries policies initiated in the 1970s, which provided generous loans and financial support to acquire new boats and equipment. The more recent history (since the early 1970s) of very low out-migration has been an important factor in the continuing strength of the culture of the island, especially since it has been combined with low rates of in-migration.

The high school yearbooks also provided some insight into the number of students who were dropping out of school prior to the completion of grade 12. By comparing the photographs of the same cohort for grades 9 and 12, it was possible to derive an approximate figure for the dropout rate. In the period before 1974, an average of 67 percent stayed in school until grade 12, whereas between 1980 and 1997, the average was 78 percent. In comparing the male-female ratios for these cohorts, it is apparent that a substantial gender difference has characterized the decision to stay in school. Even in 2003, boys told their teachers that they would be out of school as soon as they were sixteen because they could make "good money" with their fathers on lobster boats or working on salmon sites. "What could I say, that lobster might not be here when he's forty?" sighed one teacher. One year there were nine "social promotions," which involved boys who were too old to stay in grade 8 being passed up to grade 9 to "mark time" until they could leave school. Underlying these numbers is a widespread attitude that education is not important. The striking contrast with students on the mainland was made clear to one teacher who had moved to a school near Saint John. Having never experienced a different school environment before her move in 2002, she was startled by the differences between the two schools. She felt energized by students who "actually wanted to learn," who understood that "education is needed to get a job." Educators on Grand Manan acknowledged the need for new approaches, but the challenge was how to define and implement them. One important step was the decision in 2002 to hire a special education teacher who could work with a few students individually. According to at least one of these students, that decision saved his life. But the task of creating permeability within the strong youth culture was daunting.

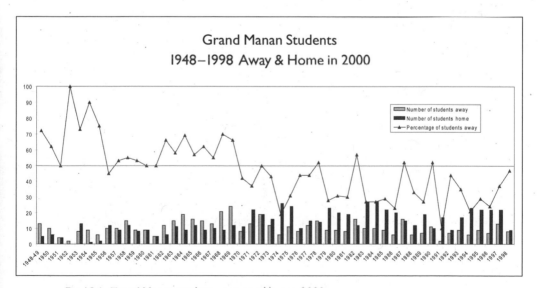

Fig. 10.1 Grand Manan students, away and home, 2000

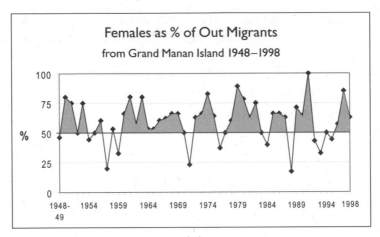

Fig. 10.2 Females as percentage of out-migrants, 1948–98

   While the higher percentage of girls graduating may not be unusual in rural communities, the figures for out-migration are quite startling in their gendered distinctiveness. Again relying upon yearbook information, it was possible to figure out the male-female split for the number of students no longer living on the island. Figure 10.2, showing the percentage of migrants who were female from 1948 to 1997, confirms what many people had reported anecdotally, that it was girls who chose to leave upon graduation, frequently marrying on the mainland and not returning. From 1948 to 1996, more than 50 percent of migrants were women in thirty-one of the forty-eight years represented by the graph. On average over the period, women account for 60 percent of the people who left Grand Manan and stayed on the mainland. Through the 1990s, the female-male difference was not as pronounced, the slight continuing disparity linked in part to lack of job opportunities for women on the island. Moreover, with their higher educational levels, the young women departed with an expectation of better job opportunities than the young men might have anticipated.
   With the closing of the smoke sheds and the fish plant, the women had fewer job opportunities and this forced them to broaden their searches. The long history of entrepreneurial spirit on Grand Manan notwithstanding, island women tended to look to tourism for business possibilities. As well, several women have successfully completed correspondence and internet-based courses to upgrade their nursing qualifications. As examples for their children, they may have contributed to at least one applying successfully for an exchange pro-

gram to Ottawa and a trip to Britain. For the children who leave the island on these excursions, because they have been more isolated than have mainland students, even short exchanges of a couple of weeks can have incalculable benefits. The implications affect both the individuals and the community.

For youth, growing up on Grand Manan was a challenge made more difficult by the limited range of options they had for friends, jobs, and recreational activities. Katz (1998, 136) argues that "the key construction of childhood and youth, in theory and in practice, is that of a time or space 'without walls,' when and where all futures seem possible, even as everyone knows they are not." While her work in Sudan explored the ways in which the importance of youth development was related to community cultures, Katz did not examine the possibility that real boundaries of isolation, such as with regard to an island culture, might constrain the evolution of new relationships. For Grand Manan youth, not only were their lives rooted in histories tied to the fishery but their futures were also constrained by a culture that did not value education, had normalized excessive drinking outside the home, and had ensured that segregation of the sexes was endemic. The social lives of teenagers were defined by a combination of factors, including being rooted in family reputations, church affiliations and village identities, and the current socio-economic realities of their peer groups. Athletic ability on the basketball court, especially for boys, was a key component for high status among peers: "Sports is the greatest barrier, especially if you're into scientific stuff. The same people are in all the sports. If you're not on a team you can still be in a group. But money doesn't help." While interviews with young people indicated a consensus about the importance of sports for peer group status, there was less unanimity related to the relevance of perceived wealth.

For the girls, while participation on basketball and volleyball teams tended to define the leaders, their peer group status was more obviously affirmed through leadership on the student council or yearbook editorial boards. A review of the yearbooks through the 1990s clearly shows that certain families are represented throughout the decade. Moreover, these same names were mentioned by younger teens during interviews as belonging to the leaders and role models. Significantly as well, these committees were dominated by girls, at a ratio of 5:1. The gender split, as in so many other areas of island life, was an entrenched part of high school committees and leadership. Despite its acknowledged importance in their lives and in the wider community, the church was not a major factor in determining peer group alliances. Asked about the role of religion, two boys said that they did not attend youth groups at church, although they did allow that they might be "good for those people." One reason cited for not attending was that there were "too many quiet people." Another teenager felt that the churches did not separate the groups and that most young Grand Mananers did "not profess to be Christians."

The self-defined social groups were an important part of youth culture on Grand Manan. While not precisely age-specific, the "troublemakers" were described as moving from a young Skunks, or Skanks, group into a group known as the Animals as they got older and became more involved in parties. Asked about names for these groups, youth were agreed that a marginalized minority were called the Skunks – a small group of at-risk kids who would hang out at the Seal Cove bridge or at the Boys and Girls Club. Mainly male, the Skunks were often the first to welcome the Newfoundlanders, which the latter saw as a mixed blessing and school teachers saw as confirmation of their suspicions that the Newfoundlanders were "backward." Many of them wore "weird clothes" and were suspected of stealing. Another group was the Partiers and its boundaries were less clearly defined than were those of the Skunks. The Partiers were variously described as being "from the lobster families" or the kids on the basketball teams. These criteria, of course, frequently overlapped. Included in this powerful group were two subgroups: the Animals and the Intimidators. The Intimidators (ages twelve to fifteen) sought power and wielded it mercilessly, according to two adults who were repeating stories told them by their children. A member of this group said: "My group is a gang. Mess with one of us and you mess with all of us. We're definitely the leaders."

A third group was variously referred to as the Quiet Group, or the Nerds. Some referred to them as "snotty" or "prissy," always having to "look good before we go to school." The members of this group often did not play sports; they might be included in some of the parties, but generally they spent most of their spare time at home, often on the computer. In any one class of thirty students, about 60 percent would be considered to be Partiers, about a quarter would be the Quiet Ones, and the remaining 10 percent to 15 percent would be the Skunks, who would become Animals in their late teens. A major dilemma for youth was that their inclusion in one of these groups was, as one teenager described it, like a "brand on my forehead." For counsellors, social workers, teachers, and parents concerned about the welfare of young people and the community, the challenge was to find a way to intervene to ensure that the boundaries between the groups would become permeable, allowing for movement between groups and the possibility for making new friends.

The Film Project, 2001–03

One program that sought to enhance the fluidity between social groups was inaugurated by the Boys and Girls Club over a period of two years (2000–02) through the support of a federal grant and, in the second year, Rogers Communications. Its goals, which were to encourage the participation of youth across all social groups and to provide a basis for social interaction skills, were framed within a film and television project. The multiple objectives of the program included: (1) to teach young people the technical aspects of film and television production; (2) to provide a vehicle for enriching their understanding of island life, its history, and its people; and, (3) to contribute to their education and social skills with respect to creating a story line, interviewing, and understanding the context of news events. As a non-athletic endeavour it was hoped that it might appeal to youth who otherwise were seen as outsiders within the mainstream youth culture, helping to boost their self-esteem and increase their status within the community.

A professional journalist and filmmaker was hired in each of the two years to train youth in the basics of making short videos for the local cable channel in collaboration with the high school, which offered the project as course credits. The students, who volunteered for the project through the school, were mainly fourteen or fifteen years old. Both years their numbers were small, ranging from four to nine, mainly girls. They were encouraged to choose topics that would be of interest to islanders and to produce short films for a program that was broadcast a couple of times throughout the week. They learned basic filming techniques in covering the always popular basketball games live; they interviewed older-generation islanders who had special craft skills or who were known as exceptional storytellers; they conducted impromptu survey-interviews when there was an issue (such as the proposal to stop broadcasting the monthly council meetings); and they covered Remembrance Day by interviewing those who had participated in the Second World War. The project enabled the youth to learn and practise a broad range of skills, including video editing, writing scripts, planning film stories, and interviewing. One of the participants went on to lead her graduating class in academic achievement and was the valedictorian in 2006. Two of the other students continued with film studies for a short time after leaving high school, attributing their interest to the project. Ultimately, while the project was enormously popular among islanders, who enjoyed the local television programs and seeing themselves reflected through them, it had to be discontinued because of lack of funding.

Many worthwhile federal programs have two key constraints built into them: the necessity to reapply every year, a process that requires both expertise and enormous time, and the necessity for the programs to become self-sustaining (i.e., self-funded). In a small rural community that lacks educational resources, the volunteer expertise that originally initiated the project, which involved several weeks of full-time work to develop the proposal and to submit it, cannot be relied upon over the long term. And, over the long term, the community simply cannot cover the cost of hiring an outside professional. These are realities. As for the success of the project itself, the youth who participated were enthusiastic, and they especially appreciated the feedback they received within the community. There was evidence that the project boosted student self-esteem and provided activities for some who were not involved in sports. That the goal of enhancing group fluidity did not seem to have been met probably had more to do with the short duration of the project rather than with the nature of the project itself.

The topics that students chose to present reflected to an important degree their perceptions of how the community should be viewed and how they had heard it described by others. Their willingness to question or confront problematic issues was never tested, mainly due to the short duration of the project. Indeed, one of the filmmakers who arrived in April 2001 expressed frustration with the need to carefully edit material that might appear to take sides when the aquaculture expansion in that year resulted in major protests by the lobster fishers. Nevertheless, even the discussion related to the necessity for non-controversial presentation required the students to consider the meaning of this conflict for the community. As Cohen (1982) has shown in his studies in Scotland, people may not be aware of their culture (except through their assumed identities) unless they stand at its boundaries. One of the values of this film project for youth was that it deepened their understanding of their "culture by trying to capture its experiential sense" (Cohen 1982, 4). In explicitly presenting island news and events, the students were being introduced to the experiential learning of the roots of their own identities. Community television allows for the situated, contextually rich reflection of identities and cultural forms, and this can encourage negotiated processes of change.

Two distinct strands of media-culture research are: (1) in North America, a focus on individual attitudes connected to audience studies; and (2) in Europe, an emphasis on the production of meaning and the relationship of media to other cultural forms (Burgess and Gold 1985, 4). In Canada, there is a historically significant relationship between communication media, cultural representation, and national identity. The intellectual contributions of Harold Innis (1951) and Marshall McLuhan (1964, 1962) were conceived in a Canadian context: the need to rely upon new communication technologies due to the vast spaces to be negotiated. Collins (1986) has suggested that neither of them would have developed their theories of the importance of the mass media without the experience of the Canadian landscape and vast spatial distances. Indeed, their scholarly ideas underpinned the subsequent establishment of the Royal Commission on the State of Arts, Letters and Sciences, which led to the Massey Report, which, in turn, "laid or rebuilt the foundations of many of Canada's cultural institutions," including the CBC and the National Film Board (NFB) (Friesen 2000, 196). In the 1960s the NFB developed a new concept of filmmaking that was grounded in the philosophy that *process* was more important than the final product. Founded as the "Challenge for Change" series, the film board sought to illuminate social problems and suggest ways in which they might be resolved through filmmaking. In so doing, it was asking questions such as: Can a film project serve as a cohesive agent and catalyst for change within a community? What role can film play in participatory democracy? The most famous of the films made during this period were those of Fogo Island, a series that became an iterative process of filming, discussion, and editing within the community. The strategy itself became known as "the Fogo process" (Bredin 1995, 180).

A related initiative in Quebec a decade later was linked to the nationalist sentiments in that province. Started in 1973 and entitled *Programme d'aide au developpement des medias communautaire,* its aim was to generate solutions to social problems while encouraging a sense of collective cultural identity among Quebecois (Stiles 1988, cited in Bredin 1995, 177). A third major initiative in Canada was the Northern Pilot Project sponsored by the Department of Communications. Designed to engage communities in conversations among themselves about regional communication priorities, and to contribute to the knowledge base related to technological infrastructure, this project was much more limited in its reach. But, in a subsequent attempt by the NFB to engage the northern Inuit in the production of locally relevant programming, in the late 1970s Wolf Koenig introduced film production to Inuit youth ages ten to twenty-five. The Northern Pilot Project was very similar in concept and results to the Grand Manan project, and Koenig commented that, even when they did not complete an entire film, they did experience a new way of seeing and of "framing reality" (Roth 1983, 109). Significantly, I think that, as on Grand Manan, the project chose not to examine socially or politically sensitive issues, and this suggests an interesting research possibility. Certainly all of these projects point to a tradition within Canada of exploring cultural and historical realities through film and through engaging local communities. It is, in my opinion, unfortunate that the lack of continued funding for the project on Grand Manan did not allow for more sustained experiential learning, which might have had greater long-term benefits for the community.

Until the advent of the film project in 2001, the local cable channel on Grand Manan had been used exclusively for bingo, advertisements, and, since 2000, the monthly village council meetings, which were broadcast live. Occasional tapings of basketball games were the only other transmissions. Starting out with a team of nine students, and in collaboration with the high school, the "media team" gradually lost members until, by Christmas, there were only four enthusiastic participants in the project. One student in particular was especially engaged with learning about developing a story, conducting interviews, editing, and basic techniques of television production. Naomi Parker enjoyed the challenge of the new technology and her new-found status within the community. The team tried to cover local stories, such as the confrontation between aquaculture and the lobster fishers, as news; however, for the most part, their focus was on developing series such as "The Great Outdoors" (later "Island Focus") and a cooking show, all of which were very popular with islanders. Some of the most successful programs were those that filmed interviews for Remembrance Day, providing opportunities for the young people to meet Second World War veterans and to hear their stories first-hand. Said one islander who had seen the interviews, "We never really knew him like that. I never realized Harry had suffered so much." One girl described meeting a bird carver, with whom she had never spoken, and enjoying sharing his pride. "He seemed really proud of his work and was glad to be able to show it to people. Most islanders did not even know what he was doing," she said. One of the episodes filmed the women working in the fish plant, their hands working quickly with scissors, snipping heads and tails as the tins moved past. This work is dangerous and physically demanding, and for many islanders this short program, during which the plant manager was interviewed, provided their first glimpse into the working experience of these piece-work-paid women. Through this film, for the first time the women's work was publicly validated. One of the women expressed pleasure that islanders might be interested in them. Another islander noted that part of "island history is being preserved." This comment was more prescient than he realized since, less than two years later, the plant was closed.

Another aspect of the film project, which related to the social and personal lives of youth, was explored during an interview with the second instructor, who arrived in May 2002. Her encounters with the young girls who were involved with the project revealed a deep well of longing for validation and, quite simply, for someone to listen to them. She soon understood that many of the youth were desperately searching for affirmation of, and outlets for, their creativity, and she expressed concern that the small-town nature of relationships, having to "tiptoe around everybody" to avoid controversy and the risk of "offending anybody," had enormous implications for the success of any local news television. But she was enthusiastic about the framework of the project, and she estimated that, had it continued, the number of participants would have doubled. Having received phone calls from teenagers, often in the middle of the night, she became aware of their need for greater support from parents and more outlets for their spare time. The key, she felt, to breaking the cycle of substance abuse was to ensure that more young people could become role models for others: "The kids here cling. They need to eat lunch with me, in my room. And they call all the time, even at home. They want to talk, not necessarily about problems, just to talk."

This view was corroborated by one of the speakers during the 4 June 2001 community meeting, which was called to consider the problems of drugs and alcohol on the island. The woman talked about the "kids who come to hang out" at her house and how they "need someone to talk to." She also said that island gossip held that she was giving the kids alcohol, but, she emphasized, "That isn't true." In her quiet, articulate message she conveyed the complex undercurrents of island culture, which assumes that it "knows everything" but that in fact often misses basic truths, thereby confounding the ability to understanding problems.

In assessing the potential for youth to develop secure identities in the context of rapid change, there seem to be several key factors. One is the family history of contact with the mainland. In other words, those families in which one parent is from-away, or at least has lived away for more than two years, seem to have a greater potential for avoiding the extreme problems of substance abuse than do other families. It is also these families that are more disposed to take advantage of student exchange programs that enable children to live in a different province or country for a period of about three months. Asked about the success of these programs, the consensus was that young people were afraid of being homesick and that parents were not very supportive.

Problems and Solutions

Even as a community discussion opened up after the events of 1999–2001, the use of more potent, injected drugs was increasing. In 2003, a high school teacher estimated that about 50 percent of the children in grade 8 had tried drugs, while probably one-third of the students in grade 11 were regular users. At the 4 June 2001 meeting, one of the emergency responders pointed out that a phone call was all that was needed to have drugs personally delivered to your home. In 2004, the addictions counsellor reported approximately ten intravenous users on Grand Manan. At the same time, break-ins at local stores and the pharmacy indicated an unprecedented desperation for money. In the summer of 2004, three break-ins in a two-week period were reportedly for money and prescription drugs. Despite the painful events that had precipitated the 4 June 2001 meeting, little seemed to have changed.

While some argued that the core problem was with a small group of young people, the effects rippled through the community. As the parents of one of the dead youths said, "It's not just the kids. It's their parents too": a situation that had been referred to at the 4 June meeting. It was becoming widely acknowledged that the problem of addictions was related to the use of drugs and alcohol on the part of parents, for whom the public consequences (i.e., vehicular deaths) were negligible because they were in their homes and camps. The youth, on the other hand, were out on the roads. So while the data represent alcohol and drug abuse among youth, there is every reason to suspect that the problem may in fact be no less serious among the older generation. In an incident reported at the high school in 2003, a boy collected $300 from children in his class, which he delivered to his father, who provided them with drugs.

Woven into these tragedies, and certainly a related effect of substance abuse, is the problem of suicide. While one case in 1998 was clearly related to chronic bipolar depression, the others all involved young adults who had problems with drugs and alcoholism. Two suicide deaths in 2003 and 2004, and several attempts (1999, 2000, 2002, 2005), reflect the malaise, tensions, and insecurities that were permeating this small community. In addition to the widely known episodes, however, there were reportedly others that remained hidden, partly because of the strictures imposed by the church. Clients had told the island social worker that they had tried to "drink [themselves] to death". As she pointed out, "self-harming behaviours are not captured in official statistics."

Perhaps the most dramatic and potentially important impact of these crises on the community was the immediate departure of several young families. Not only did their leaving represent losses for their friends but also, and more significantly, these families were all professionals – potential leaders and role models for the community. Among the group that left in the period immediately following 2001 were three teachers, a registered nurse, a small business owner, a pilot, and the executive director of the Boys and Girls Club. These were educated articulate adults, whose young children were being raised in stable family situations. Their departures were important losses for the community-at-large and, especially, for the "tomorrows" of island culture.

However, for those who stayed there seemed to be a growing determination to find solutions. A tree planting at the high school, with tags bearing the names of each of the dead students, was one gesture by Mothers Against Drunk Driving, the purpose of which was to provide a tangible reminder of what had been lost. One of the less obvious impacts of the deaths was wider community participation in issues related to parenting. A few years earlier, in 1997, the creation of Community Reach in response to widespread public concerns about youth and problems related to alcohol and drugs had been a pivotal moment for the community. At its inaugural meeting, Community Reach attracted over 125 people, representing all ages and social groups. There was consensus when someone said: "Until we as a community demand action and take action, nothing will change." A few days later, a follow-up meeting of thirty people formed what became known as Community Reach, a dedicated group of volunteers that was facilitated and animated by a professional worker in the public health sector.

Nevertheless, in 2002 the counsellor was concerned that the programs at the Boys and Girls Club were still not attracting the teenagers from the stable families and that the at-risk kids were not receiving the parental and peer group support they needed. Asked specifically about changes in behaviours and attitudes since the deaths, the social worker gave a mixed review. She did not feel that there had been any change in the level of partying, but she did point to greater efforts by the RCMP to apprehend dangerously driven vehicles. However, even given the efforts of police, medical, and emergency staff, especially following the death of Haydn Cronk in May 2001, she was critical of the failure to take advantage of formal debriefing programs that could have contributed to more effective support for families and individuals in the community. She argued that the roots of real change had to involve peer counselling. Even a year after Haydn's death, there was no closure for the family or community. According to his parents, the truth about the details of how he had died and who else might have known about the events leading up to his death were never properly investigated or revealed. Both the parents and the community felt betrayed by the media and the police, whose reporting and investigation were seen to be ineffectual.

For young people who have grown up in a culture of parties and limited choices with regard to friends and activities, the possibility of suddenly adopting new goals and attitudes means changing their behaviours when they leave school. Expecting this sudden conversion is not only unrealistic but also obviously dangerous to the health of the community. By the time these young adults reach their late twenties, many are in serious trouble; a few will acknowledge that they need help. In 2002, according to the social worker, five individuals who had serious substance abuse problems had sought professional help to "get onto a new path." They were all "partied out" and were trying to figure out how to redirect their lives. One of the problems she identified was the tendency for service providers to be from-away, and for their positions on the island to be either of short duration (less than three years) or part-time (one or two days a week, with them being based on the mainland). The lack of continuity among service providers jeopardizes the client's ability to recover. The social worker described three categories of clients: (1) the long term "lifers," for whom rehabilitation or cure is extremely problematic; (2) adults, seen every two weeks, whose treatment usually extends over about six months; and (3) children and families whose problems are especially complex because of the need to coordinate with the school, which was not always very helpful.[3] Whereas her caseload, when she arrived in 1997, had been about half adults and half young people, by the time of her departure in 2002, about 75 percent of her clients were adults, many of whom had suffered sexual abuse. In any three-month period she would see about fifty different people. Of these fifty, about twelve would be youth; a few would be adult men, mainly with addiction issues; and the majority would be women.

Rather than depending upon individual counselling, after 2001 the community began to focus on more inclusive specific programs as well as projects within the school. There was a recognition that exposure to other ways of living could open windows to new points of view. Exchange programs had not generally been very popular on the island, although a few students had taken advantage of them in the past. Even a ten-day trip in 2002 to Baffin Island had received mixed responses. While for one girl it was the "experience of a lifetime," which she described in her letters of application to university three years later, for several others the short exchange trip was a less successful venture. One somewhat disenchanted father exclaimed: "All those kids got a shock! One girl was in a house where no English was spoken; they had to get a translator."

As in so many other areas, it is the girls who tend to take advantage of student exchange opportunities; and it is these same girls who, as young women, tend to leave for university and find lives away from the island. Families who have connections with people from-away or who have lived away themselves seem able to move off-island, either to university or for exchange trips, more easily than do families who have always been on the island. While school marks were one criterion for participation in the Baffin Island exchange, the male/female ratio of three to eight also reflects a gender bias regarding who chose to apply.

A notable exception to this tendency towards gender bias in these programs was the selection of two students from Grand Manan (one male, one female) to participate in a competitive international exchange in 2004–05 with students from the Netherlands. In a group of only twelve New Brunswick students, that

Students on exchange trip, 2002

two were selected from Grand Manan was a signal honour, of which islanders were justifiably proud. The two selected, Carrie Fleet and Duncan Cronk, received students from the Netherlands into their homes for three months in the fall of 2003, and then they spent three months in Holland. For the two students it represented an enormous leap of faith and trust in the possibility of establishing new relationships. For Carrie there were new horizons of opportunity; for Duncan, whose earlier school career, especially following the deaths of his two brothers, had been fraught with tension and resistance, the exchange confirmed a significant change of direction that was a prelude to the many honours he received a year later at his graduation. There was an outpouring of community support for him when he received his graduation certificate in 2006.

While the exchange program was a life-marker for one boy, it was the crucial role played by several young teachers and school programs after 2001 that enabled such turnarounds to happen. The continuity of purpose over several years and the positive reinforcement that students were given were important factors in the hopeful environment that seemed to be emerging in 2004.

The establishment of a program called Parents Supporting Parents (PSP) in 2002 was an attempt to address some of the deeply rooted problems associated with substance abuse. According to the island social worker, the deaths provoked some parents to seek counselling for their children. But she was concerned that they were asking for "someone to talk to the kids" rather than engaging with them themselves. There was still "a link missing between the kids and the parents." Identifying two key problems – lack of parental guidance and the need to overcome individual and community denial of substance abuse – the PSP project was designed to recruit at-risk parents. Its goal was to alleviate the ongoing requirements for professional support services related to drug and alcohol abuse. To an important degree, the PSP initiative was responding to the same concerns expressed in 1997, but with the powerful added incentive of five traffic deaths.

Funded by the federal government (Health Canada), the PSP initiative had some tangible benefits, including:

- fourteen parents involved in monthly meetings, beginning in August 2001;
- the continuation after January 2002 of a core group of seven to ten parents, which met both to fundraise and to organize an evening with a motivational speaker, scheduled for August 2002;
- the inauguration of a program for new parents of young children called Nobody's Perfect, with the participation of five parents;

- increased sensitivity within the community to the urgent need to change family relations; and
- a sense that "help" is there (from national and provincial institutions) when urgently needed.

Unfortunately, the overall success of the program in changing behaviours and expectations was limited. Originally seen as a support group for parents at risk, in fact the parents who participated represented a very engaged and concerned group whose teenage children were among the least likely to have major problems. One of the parents said, "My kids were facing things, and I needed to feel I was doing something." She compared her children's environment to what she had experienced twenty-five years earlier as a teenager on the island: "It's so extreme now compared to when we grew up. We need to give parents strength, to give them training." The second characteristic of the group, which participants commented upon, was the predominance of women. Of the ten most active parents, only two were men. This was felt to be a problem since many believed that men needed to be more involved with their children and that there were too few male role models on the island.

One service provider who was interviewed at the time of the deaths felt that an important benefit of the PSP program involved bringing parents together (an apparent irony in a small community), which challenges outside perceptions as so many parents feel that there "is no other place to go." The community networking the program provided was an extremely important side-benefit that some felt might continue even after the formal group was dissolved. Nonetheless, she said that there was room for broader participation, especially on the part of parents of elementary school children, who were not being reached by any of the programs. She argued that many parents did not understand the problem until their children reached high school and, especially, as they approach graduation. Her comment that the "dads and kids are not onside" pointed to the widely accepted knowledge that, although the mothers were actively seeking solutions, the rest of the family had not acknowledged there were any problems. As well, she pointed out that mothers seemed to want to be talked to, and to have their children talked to, but that they were not communicating within the family. Even immediately after the deaths, when debriefing was offered, only two teenagers arrived, and they "stayed only twenty minutes." She noted that even the programs at the Boys and Girls Club tended to attract only the at-risk youth and that social groups were very segregated. A retired elementary school teacher described a memorable moment in

her teaching days, when a "little blond fellow in grade 4 was eating his sandwich and said to her: 'I have a problem. My mom's name is Malloy; my Dad's
name is Bennett; and my name is Smith. And I can't figure out why.'" Part of
the difficulty in trying to encourage improved parental communication and
participation is quite simply rooted in separations, divorces, and single-parent
families, all of which involve adults in their own struggles, to the detriment of
their children.

Nobody's Perfect, another initiative that occurred during this period, involved a short series of meetings, in which young parents met with a facilitator
to discuss issues related to parenting. Karey Ingalls, the facilitator, worked in
the community to publicize these discussion groups. She also helped to organize the visit of a motivational speaker; contacted businesses on Grand Manan
to try to get them to ensure a drug-free environment through a zero-tolerance
policy;[4] helped in organizing an RCMP evening about drugs in October 2001;
and contributed to a workshop called "Parents Survival Training," which was
offered through the mainland drug rehabilitation centre. In other words, she
made considerable efforts to educate, sensitize, and involve the entire community in solving the drug/alcohol and family abuse problems.

With the unfortunate departure of the executive director of the Boys and
Girls Club in 2002, the volunteer board was forced to participate more actively in program design and planning. The new director, who had lived on the
island for two years as a teacher at the high school, brought energy and enthusiasm to the position but left after only two years to return to the mainland.
Whereas the club had benefited greatly by the long tenure of the previous director, as with other professional positions on the island, assuring long-term commitments was a difficult issue. The next executive director was the wife of an
RCMP constable, another short-term appointment because of her husband's
request for a reassignment following the riot in July 2006 (see above). The frequent turnover in executive directors had immediate impacts upon the programs that could be offered at the club and also resulted in a less concerted
effort to address the problems of at-risk youth. The focus shifted slightly to
developing leadership skills among proven student leaders. But there was also
an effort to encourage the development of role models, with the hiring of several summer students in 2004–05. These students took on leadership roles and
acted as role models even as they developed activities for the children. For
adults involved in the Boys and Girls Club, these student leaders were important signs of a growing awareness of the need to contribute to the island while
at the same time gaining important social skills that would boost confidence
and self-esteem.

Another positive step in community efforts to provide more support for youth was the building of a hockey rink through the volunteer efforts of a small group of fathers. Although the island weather meant a very short season, it was seen as a big success because it encouraged family activities. As well, there were hints of changing attitudes towards education over the long term, linked to the gradual effectiveness of school-based self-esteem building programs. These seemed to offer significant hope for the future, with evidence for progress apparent during the graduation exercises at the high school. In June 2006, eighteen out of twenty-five students had achieved averages of over 80 percent; while six had achieved averages of over 90 percent. Compared to earlier years, when less than half the class would go on to postsecondary education, in 2006 about twenty of these students had plans to attend college or university on the mainland. Entering faculties of science, arts, business, and nursing, all of them had a sense of direction. In the past, of those who went away for postsecondary education, almost half would be back by Christmas, and another half would be back before the end of the academic year. In 2006, there was no expectation that this pattern would persist. With scholarships of $181,000 (including the equivalent of $100,000 from the Royal Military College for a young man entering military college),[5] these students seemed to be motivated and goal-oriented. In the comparative scale of New Brunswick schools, Grand Manan High School had gone from twenty-sixth a few years earlier to second in 2006. So what had happened?

For some people nothing had really changed. Drugs and alcohol and the parties that went with them continued to attract a large number of youth and young adults. However, the 2006 graduation did not seem to be a mere transitory bubble. Talking with a young, enthusiastic, and energetic teacher in late spring of that year, I learned of the many initiatives that the school and the community worker had undertaken over the years since 2001. A native of Newfoundland working part-time in Nova Scotia, she had received an urgent call to fill in for a teacher at the Grand Manan High School in December 2000. She "came for three weeks and then just stayed," she said, marrying an islander in 2005. At about the same time several other from-away teachers arrived, and together they became a source of energy and new ideas that created a new environment for learning. When she first arrived she was told by the principal to concentrate on boosting morale and self-esteem. "Whatever it takes, even if you don't get in all the English and math, try to build up their confidence," she was advised. Her first project in late spring 2001 was to create a "Survivor" event, designed around the series on television. The students would go to an unannounced place (Ross Island) and have to undergo a series of tests and games that would determine who was to be voted "off the island." It was a huge success, talked about throughout the community and providing the students who won not only with a sense of accomplishment but also with a major prize, which included a trip to Toronto with their parents.

Meantime, the teachers were organizing new activities such as badminton and cross-country running; and a new annual activities banquet provided the opportunity for public recognition of student achievements. The new teachers had also consulted other school districts for ideas, and this resulted in the Positive Participation Program. The students were encouraged to offer volunteer help in the community (e.g., shovelling walks for the elderly, making cookies, or cleaning ditches). For each of their own volunteer community initiatives they would receive a ballot, which would go towards a monthly award given out at a general assembly. This rewards program quickly became a key motivator to involve youth in the community and to provide them with positive feedback. A similar program rewarded their academic accomplishments. Called Way-to-Go, this program provided students with accolades for such things as high marks, handing assignments in on time, or voluntarily asking for extra help. Again, they would receive certificates that acknowledged their accomplishments, which would then be filed in their individual portfolios. These portfolios gradually accumulated through the year, to be exhibited upon graduation in binders that included their extracurricular activities, sports awards, and job references. They were learning the importance of a curriculum vitae and how such a document could be prepared for job applications.

Another school initiative was the Alternative Education Class, which focused on at-risk youth who might drop out of school. It included programs such as boat building (provided by a school volunteer), and it, too, offered awards to those who showed evidence of changed behaviours. The Turnaround Award program was a provincial initiative begun in 2004–05, and the award was immediately won by a Grand Manan student, who received it from the lieutenant-governor of New Brunswick. As well as teacher initiatives, the community worker brought in programs such as Roots of Empathy, Girls' Circle, and Awareness Day Fairs, which addressed a variety of issues, such as caring for others, date rape, and teenage pregnancy. Asked what she thought was the most important change among Grand Manan youth since her arrival in 2000, teacher Angie Russell thought briefly and said: "Leadership; being involved; taking responsibility and giving back. There are kids now who coach Little League Baseball and who helped repair the market area." For the first time, she pointed out, there were fifty students enrolled in the Duke of Edinburgh Award program. There were several boys, whose future looked bleak in 2001, who were at the top of their graduating class in 2006. One of these had received a full scholarship to the Royal Military College and had graduated with an average of over 90 percent. Even as there was growing apprehension about the economy of the island in 2006, there was optimism about the youth. In their perseverance and successes, these young people seemed to be offering hope for the future of the island.

The importance of leadership capacity and trust as contributors to resilience and adaptive change is linked to the possibility for consensus regarding leaders and role models on Grand Manan. As pointed out earlier, there is little agreement about who speaks for the community. It is as though the history of fishers as individualists and, until recently, the relative isolation and separateness of villages have together inhibited the emergence of an effective leadership structure. While the church is a crucial basis for leadership development, the distinctiveness of the many denominations serves to undermine the emergence of a cohesive structure. It seems that the school has a crucial role to play, especially in an environment in which the "most confusing day of the year is Father's Day," and family structures are threatened as never before. In planning for the future, it is crucial that "the last shall be first." As the film instructor had learned, youth need affirmation and encouragement. They need someone to talk to and someone who will listen. The school programs and activities that have been introduced since 2001 seem to be gradually having an impact. There are no immediate solutions, but the long-term possibilities seem to be brighter than they were half a decade ago.

As the economy becomes increasingly tied to the globalized networks of aquaculture and tourism, there will be greater connectivity between new perspectives and different value systems. The culture of Grand Manan is strong, rooted in family, religion, and patriarchal values that continue to guide the lives of youth. But the family structures have disintegrated to an alarming degree. Perhaps the events that occurred between 1999 and 2002 provided the key to unlocking the potential for adaptive change. The very strengths that a strong culture can impart can also be impediments to the flexibility needed in order to successfully negotiate the turbulent years of rapid change. Grand Manan seems to be on the cusp of new transitions that will open its borders, encouraging the emergence of creative and dynamic new cultures of being.

July 2006

The final event in this story that galvanized the community around the issue of drugs occurred on 22 July 2006. Simmering tensions associated with drugs, dealers, addicts, and frustrated families who were seeing loved ones in the throes of suicidal behaviours came to a tragic peak that night when the house of an alleged drug dealer was torched and his vehicle damaged by a crowd of islanders that, according to news reports, numbered between forty and fifty. An apparently powerless RCMP contingent watched as, over a period of several hours, enraged islanders shot guns in the air, threw rocks, and shouted obscenities: "Participants on both sides brandished guns, baseball bats and knives. The battle rumbled over four hours into the early morning. Three of the four RCMP officers were on duty and tried to break up what turned into a raging street riot, but they could not stop the gang from setting fire to the suspected drug den. As volunteer firefighters worked to douse the flames, the vigilantes hurled rocks at them; the house burned to the ground and more than a dozen people suffered minor injuries during the fights. Gunshots were also fired, though no one was hit" (*Globe and Mail*, 26 July 2006, A7). News reports that spread across the country in the national media described a situation that was totally out of control. Headlines such as "Vigilante Violence Shakes Serene N.B. Tourist Town" (*Globe and Mail*, 26 July 2006, A4) and "Island Tension Running High with Riot Fallout" (*Quoddy Tides*, 11 August 2006, 1), were followed within days by headlines such as "N.B. Police Appeal for Calm on Grand Manan Island before Public Meeting" (Canadian Press, 2 August 2006) and "Alleged Arsonists Hailed as Heroes, Not Criminals" (CBC news, 3 August 2006). More than seventy RCMP officers were dispatched to the island in the days immediately following the riot, which islanders greeted with both disgust and bemused sarcasm. They had been arguing for years that the police were not taking the drug problem seriously enough, and now the police seemed to be overreacting. At the public meeting attended by over five hundred people and chaired by Constable Ron Smith, islanders had two messages: "get our boys out of jail" and "do something about drugs!" Although a few dissenters such as Eric Allaby (islander and elected MLA) expressed concern that "we don't know what to do next," others were adamant that the police needed to crack down with tougher enforcement. Eventually, there were eight arrests of young men, ages twenty-four to thirty-five, while the alleged drug dealer escaped the island before he could be apprehended for questioning.

Over the coming months the five who were held in custody were moved between jails for their own safety, according to news reports, finally appearing in court in November. In the meantime, a community fundraising effort had raised about $20,000 for their legal defence, arguing that their apparently criminal behaviour had to be seen in the context of extreme frustration at the lack of action against drugs on the island. While there was no unambiguous community consensus about the "rights and wrongs" of what happened, there was agreement that the young men who had participated had done so because of a long history of frustration regarding the drug situation on Grand Manan – a frustration shared by every islander. However, as one police officer pointed out at the meeting, some of those cheering for "justice" had themselves been known to buy cocaine. There were no clear boundaries between truth and lie, hero and criminal. One fact was incontestable however: not a single family felt untouched by the events. Everyone was implicated, and everyone experienced a mixture of embarrassment at being in the national media and anger that their situation no doubt reflected the realities of many other rural communities across the country. Indeed, in e-mails and letters-to-the-editor from across Canada, islanders received confirmation that they were not alone in their struggles to mitigate the ravages of drug addiction.

Eventually, two of the men were given sentences that could be served at home under conditions of house arrest, while the other three were given probation. Three of the initial eight had been released prior to the week-long trial. There is no question that, coming after the events of 1999–02, the July arson and riot in 2006 confirmed for the community that solutions had not really been found to the alienation being experienced by so many youth and young people on Grand Manan. Nonetheless, the "silver lining" that everyone seeks during times of tragedy did in fact materialize within months of the trial. After years of trying to obtain support for an expansion to the Boys and Girls Club, suddenly all three levels of government (federal, provincial, and municipal) were prepared to agree to an equal cost-sharing program that would build a new facility on the island. Whereas just four years earlier the executive director could not get any tangible support from the village council or find an islander willing to chair the fundraising committee for an expansion project, now there was the promise of funding for up to $4 million for a new community centre. This represented a major contribution to island institutional and activity potential for the future. Despite the unfortunate loss of a successful new director of the club due to her RCMP husband's decision not to stay on the island following the riot, the club quickly recruited a new director in the spring of 2007, and the promise of a new expanded facility seemed to suggest hope for the future.

The possibility for resilience and new directions for islanders as they seek new forms of employment within stable social milieux remains a challenge. While the underpinnings of historical values that depend upon independence and self-containment are being eroded, they continue to dominate the way islanders respond to problems. Volunteer committees continue to have difficulty encouraging people to take on leadership roles; even positions as board members are not easily filled. The uncertain future of the traditional fishery and of salmon aquaculture contributes to a strong undercurrent of apprehension among islanders as they consider their options. The rays of hope offered in the high ranking of Grand Manan students in 2006 is set against the ongoing drug problems that everyone acknowledges continue to plague many families. At the same time, increasing numbers of young people who graduate from the high school and go away to college are not returning to the island, as they have in recent decades. There are signs of a greater willingness to risk the social struggles of adapting and integrating into mainland structures as some young people attend New Brunswick universities and others find jobs in the RCMP or trades in Halifax or Saint John. While some will eventually return, there is a sense that moving away is one way to strengthen one's chances and to broaden one's options. A few have travelled abroad to Australia and Europe, and more are completing the university degrees they start. In the spring of 2007, several young women from Grand Manan graduated from the nursing program in Saint John, and none had plans to return to the island. Although some islanders had a sense that young people's leaving for jobs elsewhere meant that "they're lost," as an eighty-three-year-old woman said sadly, many islanders could see the possibilities being opened up for these young people. Adaptation to the forces of globalization on the island seems to require some time away in order to gain both technical and social skills as well as an understanding of new ways of connecting and communicating within a wider social context. In other words, the creation of social capital through being educated and trained away from the island may be the long-term safety net that allows for resilience and a renewal of community spirit and confidence. The caveat, however, is the concern being voiced by some islanders in the spring of 2008 that young families are making plans to move away and that overall student enrolments are showing signs of decline. The challenge will be to retain and nurture youth and young families as they seek new directions for growth. Grand Manan has always adapted to the vagaries of the fishery; now it must find the resilience to adapt to the challenges of globalization. It is very much a community between history and tomorrow.

11

Reimagining Grand Manan

As the flood tide of change began to make itself known, such activity seemed like innocent child's play. With only the most subtle warnings, a dozen simultaneous pressures abruptly appeared. ~ (Simon 1973, 3)

The changes on Martha's Vineyard described by Anne Simon in 1973 altered forever the landscape and identity of that island. For the community of Grand Manan there is a growing sense that they are confronting a fundamentally different new reality. As one woman said, "I don't know exactly how it has changed, but I have this overwhelming feeling of everything having changed and not knowing where we're going." For Martha's Vineyard, thirty-five years earlier, there had been a "struggle to abandon habits of thought which had persisted for centuries" (Simon 1973, 3). There are no givens, no unassailable boundaries that protect and sustain the way things have been. For Grand Mananers the historic affirmation of family roots and lives defined by the wild fishery and their island isolation are no longer realities that can guide choices and values. The incursions of globalization through restructuring in the agro-food industry and salmon aquaculture; the loss of a reliable diverse year-round fishery through inappropriate regulatory regimes and government interventions; and growing connections to the mainland through tourism, the internet, and outsiders moving to the island have all contributed to changes that have forever transformed the meaning of belonging to the community of Grand Manan. The boundaries and edges that have both constrained lives and defined new possibilities have opened up significantly over the past decade to become zones and spaces of transformation. Nevertheless, the adaptability of generations, still apparent in the details of family lives, seems to be threatened.

In 1999 I sat talking with Peter Cronk, who was picking strawberries from his organic garden, happy that they were two weeks early and would represent some unexpected income at this in-between season of the year. The garden was also planted with lettuce, beans, peas, broccoli, carrots, beets, and potatoes, as he was preparing for the weekly markets in the village that would begin on the first of July. His wife was still teaching, although classes would soon be over, allowing her time to help out in their half-acre garden. Peter had bought a share in a Whale Cove herring weir (Winner) in 1984, when a number of men wanted to sell because of several bad years. "Last year [1998]," he said, "had been the best since 1986," and he had been able to clear $7000 (pre-tax) during the July to September season. A former customs officer who had lost his job when the government office closed, he was typical of many islanders, perpetually looking for opportunities and adapting to seasons and niches of part-time work. Together, the Cronks had built two small cottages on their land, which were being rented out to tourists for about $700 per week, providing income (and a lot of work) for about four months of the year.

When an American summer resident wanted a new stone house atop a cliff overlooking North Head, Peter was asked to build it, and this included finding the field stones. The superb result is another example of the unheralded skills of so many Grand Mananers. He had also experimented with an orchard of apples and other small fruits, which proved to be a drain on family resources and was eventually abandoned. When the wharves were being reconstructed there was a need for major amounts of gravel. The company that had been contracted for the work negotiated with the Cronks for access to a suitable area of their land in order to develop a quarry, providing yet another source of income.

Peter's wife Marilyn, originally from Nova Scotia, had met her husband at university and followed him back to the island. When they were first married they lived in a trailer without electricity and subsidized their food budget through dulsing. It was not an easy life. Eventually, Marilyn was able to get a teaching position that she maintained over the next twenty years as she and her husband brought up their four boys. She gardened, sold at market, taught, and kept house and raised their children, while her husband roofed the house, tended weir, and cut wood for sale to islanders. These activities and more defined the worklife of this energetic and intelligent young couple. That it was two of their sons who, in separate incidents eight months apart, should have been killed in drug-related traffic accidents seemed incomprehensible to an island where families with deeply rooted abuse problems were not unusual.

This was an apparently stable family whose members took holidays together, in which parents worked and played with their boys, helped them to prepare résumés, and helped them move to new apartments on the mainland. Their children were loved and nurtured.

The events that occurred between 1999 and 2002 changed forever this couple's feelings about the island as well as the island's complacent attitude that alcohol and drug-related issues were the problem of someone else. That period created a determination to change, although the Cronks to this day question whether the deep attitudinal changes that are required have actually happened. Nonetheless, as described in chapter 10, led by the school, a combination of energetic new teachers and new programs seems to have helped people begin to acknowledge the deeply rooted problems that have permeated so many lives in the community. Virtually every family on the island had been touched in some way by the events of that period. When a story in the *Globe and Mail* (19 May 2007, F4-5) described the riot of July 2006 as "big-city problems" arriving in rural areas, it could not have provided an analysis that was further from the truth: "Counterintuitive as it may seem, given that crack cocaine is so associated with inner-city slums, Grand Manan was – and but for some intervening circumstances, might still be – a drug dealer's dream." Urban Canada's perception that drugs are an urban problem is not only wrong, it is dangerously misguided because it informs social policy and financial commitments that ignore the deep-rooted and long-standing issues pervading rural communities. It is not merely about unemployment or poverty. Rural communities need ongoing connections to educational, cultural, and political institutions that broaden their perspectives and encourage dreams for the future. New realities have to be nurtured through hope and belief in new possibilities. The entrepreneurial spirit of Grand Mananers can be a significant basis for retrenchment, but it needs to have the support of broadly based initiatives that include education and experiences that nurture self-esteem. These begin with youth. Furthermore, insofar as social capital has a role to play in supporting resilience, there are important issues that the community needs to address in areas of leadership and consensus building.

The entrepreneurial spirit of islanders is one of their most outstanding characteristics, and it defines attitudes, values, and priorities to an extent that often leads to disappointment but that also leads to remarkable successes. In 1980, Glen McLaughlin and John L'Aventure's pioneer establishment in Dark Harbour of one of the first Bay of Fundy aquaculture sites was a special milestone in terms of the scale of business investment that would eventually be involved.

Of truly global proportions, the initial investment barely reflected the magnitude of what was to come. The perceptive enterprise of these two men was combined with L'Aventure's long-term determination to create a technologically advanced company – Fundy Aquaculture – that was vertically integrated and, for a few short years, able to withstand highly competitive international pressures. He experimented with raising smolts, an egg stage hatchery, processing and shipping, feed blowers, and million-dollar scows. What he was unable to control was global prices, which plummeted after 2003, partly in response to a damning scientific report that described high levels of PCBs in farmed salmon and partly in response to increasing world supplies. Given long-term equity requirements, his business acumen in establishing vertical integration was not enough. When the bank called in loans, he had to negotiate a below-asset cost sale rather than go into bankruptcy. The high stakes global environment of salmon farming, which depended upon government support, ultimately proved to have too many unpredictable and uncontrollable factors for a relatively small investor to sustain. It was an industry that was suitable only for large companies.

I have already described the inventiveness of islanders, but it is worth noting that their ingenuity carries forward into every aspect of their working and social lives. Engine problems, leaking wood stoves, air circulation, and how to protect dulse from inclement weather are only a few of the puzzles that islanders have solved. And, historically, there is evidence of islanders' problem-solving inventiveness. As reported in the *Island Times*, the mechanical dulse shaker was only invented in the 1960s. Before this time the spreading of dulse for drying had involved separating the seaweed by hand, which was extremely time-consuming. Dulser Harry Greene described the earlier process, which involved "a shaking table, a big thing made out of wood that you'd put your dulse on and shake it up a little bit at a time and put it over in the corner. After we finally got enough for a basket we'd spread it. The dulse shaker sure made it a lot easier" (Jessica [Fitzsimmons] Newell, "An Innovative Island Invention: The Mechanical Dulse Shaker," *Island Times*, 10 January 2006, 1). Roland Flagg pointed out the time saved, with the dulse shaker taking about thirty seconds for six pounds and the hand method taking fifteen minutes. The inventor of this device was Sam Cary, a dulser who persevered through hilarious trial runs, which involved dulse being spewed in all directions, until he finally devised a workable design that has since been copied across the island. Other inventions that Cary developed (such as a heated clothesline and a motorized ceiling washer) were not quite so successful.

Fundy Aquaculture was not the only enterprise with a global reach. For islanders the new reality is that the ownership of almost 80 percent of the aquaculture sites around Grand Manan lies with the New Brunswick company Cooke Aquaculture. Regardless of which mainland or international company controls the industry, however, the crucial issue for Grand Manan is the loss of control by islanders, who now work as hourly paid labour. This global industry depends upon a pool of local unskilled labour, with only a few positions demanding the education and technical skills that allow for upward mobility. The hope is that, as the industry regains strength and is able to restructure its production sustainably, there will be opportunities for island entrepreneurship to make more substantial contributions.

Independence and the ability to apply multiple skills, which have traditionally defined fishers involved in the wild fishery, are significant and probably irreplaceable underpinnings of island culture, and they are being lost. The source of personal and collective identities for at least one-third of the residents of this small community has been seriously undermined by the introduction and growth of salmon aquaculture. On the one hand, aquaculture has been described as the economic saviour of the island; on the other, it is seen as a sign of the eventual demise of this strong culture. Islanders themselves are split regarding what the changes mean. Some believe that what is happening is only a temporary downturn and that "things have always gone in cycles and this is no different." Others feel that the changes this time are qualitatively different from those that have gone before and that the safety nets of alternative sources of income (such as picking dulse) are no longer viable. Whereas in the past dulse would get a family through a few months until a seasonal fishery opened up, now those fisheries are dead, with no sign that they can be revived. For many islanders, the experience in aquaculture has illuminated the threats to secure livelihoods in that industry as well. Their apprehension is well-founded.

When the reality of the high-stakes risk and low profit margins associated with aquaculture became increasingly apparent after 2004, many Grand Mananers began to detach themselves from the industry. While some, such as Harley Griffin, had to suffer the ignominy of bankruptcy, others were able to liquidate more gracefully. Wayne Green was able to sell his interest in his site near Wood Island to Cooke's in 2005, and almost immediately he was conjuring up new ideas for a new business venture. The purchase of two sheds on the Woodwards Cove breakwater established his "beach head" for a new enterprise, which he hoped might involve dulse. Then, in 2006, he invested in the restoration of the old weir, the North Air, behind the seawall at Dark Harbour;

and in 2007 he was considering investing in two new weirs along the back-of-the-island. The sense that many islanders were trying to rebalance their work commitments and investments was palpable in 2006–07. Conversations everywhere reflected apprehension. Despite such constant refrains as "there's always work if you want it" or "there's always something else that comes along," this time people were not so sure. Said one former fish plant worker, "Since the plant closed [in 2004] everything has changed." She went on to give the example of a cancelled dinner and auction event for Ducks Unlimited: "They couldn't sell enough tickets; nobody has the money." Islanders had become accustomed to higher standards of living, and the combined effects of closing the fish plant and restructuring within the aquaculture industry, plus severe constraints on the traditional flexibility of the fishery, were all clear signs of long-term problems. Questions were being asked about the viability of salmon aquaculture as a reliable and broadly based means of obtaining a livelihood. Reflecting uncertainty and worries about the future, houses were put on the market as some islanders tried to retrench, to restore financial security in the face of concerns for the long term.

Again, the forecasts were mixed. Quoted in a business magazine, the local MLA described the entrepreneurial spirit and self-sufficiency of Grand Manan as the island's biggest asset, indicating that when one thing runs its course, something else always comes along. It is not only the ability to see opportunities but also, and foremost, the capacity for flexibility and taking risk that seems to characterize the Grand Manan enterprise. While Bev Fleet's successful company, Dutchmen, is involved in trucking and heavy machinery, he has also worked as a dulse picker, school bus driver, and builder as well as running a deer farm. Many talents, the awareness of global opportunities, and a willingness to risk financial investments have all contributed to remarkable successes on this small island.

Meantime, caught up in the economic changes were gender relations and the futures for youth. Some curious contradictory initiatives were introduced in the period after 2001, when it seemed the future was all positive. Choices being made for new businesses were predicated upon money and more leisure and holiday time. For women, divorce can lead to creative solutions to their predicament. For example, a woman from-away became a fitness instructor and then established a spa. Two women bought into a national fitness franchise (Sisters) that operated for two years (2004 and 2005) before closing because business had declined and then reopening in 2007. Other women incorporated tanning beds into their hair salons so that Grand Mananers could better tolerate the cold, dreary winter months and/or be prepared for their Caribbean March holidays, which became fewer with the downturns after 2004. Two women invested in a condominium project that puzzled most islanders as they wondered who would ever buy into such a venture. "It's not how we are," was the common refrain. Three years after the gala opening in 2003, only one of the four units was being lived in. The opening of spas was another hazardous venture that long-established cottage owners and tourism operators worried might represent a new direction for the island, a move away from the more "natural" hiking and birdwatching activities.

Implicated in all of these social and economic changes were the evolving patterns of spatial relations that affected landscapes and lives on the water, along the shores, and in the woods and villages. Women ventured out into spaces that had earlier been unknown to them in their own ATVs and in their own trucks. They enjoyed weekends at the camps in the woods and made their own trips to the liquor store. Whereas walking alone along the road might have been viewed with suspicion in 1995, by 2006 young women were regularly jogging or walking daily, sometimes in pairs but just as often alone. Despite older women continuing to feel uneasy about public wanderings, the younger generation has claimed spatial equality across island spaces in ways unprecedented in Grand Manan history. While they still do not "invade" the men's crib games in the sheds, there is a sense that gendered space is not as easily defined or exclusively protected as it was in the past.

The details of the landscape and home designs provide clues to the transformation of Grand Manan culture. The civic numbers that suddenly appeared in 1996 are now taken for granted; new street names appear without comment. In 1995, few homes were surrounded by gardens, fences, or paved driveways. Today, decorative and privacy fences, lawn ornaments and lawn furniture, elaborate gardens with borders of spring bulbs (peonies, iris, and cornflower), and neat driveways all testify to both more time and money and changing aesthetic values. Houses have been "suburbanized," spread out between the villages on hilltops or along quiet roads, thereby loosening up the spatial fabric of residences. Even as island services were being centralized, the dispersal of homes was weakening the strong village-centred nature of identities and daily lives. Houses are larger, and while most continue to be built with the kitchen as the entrance room, they also incorporate front doors and entrance hallways as well as door bells. The sense that the home is for women while the outdoors is for men is no longer a pervasive theme. The "work week" is much closer to a Monday-to-Friday, 6:00-AM-to-6:00-PM schedule than it was even a decade ago, when many boats were out for days at a time. The daily rhythms of lives on Grand Manan now allow for the tending of flowers, along with the traditional vegetable gardens. Unlike 1995, when everyone on the island was assumed to have been invited to a wedding, now invitations are sent. Doorbells announce visitors. Voice messages on the phone and internet usage define communications strategies that didn't exist ten years ago. In 2007, describing a tension-filled argument with a neighbour and friend, a young woman told of their e-mail exchanges rather than of their telephone calls. Lives have been changed even in the smallest details of daily routines.

Through these changing relationships with the "outside world," both in terms of economic restructuring and social and cultural values, there have been turbulent times for the community. The arrival of the Newfoundlanders and Aboriginals, the loss of fishery licences through the government program on behalf of Aboriginal reserves, the rapid growth of aquaculture, the closing of the Connors fish plant in 2004, and the tragic events that occurred between 1999 and 2006 have all changed the island forever. Views on the church, alcohol, and education are changing; more islanders are travelling; and tastes in food, art, music, and books have evolved. Change is everywhere. For islanders the twelve years between 1995 and 2006 have been years of enormous growth and sudden stall, with the future very uncertain. What is inevitable, however, is that there will be no going back to 1995. The wild fishery has been depleted to an extent that is probably not reversible. Salmon farming offers a possibility for secure employment, although it may still be a few years before stability is assured. The change to wage employment and dependence upon a global agro-food sector has implications for social relations that are only beginning to be experienced. For a community that has long depended upon the independence of operating its own fishing enterprises and the flexible seasonal rhythms of a resource-based economy, the adjustments will take years. Adapting to the rhythms of nature is not the same as adapting to currency markets and corporate strategies.

The nature of tourism itself is changing, from a focus on rustic cottages to a focus on luxurious summer homes for wealthy mainlanders. The services that these new incomers expect will be more sophisticated, and the willingness of Grand Mananers to cater to them will be tested. While Grand Mananers have always had a strong interest in family lineage and are proud of their material and photographic records of their historical ties to the island, they have been reticent to share their history in a public way. The protection of heritage sites such as the smoke sheds at Seal Cove or the Swallowtail lighthouse were left to people from-away. Even as a few islanders begin to consider the possibility of broadening their tourism base, there are difficult challenges related to the short season and to a culture of not valuing the material landscapes of history.

For islanders, the loss of many young families who moved to the mainland with their children after 2001 had implications for the future. Not only young families were leaving; women whose spouses had died were also leaving, moving closer to children and grandchildren on the mainland. Even as the school worked to improve its programs there were concerns that young people did not have the range of opportunities that were offered youth who lived closer to urban areas. Summer jobs for young people at local kiosks or on a salmon site did not prepare them for a wide range of positions, not even in tourism or in offices. The self-assurance that the film instructor (see chapter 10) said was lacking among young people continued to be an issue in 2007. While this problem is being addressed through school programs, it needs to be given more broadly based consideration throughout the recreational lives of youth. While a new Boys and Girls Club facility certainly provides the potential for offering rich new opportunities, there will be an urgent need for ongoing operational support in the form of staff and equipment that can contribute to the development of Grand Manan's youth. There are signs that, following graduation from school, young people are making plans to leave the island at a rate that has not been seen since the 1960s. While this can be seen as a positive step for individuals, it does not necessarily bode well for an island faced with an aging population whose experiences have been framed by the traditional fishing economy. The future of Grand Manan could be made brighter by the possibility of youth seeking off-island education and work experience and then, in a few years, returning and bringing with them new skills and ideas that could build upon the strengths of Grand Manan's culture.

Islanders, both consciously and unconsciously, resist the idea that "rural" and "progress" are mutually exclusive. The myths of rurality that are impediments to change can, in fact, become a treasure trove for coping with the future (Walbert 2002; Lasch 1991). Family lineage, generations of struggle with the sea, independent spirits that resist bureaucratic solutions, and the shared experiences of a collective community all provide potential core strengths upon which to build. While it may be necessary to loosen the ties to the past, which are seen as defining features of social and economic relationships, islanders will discover that, more than simply living "in" their community they will have to acknowledge the importance of being "of" their place – Grand Manan Island. This means participating in and giving to the collectivity in a way that has not previously characterized their attitudes and understandings of what it is to be a Grand Mananer. As Annette Harland articulated at the opening of the public meeting in June 2001, the process will not be about anyone telling the island what the solutions are or what islanders need. Islanders must define those for themselves. Change will be difficult and it will be uncomfortable.

Ferry passing Swallowtail

Leadership and volunteerism will be part of this new dynamic.[1] As well, there are opportunities for businesses related to the agro-food industry (such as transportation and processing), tourism, and the arts that require only imagination and perseverance – but within the context of mainland experience and understandings. To a significant extent, governments need to recognize and support initiatives that encourage such developments. The federal government's cancellation of summer job program subsidies for youth in the spring of 2007 was precisely the opposite of what should have happened. If there is one group that governments need to support it is rural youth. The out-migration that has threatened so many communities, and, in Grand Manan's case, the isolated, insular culture that has defined it for generations, are causes of regional disparities – the urban/rural split – that are becoming stronger and more difficult to solve with each year. Programs such as the film project described in chapter 10 can provide hope for sustaining rural communities. But they will need the understanding of all Canadians who value our regional differences and local cultures, and they need the support of the government.

Grand Manan has a strong culture that has persevered through hundreds of years. But its rurality and relative geographic isolation, which have created a tradition of valuing the past, cannot continue. The current period demands new perceptions and views. The boundaries are no longer clear. The edges of transition are rich in possibilities but hazardous for those who cannot change. Grand Manan is an island that will need to call upon all its historic resources of determination and creativity, one hopes with the support of others, if it is to be able to regain control over its future and a renewed sense of pride and hope.

Epilogue

Let the preachers learn,
Hell may be fiery,
Hell may be firey, but the
Let the scholars learn,
Shielding the sin, they
Share, they share the
Floods will ebb, the tide
The tide will, floods will ebb,
The tide, the
~ (Britten 1963, 65)

Thanksgiving Sunday, October 2007, was a cool but beautiful day on Grand Manan. I had arrived for a week of revisions and personal renewal prior to winter's setting in. That day became a veritable day of thanksgiving and reflection. It had been eighteen years since I had first visited Grand Manan, and many of life's passages had been confronted along the way. In the last four alone I had struggled with breast cancer and grieved the deaths of both my parents and my husband. But as well, I had taken delight in the birth of my first grandchild and was looking forward to the birth of a second one. My passionate engagement with Grand Manan had not precluded personal challenges at home. Not long after the early morning sun's purple and orange hues were streaking the sky, I was walking along the beach at the Anchorage, reflecting upon the manuscript that was soon to become a book, and wondering about the community of people who had become so important to me. The full moon that had brought the extremely high and low tides of a week earlier had waned, and the beach seemed relatively clean of the dramatic debris that often litters it following the high tides and storms of autumn months. On the one hand, it seemed like a short week from full moon to the sliver that was in the sky now, a sudden transformation that was at odds with the feeling of the daily experience of change being slow, inexorable, and predictable. The tide metaphor seemed especially appropriate, a reality defining so many issues of change on the island: the inevitable change of globalization appears as a creeping new way of being, seeping into island consciousness. And yet its ultimate impacts had been so sudden, so abrupt, and so widespread, and certainly not predictable. From the spirit of optimism that the island experienced in the late 1990s to the mood of devastation and apprehension in 2006, it seemed, in 2007, that there was a gradually evolving acceptance of new realities. Most significantly, perhaps, Grand Manan's youth were noticeably more willing to move away for education and job opportunities, a situation that would unquestionably have profound implications for the future.

At mid-day I drove to a cemetery in the middle of the island to attend a funeral service for a woman whose special strengths, generosity, wisdom, and artistic talents had brought unique dimensions to the island thirty-five years earlier. She had arrived on the island as a single mother of four, without a high school diploma and as a dedicated member of the Baha'i community. Within a few short years she had achieved a university degree, was teaching, and later bought a small inn that she operated for tourists. Through her life she personified so magnificently the possibilities for Grand Manan to overcome terrible hardships with a spirit of generosity, creativity, and fortitude. As described in the eulogy, she was able to "reinvent" herself, as surely as will Grand Manan. As an entire generation of elders who have defined island culture through the wild fishery gradually pass on, their children, now in their fifties and sixties, acknowledge that new economic realities, driven largely by technologies such as salmon farming and communications, will transform the culture and community of the island. The edge of tourism development is becoming more deeply embedded through land sales and summer homes, another potential avenue of ongoing change that will define a different Grand Manan.

Later the same afternoon I dropped in to visit with an island friend, who happened to have two of her children and their families visiting from the mainland for the weekend. As we chatted we touched on many issues, some of which will always be part of any small community, and others that seem specific to Grand Manan. I was shown the preliminary architectural drawing for the approved expansion project for the Boys and Girls Club, which drew comments related to youth activities, parental participation, and the role of the village council. Evidence for these young couples of new attitudes and goals on the island was tempered by their satisfaction in having chosen to move away with their young families to ensure their exposure to a broad range of ideas and people. Discussion moved to the ongoing challenges that islanders will need to address, especially those related to youth.

That evening I had been invited for Thanksgiving dinner at the home of a retired couple who had moved permanently to the island from Toronto twelve years earlier. Also at the table was a man who had moved to the island with his wife three years earlier and a couple from British Columbia who were spending a week in their summer home, bought in 2003. The six of us, all from-away, reminisced about our first visits to Grand Manan and what it was that had captivated our spirits. Acknowledging the difficulties of negotiating the teenage years, these "Choosies" were all optimistic about the future of the island. Active in various youth-related projects, they were convinced of the gradual change in direction that was taking root, indicative of greater contacts with mainland institutions, the internet, travel, and receptivity to the opportunities for advanced education. Isolation, for so long a defining feature of Grand Manan, was gradually losing its power on socio-economic spaces and relationships

As I watched the soaring gulls and tiny shorebirds skittering across the wet sand along the beach, I felt the full gamut of emotions as an outsider that had been privileged to share a small part of other lives within their community culture. While I am saddened by their struggles and fears for the future, I also rejoice in their pride, their perseverance, and their determination to redefine their lives. Grand Manan is very much a part of me, and I feel specially honoured when someone asks me, after I have arrived back on the island, "When did you come home?"

Appendix A

Aquaculture sites – June 1999

SITE LOCATION	LOCAL LESSOR / COMPANY	(MANAGER, LOCAL)
Long Id.	S. Kinghorne/Grand Isle, J. Ross; **Cooke**	(Mgr. Jamie Ellis)
	D. Outhouse, R. Urquhart/**Ocean Salmon**-C. Saulnier	
	A. and B. Brown, T. Lyons/**Bothwick-Limekiln**	
	Neil Morse/**AquaFish** (Saulnier and Hamilton)	(Mgr. Nathan Bass)
	R. and H. Lambert/**AquaFish** (Saulnier and Hamilton)	(Mgr. Nathan Bass)
Nantucket	Harley Griffin and Bev Frost/Stolt	(Mgr. H. Griffin)
	Darcy, Dale, and Frank Russell/Heritage	(Mgr. D. Russell)
Andy's Ledge.	M. and M. Ingersoll (Gary Guptill)/**Cooke**	(Mgr. Michael Ingersoll)
Duck Id.	J. Ingersoll, B. Russell, M. Guptill/Grand Isle; **Cooke**	(Mgr. Matthew Ingersoll)
White Head	B. Ossinger and B. Russell/**Jail Id.**, later **Cooke**	(Mgr. B. Ossinger)
	Jamie Small/**Stolt**	(Mgr. Joe Middleton)
Ross Id.	Kenny Brown/**Stolt**	(Mgr. Stacy Brown)

SITE LOCATION	LOCAL LESSOR/COMPANY	(MANAGER, LOCAL)
Seal Cove	Long Pond: J. l'Aventure/**Fundy**	(Mgr. Chris Rayner)
	S.C. J. l'Aventure/**Fundy**	(Mgr. Chris Rayner)
	G. and G. Brown/**Cooke**	(Mgr. self)
	Wayne Green/**Stolt**	(Mgr. self)
	Ron Benson/**Stolt**	(Mgr. R. Zwicker)
Dark Harbour	E. Carpenter, Deer Id.	(Mgr. Artie Neves)

New Sites – May 2001

SITE LOCATION	LOCAL LESSOR/COMPANY	(MANAGER, LOCAL)
Pat's Cove	Bradley and Ann Small/**Heritage**	(Mgr. A. Small)
Pond Point	R. Lambert, D. Russell, and Harley Griffin/**Heritage and Stolt**	
	*split site with Nantucket	(Mgr. self)
Outer Wood Id.	**Ocean Salmon and AquaFish**	
	(Saulnier and Hamilton)	(Mgr. K. Brooks)
Lower Wood	Mainland "exchange" site/**Stolt**	(Mgr. D. Greenlaw)
White Head	Stolt (not developed)	
Nantucket Id.	Harley Griffin and Darcy, Blain & Russell/**Heritage**	
	*split site with Pond Point	

2001 Sites Summary: Stolt – 7 sites; Ocean Salmon – 2 sites; AquaFish 3 sites; Fundy – 2 sites; Heritage – 4 sites; Bothwick – 1 site; Cooke – 5 sites; Deer Id. – 1 site (= 23 working total, with overlap between Ocean Salmon and AquaFish in 2, and split sites at Pond Point and Nantucket)

Aquaculture sites – Summer 2006

SITE LOCATION   LOCAL LESSOR/COMPANY   (MANAGER, LOCAL)

* bold – change of ownership

SITE LOCATION	LOCAL LESSOR/COMPANY	(MANAGER, LOCAL)
Long Id.	Sherman Kinghorne/**Cooke**	Ryan Green
	David Outhouse, Ray Urquhart/**Cooke**	Jen Hutchison
	No local ownership/**Cooke**	Ann Small
	Neil Morse/AquaFish (Saulnier and Hamilton)	Nathan Bass
	R. and H. Lambert/AquaFish (Saulnier and Hamilton)	Nathan Bass
Nantucket	**all sites closed until autumn 2007 or 2010**	
Duck Id.	M. and M. Ingersoll, G. Guptill/**Cooke**	M. Ingersoll
	J. Ingersoll, B. Russell, G. Guptill/**Cooke**	D. Greenlaw
White Head	no local ownership/**Cooke**	
	Jamie Small/Cooke (contract lease)	Jamie Small
	White Head (not developed)	
Ross Id.	No local ownership/**Cooke**	Todd Clinch
Seal Cove	Long Pond: Brown/**Admiral**	Chris Rayner
	Billy and Glendon Brown/**Admiral**	Silas Daggett
	Billy and Glendon Brown/**Admiral**	Silas Daggett
	No local ownership/**Cooke**	
	Ron Benson/Cooke (contract lease)	
Dark Harbour	M. Green/**MG Fisheries** (closed 2007)	Artie Neves
Pat's Cove	Bradley and Ann Small/**Cooke**	A. Small
Pond Point	No local ownership/**Cooke**	R. Lambert
Wood Id.	No local ownership/**Cooke**	

Possible future site:

Cheney's Head   Ron Benson

Summary of Sites, 2007: Cooke – 13 sites; Admiral – 3 sites; AquaFish – 2 sites; mg Fishery – 1 site; Benson – 1 site.
(= 20 sites, of which 8 had fish in June 2007)
Fish: Cooke – approx. 2.1 m. fish + others, approx. .5 m fish
Value: est. @ $3.50/lb. $60 mill.

Appendix B

Grand Manan youth interviews value rankings, 1999–2000
(Highest = 10)

	F	F	F	F	F	F	M	M	M	M	M	M	M	M	M	M
Fa	10	9	9	9	10	9	10	10	9	10			10	10	10	7
Fr	9	8	8	8	9	7	8	9	7	9	7	10	9	7	8	9
Ed	8	7	10	8	8	8	9	8	8	8	8	8	7	8	9	3
Re		10			10	2	10	2	5	10	3		8	9	5	10
Mo	6	5		6.5	7	1	7	6.5	2	2	9	9	6.5	5	6	8
Fs											5	5			7	4
Jb	5	6		8			6.5	5	4				6.5	4	7	6
bl	7	3	7	6.5					4	1		7				5
Sp					5	4	4	6	7	10				6	4	
Mr	5	4					1	6						3		2

(In order, the categories are as follows: "Fa" family; "Fr" friends;
"Ed" education; "Re" religion/church; "Mo" money; "Fs" fishing;
"Jb" job; "bl" being my own boss/doing what I like; "Sp" sports; "Mr" marriage)

Notes

1 The officially designated areas for fishing are drawn by the Department of Fisheries and Oceans Canada (DFO) and are known as Areas. The term "sector" refers to different species and harvest techniques. Thus, fishers may have licences for the lobster, groundfish, or scallop sectors. But it is not a "herring sector"; at association meetings they refer to the "weir sector" for herring, differentiating it from the seiner sectors.

Prologue

1 The study formally extends over twelve years, from September 2005 to July 2006. However, I have included some references to personal visits in 1989, 1991, and 1993 as well as to the year of manuscript revisions, 2007–08. The reality is that, even as I was writing and revising, I could not resist updating when it seemed to be especially relevant to the main themes and issues in the book.

Chapter 1

1 Grand Manan Historical Society, *Grand Manan Historian* 11 (1967 [2nd ed. 1973]): 13.

2 The term "Indian" is used here because that is how they are referred to in all of the historical literature. However, later in the book I discuss the introduction of the terms "First Nations," "Natives," and "Aboriginal" to Grand Manan as well as the impact of Aboriginal peoples on various aspects of the island fishery

and social milieu. While Canadian usage of these three terms is interchangeable, and I have spoken with First Nations people about this, I have mainly used the term "Aboriginal," with the understanding that I am referring to the New Brunswick Maliseet and the Maine Passamaquoddy.

3 Report of 1853, quoted in Grand Manan Historical Society, *Grand Manan Historian* 11 (1967): 14.

4 "Of fifty individuals, heads of households claiming the 54 surveyed lots, twenty-four were shown as 'Emigrants from the United States,' which merely means they came from the country following the Treaty of 1783 at the end of the Revolutionary War; thirteen were listed as 'United Empire Loyalists,' most of whom had arrived here after temporary residence elsewhere in British territory; three were 'brought up on the island,' two were natives of Campobello; five can be shown as coming from Great Britain and were probably discharged servicemen who had fought in the late war; one was born in Charlotte County, one in New Brunswick, and one in Nova Scotia" (Grand Manan Historical Society, *Grand Manan Historian* 11 [1967 (2nd ed. 1973)]: 21).

5 Ibid., 1 (1967 [2nd ed. 1973]): 17.

6 Ibid., 11 (1967 [2nd ed. 1973]): 17.

7 Ibid., 22 (1980): 1.

8 Ibid., 22 (1980): 25. The text actually refers to "one decade," despite referring explicitly to the years 1861 and 1881.

9 Ibid., 21 (1979): 32

10 *Report on Grand Manan*, unpublished report, 1995, FGA Consultants, Fredericton, NB, 1995.

11 Study of Grand Manan Ferry Service-Update, unpublished report, FGA Consultants, Fredericton, NB, 1999, 7.

Chapter 2

1 In 1882, production was between 300,000 and 400,000 boxes of smoked herring, representing a harvest of five thousand to six thousand hogsheads in the local weirs (Gilman 2003, 112).

2 A hogshead is equivalent to 1,200 pounds.

3 Interview, 17 June 2007. On Campobello Island, fishers developed a sleeve to join two stakes, allowing for weirs to be placed further offshore as well as for the purchase of less expensive stakes.

4 In 2005, Stephen Bass considered rebuilding the Challenge, located off Indian Beach, but decided against it because of the enormous costs involved.

5 In 2003, according to the Department of Fisheries and Oceans, there were 20,000 tons taken from weirs and 12,000 tons taken from purse seiners. This is cited in D. Wilbur, "Fisheries File," *Courier Weekend*, 13 May 2005, A-4.

6 Weirs were actually licensed by the DFO as "privileges," which carried rules related to usage. In order for a family to retain the privilege, weirs had to be "built" and "tied" every year in preparation for harvest. If, after three years, the family did not execise its privilege, then the DFO could offer the weir for sale, with the existing "owner" having right of first refusal. The family could sell the assets, notably the two sets of twine (nets) that would have been precisely measured to fit the design and size of the weir.

7 The problem of inadequate diets and obesity is only beginning to be explored. An unpublished study by Melissa Legrand, which focused on mercury levels, found that fewer people are eating fish as a regular part of their diet now than was the case several years ago.

8 The official designation for the lobster areas within which fishers may harvest is established by the DFO. Occasionally, the men refer to "District 38." The "sector" refers to the species, such as the "weir" sector for herring or the groundfish sector. For Grand Manan, Area 38 is the designated area. But there is also a "buffer" zone, Area 37, which may be fished only during the season defined for Deer Island and Campobello Island (until January). In 2002, the opening of the "Grey Zone" for lobster fishing during the summer months until the end of October created a third (overlapping) area – 38B.

Chapter 4

1 See *Sustaining America's Fisheries and Fishing Communities: An Evaluation of Incentive-based Management* at environmentdefense.org. Viewed 9 April 2007.

2 The high values were explicitly for those weirs on the "front" of the island, which were seen as potential aquaculture sites. On the "back," or western side, of the island, where weirs had taken increasing harvests through the years after 2000, the market values were lower, reportedly in the range of $75,000 to $100,000. Connors' interest in weirs at Dark Harbour and Indian Beach was prompted by the growing market for herring as bait (lobster) and feed (salmon). The most recent sale of a weir privilege, in the summer of 2007, was for the Challenge weir at Indian Beach, which was bought by a group from Campobello Island. Speculation was that the price was low as the owner had not built the weir for five years, with the result that the stakes and top poles had deteriorated.

3 See appendix for list of aquaculture sites in 1999 and 2001.

4 In 2005 the restructuring of the industry and another epidemic of ISA led the government to change to a three-year cycle that included one fallow year. See p. 206.

5 This calculation is based upon 136 licences for Sector 38 and a harvest in 2004 of 700 tons and an average price of five dollars per pound.

6 Notes from interview with Wendy Dathan, June 2007.

7 Atlantic Policy Congress response to the Marshall decision. Available at http://www.apcfnc.ca/pscf_response1.htm.

8 It was reported in spring 2005 that one young fisher had just acquired a boat and lobster licence, worth over $700,000, with the backing of one of these groups.

9 Some of the boats, such as *Against the Wind* (Tobique), had been purchased from island fishers, while others, such as *Dr Peter Paul* (Woodstock) and the *Charles Paul* (Tobique) were specially built. Altogether, Tobique had five boats, Kingsclear three, Woodstock four, and the Native Council of New Brunswick one.

10 The only interview granted was one with students who were on the island for three days as part of a university course.(McGill University, Autumn 2001, field course held at the Huntsman Centre in St.Andrew's). They noted the terse response to questions related to financial structures.

11 A fisher suggested that the patrol boat *Cumella* was not being seen so often because it was needed in other regions of the Atlantic provinces for problems related to the new Aboriginal fishery. In other words, a confluence of factors had combined to create an increasingly difficult situation in the Grey Zone.

Chapter 5

1 DFO and DFA NB, Rockweed Management and Development Strategy for the Bay of Fundy (March 1992, draft); and DFO and DFA NB, *Canada-New Brunswick Development Initiative on Rockweed,* Ascophyllum Nodsum: *Harvesting in the Bay of Fundy* (draft, 1994).

2 This consisted of a box of matches, the traditional uncomfortable warning.

Chapter 6

1 This is a paraphrase of Whatmore's argument that the OECD definition of globalized food networks applies to the activities and relationships within food production generally.

2 Sales in 2003 for Bumble Bee were approximately $750 million; for Connors Brothers, they were $150 million.

3 20 June 2007. The site manager's estimate was that less than 5 percent of the feed reached the ocean bottom unconsumed by the salmon. Cameras that monitor the flow are one indicator. But there is also the economic encouragement to reduce feed to its optimal amount.

4 Usually referred to as fifty-, seventy-, and one hundred-metre cages respectively.

5 Usually referred to as 10 grams.

6 In 1995 the Canadian dollar was only US$0.63, as compared to US$0.90 in April 2007 and was at parity by 2008. In effect, the rise of the Canadian dollar against the American dollar, by 60 percent since 2002 (*Globe and Mail*, 6 May 2008, A8), meant that the prices for Grand Manan lobster could not increase. Indeed, at the beginning of the winter season in 2006, the price, at $5.50 per pound, was only 20 percent higher than it was in 1995, not taking inflation or the exchange rate into account.

7 The author was a young visitor from Stoney Creek, Ontario.

8 "Shoulder months" is a term commonly used by tourist operators to describe the period at the margin of peak season.

9 Tourist survey conducted as exit survey at the ferry terminal, 26 June–6 July 1997, 285 returns (approx. 748 individuals).

Chapter 8

1 Conversation, October 2007, concerning the possibility that the village office will take up space in the proposed new Boys' and Girls' Club.

2 Although in 2006 it seemed that one of the North Head tourist businesses would be moving to Seal Cove, in the end Kevin Sampson's kayak business remained in North Head.

3 The young nineteen-year-old was drafted by the Milwaukee Brewers but subsequently received an athletic scholarship to play at the University of Illinois. With his brother also playing with a university baseball team in Alberta, the family has never regretted its decision to move to the mainland.

4 Named by the fisher owner after Lobster Area 38, because he felt that the area had been "so good to him."

5 The informant noted that Gallaway's had been criticized for its noisy bar.

Chapter 9

1 Based on a count of telephone-book residents in 2006, with the help of Joan Small.

Chapter 10

1 Taken from the provocative title of Gerald Sider's *Between History and Tomorrow: Making and Breaking Everyday Life in Rural Newfoundland.* This book was based on his 1986 *Culture and Class in Anthropology and History.* See Sider (2003) and Sider (1986).

2 The data for these graphs are based upon analysis of the high school yearbooks and the information from one of the islander school teachers, who was involved in collating information for a reunion in 2000.

3 At least three interviewees noted problems associated with collaboration with the school.

4 The effort was unsuccessful. The aquaculture companies said that it would be "unrealistic" to enact zero tolerance as it could not be enforced.

5 In the end the student decided not to attend the military college, thus foregoing his scholarship.

Chapter 11

1 "As a report on the future of rural areas makes clear, the survival of small communities depends upon local elites, leadership within the community" (Crossing Boundaries National Council, quoted by Ibbitson, *Globe and Mail,* 7 October 2005, A4).

References

Acheson, J.M. 1987. "The Lobster Fiefs Revisited: Economic and Ecological Effects of Territoriality in Maine Lobster Fishing." In *The Question of the Commons: The Culture and Ecology of Communal Resources*, ed., B.J. McCay and J.M. Acheson, 37–65. Tucson: University of Arizona Press.

Adams, James Luther. 1965. *Paul Tillich's Philosophy of Culture, Science and Religion*. New York: Harper and Row.

Agnew, J. 1987. *Place and Politics: The Geographical Mediation of State and Society* London: Allen and Unwin.

Allaby, F. 1984. *Grand Manan*. Saint John: Grand Manan Museum Inc.

Andersen, R., ed. 1979. "Public and Private Access Management in Newfoundland Fishing." In *North Atlantic Maritime Cultures: Anthropological Essays on Changing Adaptations*, 299–35. The Hague, Paris, NY: Mouton.

Apostle, R., G. Barrett, P. Holm, S. Jentoft, L. Mazany, B. McCay, and K. Mikalsen. 1998. *Community, State, and Market on the North Atlantic Rim: Challenges to Modernity in the Fisheries.* Toronto: University of Toronto Press.

Apostle, R., and G. Barrett. 1992. *Emptying Their Nets: Small Capital and Rural Industrialization in the Nova Scotia Fishing Industry*. Toronto: University of Toronto Press.

Appiah, K.A. 2005. *The Ethics of Identity.* Princeton, Oxford: Princeton University Press.

Arbo, P., and B. Hersoug. 1997. "The Globalization of the Fishing Industry." *Marine Policy* 21: 121–42.

Bartlett, John. 1980. *Bartlett's Familiar Quotations*. 15th ed. Boston, Toronto: Little, Brown.

Bauman, Z. 2001. *Community: Seeking Safety in an Insecure World*. Cambridge: Polity Press.

Berton, P. 1963. *The Comfortable Pew: A Critical Look at Christianity and the Religious Establishment in the New Age*. Toronto: McClelland and Stewart.

Berrill, M., and D. Berrill. 1981. *The North Atlantic Coast*. San Francisco: Sierra Club Books.

Bottomley, G. 1992. *From Another Place: Migration and the Politics of Culture*. Cambridge: Cambridge University Press.

Bourdieu, P. 1984. *Distinction: A Social Critique of the Judgement of Taste*. Cambridge, MA: Harvard University Press.

Bredin, M. 1995. "Aboriginal Media in Canada: Cultural Politics and Communication Practices." PhD diss., Graduate Programme in Communications, McGill University.

Britten, B. 1945. *Peter Grimes: An opera in three acts and a prologue. Derived from the poem of George Crabbe; words by Montagu Slater*. 1963 ed. London, New York, Bonn, Sydney, Tokyo, Toronto:Boosey and Hawkes.

Burgess, J., and J.R. Gold, eds. 1985. *Geography, the Media and Popular Culture*. New York: St Martin's Press.

Burridge, L.E. 2007. *A Scientifc Review of the Potential Environmental Effects of Aquaculture in Aquatic Ecosystems*. Vol. 1. Department of Fisheries and Oceans website, www.dfo-mpo.gc.ca/home-acceuil_e.htm, viewed 25 May 2008.

Buttimer, A. 1980. "Home, Reach, and the Sense of Place." In *The Human Experience of Space and Place*, ed. A. Buttimer and D. Seamon, 166–87. New York: St Martin's Press.

Carey, R.A. 1999. *Against the Tide: The Fate of the New England Fisherman*. Boston/New York: Houghton Mifflin Co.

Charney, J.I., and L.M. Alexander, eds. 1993. *International Maritime Boundaries*, vol 1. Dorchecht, Boston: Martinus Nijhoff Publishers.

Clapp, R.A. 1998. "The Resource Cycle in Forestry and Fishing." *Canadian Geographer* 42, 2: 129–44.

Coates, K. 2000. *The Marshall Decision and Native Rights*. Montreal and Kingston: McGill-Queen's University Press.

Cohen, A. 1985. *The Symbolic Construction of Community*. London and New York: Tavistock.

Cohen, A., ed. 1982. *Belonging: Identity and Social Organization in British Rural Cultures*. Social and Economic Papers No. 11, Institute of Social and Economic Research, Memorial University, Newfoundland.

Collins, Richard. 1986. "Broadcasting Policy in Canada." In *New Communication Technologies and the Public Interest: Comparative Perspectives on Policy and Research*, ed. M.Ferguson, 150–63. London: Sage Publications.

Davis, Anthony. 1991. *Dire Straits: The Dilemmas of a fishery: The case of Digby Neck and the islands*. St John's, NF: ISER (Newfoundland Institute of Social and Economic Research, Social and Economic Studies no. 43).

Davis, D. 1995. Women in an Uncertain Age: Crisis and Change in a Newfoundland Community. In *Their Lives and Times: Women in Newfoundland and Labrador – a Collage*, ed. C. McGrath, B. Neis, and M. Porter, 279–95. St John's, NF: Killick Press.

Davis, D., and J.L. Bailey. 1996. "Common in Custom, Uncommon in Advantage. Common Property, Local Elites, and Alternative Approaches to Fisheries Management." *Society and Natural Resources* 9: 250–65.

De Tocqueville, Alexis. 1990 [1840]. *Democracy in America*. Vol. 2. New York: Vintage Books.

Dodgshon, R.A. 1998. *Society in Time and Space: A Geographical Perspective on Change*. Cambridge, Cambridge University Press.

Doucet, R., and R. Wilbur. 2000. *Herring Weirs: The Only Sustainable Fishery*. St. George, NB: Image Press.

Dunn, R.G. 1998. *Identity Crises: A Social Critique of Postmodernity*. Minneapolis: University of Minnesota Press.

Eliade, Mircea. 1957. *The Sacred and the Profane*. New York: Praeger Palo.

Eythorsson, E. 2000. "A Decade of ITQ Management in Icelandic Fisheries: Consolidation without Consensus." *Marine Policy* 24 (6): 483–92.

Food and Agriculture Organization. 2006. *The State of the World Fisheries and Aquaculture*, Aquaculture Production, 2004, vol. 100/2; Capture Production, 2004, vol. 100/1; Rome. Available at FAO.org/fi/eims_search/advanced_s_result

Featherstone, M. 1993. "Global and Local Cultures." In *Mapping the Futures*, ed. J. Bird, B. Curtis, T. Putnam, and G. Robertson, 169–87. New York and London: Routledge.

FGA Consultants Ltd. 1995. *Economic Report on Grand Manan*, Fredericton. N.p.

– 1999. *Study of Grand Manan Ferry Service Update.* N.p.

Fitchen, Jane. 1991. *Endangered Spaces, Enduring Places: Change, Identity and Survival in Rural America.* Boulder: Westview Press.

Forman, R.T.T. 1995. *Land Mosaics: The Ecology of Landscapes and Regions.* Cambridge, NY: Cambridge University Press.

Fox, Bonnie J. 1988. "Conceptualizing Patriarchy." *Canadian Review of Sociology and Anthropology* 25 (2): 163–82.

Frank, K.T., B. Petrie, J.S. Choi, and W.C. Leggett. 2005. "Trophic Cascades in a Formerly Cod Dominated Ecosystem." *Science* 308, 5728: 1621–3.

Friesen, G. 2000. *Citizens and Nation: An Essay on History, Communication and Canada.* Toronto: University of Toronto Press.

Fukuyama, F. 1995. *Trust: The Social Virtues and the Creation of Property.* London: Penguin.

Game, A., and A. Metcalfe. 1996. *Passionate Sociology.* London: Sage Publications.

Geertz, C. 1973. *The Interpretation of Cultures.* New York: Basic Books.

Gentilcore, R.L. ed. *Historical Atlas of Canada.* Vol. 2; *The Land Transformed, 1800–1891.* Toronto: University of Toronto Press.

Gilman, J. 2003. *CANNED: A History of the Sardine Industry.* St Stephen, NG: John Gilman.

Grand Manan Historical Society. 1967. *The Grand Manan Historian.* Nos. 8, 9, 10, 11, 14, 21, 22. Edited by L.K. Ingersoll.

Gregory, D. 1990. "Grand Maps of History: Structuration Theory and Social Change." In *Anthony Giddens: Consensus and Controversy*, ed. J. Clark, C. Modgill, and S. Modgill, 217–33. Brighton: Falmer Press.

Gupta, S., ed. 1993. *Disrupted Borders: An Intervention in Definitions of Boundaries.* London: Rivers Oram Press.

Gusfield, J.R. 1975. *Community: A Critical Response.* New York: Harper and Row.

Hansen, P.K., B.T. Lunestad, and O.B. Samuelson. 1993. "Effects of Oxytetracycline, Oxolinic Acid and Flumequine on Bacteria in an Artificial Marine Fish Farm Sediment" *Canadian Journal of Microbiology* 39: 1307–12.

Haraway, D. 1988. "Situated Knowledges: The Science Question in Feminism and the Privilege of Partial Perspective." *Feminist Studies* 14 (3): 575–99.

Hardin, G. 1968. "The Tragedy of the Commons." *Science* 162: 1243–8.

– 1974. "Living on a Lifeboat." *Bioscience* 24: 561–8.

Harrison, C.M., and J. Burgess. 1994. "Social Construction of Nature: A Case Study of Conflicts over the Rainham Marshes." *Transactions of the British Institute of Geographers* 19: 291–310.

Haya, K., L.E. Burridge, and B.D. Chang. 2001. "Environmental Impact of Chemical Wastes Produced by the Salmon Aquaculture Industry." *ICES Journal of Marine Science* 58: 492–6.

Hewitt, J.P. 1984. *Self and Society: A Symbolic Interactionist Social Psychology*. Boston: Allyn and Bacon.

– 1989. *Dilemmas of the American Self*. Philadelphia: Temple University Press.

Ingersoll, L.K. 1963. *On This Rock: An Island Anthology*. Grand Manan: Gerrish House Society.

Innis, H.A. 1951. *The Bias of Communication*. Toronto: University of Toronto Press.

Katz, C. 1998. "Disintegrating Developments: Global Economic Restructuring and the Eroding of Ecologies of Youth. In *Cool Places: Geographies of Youth Cultures*, ed. T. Skelton and G. Valentine, 130–44. London, New York: Routledge.

Katz, C., and J. Monk. 1993. "Making Connections: Space, Place and the Life Course." In *Full Circles: Geographies of Women over the Life Course*, 264–27. New York and London: Routledge.

Krkosek, M., M.A. Lewis, and J. Volpe. 2005. "Transmission Dynamics of Parasitic Sea Lice from Farm to Wild Salmon." Proceedings of the Royal Society B. published online at www.math.ualberta.ca/~mkrkosek/publications.htm 1 April 2005, 1–8.

Lasch, 1991. *The Culture of Narcissim: American Life in an Age of Diminishing Expectations*. New York: W. Norton.

Liebcap, G.D. 1989. *Contracting for Property Rights*. Cambridge and New York: Cambridge University Press.

Lippard, L. 1997. *The Lure of the Local: Senses of Place in a Multicentered Society*. New York: W.W. Norton.

Livingstone, D.N. 1992. *The Geographical Tradition*. Oxford: Blackwell.

Macdonald, S. 1997. *Reimagining Culture*. Oxford and New York: Berg.

Mannheim, Karl. 1968. *Essays on the Sociology of Knowledge*. London: Routledge and Kegan Paul.

Mansfield, B. 2004. "Rules of Privatization: Contradictions in Neoliberal Regulation of North Atlantic Fisheries." *Annals of the Association of American Geographers* 94, 3: 565–84.

Marshall, J. 1999. "Insiders and Outsiders: The Role of Insularity, Migration and Modernity on Grand Manan Island, New Brunswick." In *Small Worlds, Global Lives: Islands and Migration*, ed. R. King and J. Connell, 95–113. London and New York: Pinter.

– 2001. "Landholders, Leaseholders and Sweat Equity: Changing Property Regimes in Aquaculture." *Marine Policy* 25: 335–52.

Martin, D. 1978 *A General Theory of Secularization.* Oxford: Basil Blackwell

Massey, D. 1994. *Space, Place and Gender.* Minneapolis: University of Minnesota Press.

– 1997. "A Global Sense of Place." On *Reading Human Geography: The Poetics and Politics of Enquiry*, ed. T. Barnes and D. Gregory, 315–23. London and New York: Arnold.

– 1998. "The Spatial Construction of Youth Cultures." In *Cool Places: Geographies of Youth Cultures*, ed. T. Skelton and G. Valentine, 121–9. New York and London: Routledge.

McCay, B. 1999. "'That's Not Right': Resistance to Enclosure in a Newfoundland Crab Fishery." In *Fishing Places, Fishing People: Traditions and Issues in Small-Scale Canadian Fisheries*, 301–20. Toronto: University of Toronto Press.

McEwen, A. 2001. "The Water Boundary between Maine and Canada." Paper presented at water boundary conference, University of New Hampshire, Portsmouth. Available at www.ucalgary.ca/~amcewen/maincan.pdf/.

McLuhan, Marshall. 1962, *The Gutenburg Galaxy: The Making of Topographic Man.* Toronto: University of Toronto Press.

– 1964. *Understanding Media.* London: Routledge and Kegan Paul.

McHugh, K.T., and R.C. Mings. 1996. "The Circle of Migration: The Attachment to Place." *Aging* 86: 530–50.

Melucci, A. 1996. *The Playing Self: Person and Meaning in the Planetary Society.* Cambridge: Cambridge University Press.

Milewski, I,, J. Harvey, and B. Buerkle. 1997. *After the Gold Rush: The status and Future of Salmon Aquaculture in New Brunswick.* Fredericton: Conservation Council of New Brunswick.

Misztal, B. 1996. *Trust in Modern Societies.* Cambridge, MA: Polity Press.

Monmonier, M. 1995. *Drawing the Line: Tales of Maps and Cartocontroversy.* New York: Henry Holt and Co.

Neis, B., R. Jones, R. Ommer. 2000. "Food Security, Food Self-Sufficiency, and Ethical Fisheries Management." In *Just Fish: Ethics in Canadian Marine Fisheries*, 154–73. St John's, NF: Institute of Social and Economic Research, Social and Economic Studies no. 23.

New Brunswick. 2000. *Bay of Fundy Marine Aquaculture Site Allocation Policy.* Fredericton: New Brunswick Department of Agriculture, Fisheries and Aquaculture.

Newell, D., and R. Ommer, eds. 1999. *Fishing Places, Fishing People: Traditions and Issues in Canadian Small-Scale Fisheries.* Toronto: University of Toronto Press.

Newman, D., and A. Paasi. 1998. "Fences and Neighbours in the Postmodern World: Boundary Narratives in Political Geography." *Progress in Human Geography* 22, 2: 186–207.

Nicolson, A. 2001. *Sea Room: An Island Life in the Hebrides.* New York: North Point Press.

Ogden, P.E. 2000. Entry for "Migration" in *Dictionary of Human Geography*, ed. R.J. Johnston, D. Gregory, G. Pratt, and M. Watts, 380–1. Oxford: Blackwell.

Ommer, R. 2000. "The Ethical Implications of Property Concepts in a Fishery." In *Just Fish: Ethics and Canadian Marine Fisheries*, 117–39. St John's, NF: ISER, Social and Economic Papers No. 23, Memorial University.

Ostrom, E. 1987. "Institutional Arrangement for Resolving the Commons Dilemma." In *The Question of the Commons: The Culture and Ecology of Communal Resources*, ed. B.J. McCay and J.M. Acheson, 250–65. Tucson: University of Arizona Press.

– 1990. *Governing the Commons.* New York: Cambridge University Press.

Oxman, B.H. 1993. "Political, Strategic, and Historical Considerations." In *International Maritime Boundaries*, vol. 1, ed. J Charney and L.M. Alexander, 3–40. Dorchecht, Boston: Martinus Nijhoff Publishers.

Palmer, C., and P. Sinclair. 1997. *When the Fish Are Gone: Ecological Disaster and Fishers in Northwest Newfoundland.* Halifax: Fernwood Publishing.

Pavlakovich-Kochi, V., B. Morehouse, and D. Wastl-Walter, eds. 2004. *Challenged Borderlands: Transcending Political and Cultural Boundaries.* Burlington, VT: Ashgate.

Perley, M.H. 1852. *Reports on the Sea and River Fisheries of New Brunswick*, 2nd ed. Fredericton: J.Simpson Printer to the Queen's Most Excellent Majesty.

Pillay, T. 1999. "Resources and Constraints for Sustainable Aquaculture." In *Sustainable Aquaculture: Food for the Future?* ed. N. Svennevig, H. Reinersten, and M. New, 21–7. Rotterdam: A.A. Balkema.

Pinkerton, E. 1999. "Directions, Principles, and Practice in the Shared Governance of Canadian Marine Fisheries." In *Fishing Places, Fishing People: Traditions and Issues in Small-Scale Canadian Fisheries*, 34–354. Toronto: University of Toronto Press.

Pocius, Gerald. 1991. *A Place to Belong: Community Order and Everyday Space in Calvert, Newfoundland.* Athens and London/Montreal: University of Georgia Press/McGill-Queen's University Press.

Polanyi, K. 1944. *The Great Transformation: The Political and Economic Origins of Our Time.* New York: Rinehart and Co.

Portes, A. 1998. "Social Capital: Its Origins and Applications in Modern Sociology." *Annual Review of Sociology* 24 (1): 1–24.

Rangeley, R. 1991. *A Critique of the Proposed Rockweed Management and Development Strategy for the Bay of Fundy.* Montreal: Department of Biology, McGill University (submitted to Camille Theriault, Minister of Fisheries and Aquaculture, New Brunswick).

Recchia, M. 1997. "Catching Lobsters in a Community Mesh: The Dynamics of Local Lobster Management on the Grand Manan Archipelago." MA thesis, Dalhousie University, Halifax.

Rhee, S-M. 1980. "The Application of Equitable Principles to Resolve the United States-Canada Dispute over East Coast Fishery Resources." *Harvard International Law Journal* 21 (3): 667–83.

Rogoff, I. 2000. *Terra Infirma: Geography's Visual Culture.* London: Routledge.

Roth, L. 1983. *The Role of Communications Projects and Inuit Participation in the Formation of a Communication Policy for the North.* MA thesis, Graduate Programme in Communications, McGill University.

Said, Edward. 1984. "The Mind of Winter: Reflections on Life in Exile." *Harpers*, vol. 269, issue 1612, p. 51.

Scherman, K. 1971. *Two Islands.* New York: Ballantyne Books.

Sibley, D. 1995. *Geographies of Exclusion: Society and Difference in the West.* London: Routledge.

Sider, G. 1986. *Culture and Class in Anthropology and History: A Newfoundland Illustration.* Cambridge: Cambridge University Press.

— 2003. *Between History and Tomorrow: Making and Breaking Everyday Life in Rural Newfoundland.* Peterborough, ON: Broadview Press. (rev. ed. of Sider 1986.)

Simon, Anne. 1973. *No Island Is an Island: The Ordeal of Martha's Vineyard.* Garden City, NY: Doubleday.

Sinclair, P., H. Squires, and L. Downtown. 1999. "A Future without Fish? Constructing Social Life on Newfoundland's Bonavista Peninsula after the Cod Moratorium." In *Fishing Places, Fishing People: Traditions and Issues in Canadian Small Scale Fisheries,* ed. D. Newell and R. Ommer, 320–39. Toronto: University of Toronto Press.

Small, D. 1997. "Picturing Grand Manan: Nineteenth-Century Painting and the Representation of Place." MA thesis, Trent University, Peterborough, Ontario.

Small, Ina. and Ernie Mutimer. 1989. *Ina of Grand Manan: A Stranger from Away.* Halifax: Nimbus.

Statistics Canada. 1981. *Profile of Census Divisions and Subdivisons in New Brunswick.* Ottawa: Ministry of Industry, Science and Technology.

— 1991. *Profile of Census Divisons and Subdivisions in New Brunswick.* Ottawa: Ministry of Industry, Science and Technology, 1992.

— 2001. www45.statscan.gc.ca

— 2006. www45.statscan.gc.ca

Stich, K.P. 1989. "The Cather Connection in Alice Munro's 'Dulse.'" *Modern Language Studies* 19, 4: 102–11.

Stiles Associates Inc. 1988. *History and Status of Community Radio in Quebec.* Toronto: Ministry of Culture and Communications.

Stonich, S., and P. Vandergeest. 2001. "Violence, Environment and Industrial Shrimp Farming." In *Violent Environments,* ed. N. Peluso and M. Watts, 261–86. Ithaca, NY: Cornell University Press.

Tomlinson, John. 1999. *Globalization and Culture.* Chicago: University of Chicago Press.

Turner, N.J., I.J. Davidson-Hunt, and M. O'Flaherty. 2003. "Living on the Edge: Ecological and Cultural Edges as Sources of Diversity for Social-Ecological Resilience." *Human Ecology* 13 (3): 439–61.

Wagner, J., and A. Davis. 2004. "Property as a Social Relation: Rights of 'Kindness' and the Social Organisation of Lobster Fishing among Northeastern Nova Scotian Scottish Gaels." *Human Organisation* 63 (3): 320–33.

Walbert D.J. 2002. *Garden Spot: Lancaster County, the Old Order Amish, and the Selling of Rural America.* New York: Oxford University Press.

Watts, M., and D. Goodman. 1997. "Agrarian Questions, Global Appetite, Local Metabolism: Nature, Culture and Industry in the *fin de siecle* Agro-Food System. In *Globalizing Food: Agrarian Questions and Global Restructuring*, ed. D. Goodman and M. Watts, 1–32. London: Routledge.

Weil, P. 1993. "Geographic Considerations in Maritime Delimitation." In *International Maritime Boundaries*, vol. 1, ed. J. Charney and L.M. Alexander, 115–30. Boston, London: Martinus Nijhoff Publishers.

Whatmore, S. 1995. "From Farming to Agribusiness: The Global Agri-Food System. In *Geographies of Global Change: Remapping the World in the Late Twentieth Century*, ed. R. Johnson, P. Taylor, M. Watts, 36–49. Oxford: Blackwell.

Wilbur, R. 2002. "Fishing with the Elephant: In the Grey Zone." *Weekend Courier*, 26 July 2002.

Wilson, Bryan. 1982. *Religion in Sociological Perspective.* Oxford, New York: Oxford University Press.

Woolcock, M., and D. Narayan. 2001. "The Place of Social Capital in Understanding Social and Development Outcomes." *Canadian Journal of Policy Research* 2 (1): 11–17.

Yaeger, P. 1996. "*The Geography of Identity*, 1–38. Ann Arbor: University of Michigan Press.

Index